CASCADE OF ARMS

ANDREW J. PIERRE
Editor

CASCADE
of
ARMS

*Managing Conventional
Weapons Proliferation*

BROOKINGS INSTITUTION PRESS
Washington, D.C.

THE WORLD PEACE FOUNDATION
Cambridge, Massachusetts

Library of Congress Cataloging-in-Publication data:
Cascade of arms: controlling conventional weapons proliferation /
  Andrew J. Pierre, editor.
    p. cm.
  Includes bibliographical references and index.
  ISBN 0-8157-7064-2 (cloth).—ISBN 0-8157-7063-4 (pbk.)
    1. Defense industries. 2. Arms transfers. 3. Arms control.
  I. Pierre, Andrew J.
  HD9743.A2C33 1997   97-33822
  338.4'76233—dc21   CIP

                    9 8 7 6 5 4 3 2 1

Contents

Tables

Figures

Foreword

IN MAY 1982, when the warships and troopships of the Royal Navy completed their 12,000-mile journey and began the campaign to recapture the Falkland Islands from their Argentinian invaders, they encountered an obstacle they arguably should never have faced: state-of-the-art *Exocet* anti-shipping missiles designed and made not in the factories of Britain's Warsaw Pact adversaries but in those of one of its closest allies, France.

Moreover, the *Super Etendard* attack aircraft that launched the missiles that caused so much damage and took so many British lives were also made in France. They were flown by Argentine pilots not because of any affinity between French democracy and Argentina's military dictatorship, but because the price—the combination of financial terms and the capabilities of the weapons system—was right. Indeed, when Aerospatiale won the contract to supply the *Exocet* to the Argentines, among rival firms bidding for a share of the lucrative contract was undoubtedly British Aerospace.

Had the Falklands war occurred later there would have been many more players in the game. Enterprises from all over the former Communist world, and from a still-Communist China as well, might have bid for the money and for the desperately needed jobs that are the rewards for winning the competition for a major weapons program. Moreover, the ranks of would-be suppliers might be swollen by yet another set of new entrants into the global arms market—firms, sometimes state-owned, sometimes private, from what we used to call the Third World. Brazil is the most prominent arms merchant among this category of states, but capabilities for making all but the highest-technology weapons exist in a dozen or more others. Very few of the armed conflicts of our era are marked by the employment of high-tech weaponry. Quite the contrary: in Rwanda the weapon of choice of those intent upon genocide was the machete, a true "dual-use" weapons system.

ix

When the Trustees of the World Peace Foundation sought a guide through the labyrinths of the post-Cold-War international arms trade they turned to Andrew J. Pierre, whose 1982 book, *The Global Politics of Arms Sales,* was widely regarded as the standard work on its subject. Pierre in turn recruited a set of specialists to join him in writing the essays that make up *Cascade of Arms.* They amply document both the demand and supply side of the equation. In doing so, they demonstrate just how difficult it will be to establish an effective international regime for controlling arms sales, especially in the circumstance that the United States has long been, and will continue to be, the world's leading arms merchant.

The contributors to *Cascade of Arms* are all realists. None of their essays attempts to solve problems rhetorically or through the sudden creation of new and powerful international institutions. Pierre's concluding chapter, in which he discusses at length the formidable difficulties that lie in the path of regime-builders, nevertheless puts forward a number of practical suggestions for departing from the status quo. It is a model of clear, practical thinking. Edwin Ginn, the founder of the World Peace Foundation, was no advocate of utopian formulas. He would have been pleased, as are the present trustees, with the volume Pierre has so ably brought to conclusion.

Richard H. Ullman
Chairman, Board of Trustees
The World Peace Foundation

Acknowledgments

Richard Bloomfield, then President of the World Peace Foundation, proposed this book to me and was an early source of encouragement. His successor, Robert Rotberg, followed on with his own support and helped greatly to bring the book to publication. I warmly thank both as well as the trustees of the World Peace Foundation.

Most of all, I am indebted to my fellow authors for their contributions. Each chapter stands on its own but the authors benefited from each other's advice at two important workshops which shaped the book. We all were greatly assisted by the comments of a number of additional experts who participated in the workshops.

I would like to thank The John Merck Fund for supporting my travel related to this book.

The book was written during the time that I was a Senior Associate at the Carnegie Endowment for International Peace. I would like to thank its President, Morton Abramowitz, for his continuing support as well as that of my colleagues at the Endowment. Special mention should be made of the doyen of non-proliferation, Leonard Spector, and the Endowment's librarian, Jennifer Little and her staff. Also to be thanked are a number of successive Carnegie Junior Fellows and Interns who were always ready to undertake a great deal of fact checking and other tasks: Sahr Conway-Lanz, Glenn Hodes, Douglas Johnson, Konstantinos Karistinos, Timothy Lehmann, Jeanette Moore, Ghandi Moussa, Lanette Mumford, Aaron Sedley, Alison Smith and Ann Walsh.

Finally, I am especially grateful to two persons who labored hard to perfect the manuscript. Dana Francis, of the World Peace Foundation, assisted in many ways and meticulously made certain that any omissions were properly corrected. Whitney Watriss, my skilled editor, always cheerfully produced a solution to whatever the problem. It was a joy to work with both.

Andrew J. Pierre

ONE

Introduction

Andrew J. Pierre

ARMS KILL.

Over the past fifty years an estimated 159 wars have claimed over 25,000,000 lives.[1] The end of the Cold War has had the perverse effect of accelerating the amount of armed conflict. For decades the East-West competition served as a geopolitical straightjacket, keeping many latent conflicts in check. In the new world disorder, it is well-recognized that ethnic wars, territorial disputes, religious strife and other destabilizing conditions are on the rise. With this trend has come a world increasingly awash with arms of a *conventional* nature.

In recent years the world's attention has been concentrated on the risks of nuclear proliferation. While this problem must remain a central focus, it is important to recall that no nuclear weapon has been used in the more than half century since the close of the Second World War. And nuclear proliferation has spread much more slowly than most had predicted.

A consequence of the concentration on the nuclear issue has been to mask the risks and dangers resulting from the proliferation of conventional weapons. Within the policy and academic communities, there is an almost total preoccupation with weapons of mass destruction. Conventional arms are usually treated as a difficult, and preferably forgotten, orphan. This book is intended to help rectify this inattention.

Why Conventional Arms Are of Concern

Conventional arms proliferation requires considerably greater attention for at least a baker's dozen—thirteen—reasons.

First, the basic military forces of nations are essentially conventional, and these are the weapons used in conflict. Conventional arms remain at the

1

heart of the battlefield and military planning of nations. There is still a structure of deterrence associated with the use of nuclear, chemical and biological weapons. This was evident when Iraq proved reluctant to use most of its chemical and biological capabilities when it invaded Kuwait. But deterrence often does not exist when the use of conventional arms is contemplated. We should be acutely aware that conventional weapons do all of the killing in today's world.

Second, conventional arms are rapidly becoming far more sophisticated and lethal. Smart, computer-driven weapons have come of age. Military analysts speak of a Military-Technical Revolution that is ushering in weapons with dramatically enhanced capabilities. This is making it possible to have a high kill probability with a single salvo, to extend the range of delivery systems such as cruise missiles and conventional artillery while assuring great accuracy, to detect and identify targets deep in enemy territory through sensor systems, and to have highly intelligent command, control, communications and intelligence (C^3I) systems.

Third, as a consequence of the new military technologies, the traditional distinction between conventional weapons and those of mass destruction is eroding and becoming more artificial. A number of the new conventional weapons systems are dual use, capable of carrying chemical, biological and nuclear warheads or materials as well as classic high explosives. Indeed, a third-world atomic bomb would probably be delivered by a combat aircraft built in the developed world. Moreover, because of the enhanced power and accuracy of some of the new conventional weapons, their employment on the modern battlefield may be as devastating as tactical nuclear weapons. Hiroshima can now be achieved with *non*-nuclear arms.

Fourth, these high-technology weapons—advanced combat aircraft, electronic warfare technology, longer range and accurate missiles, smart munitions and quiet submarines—are becoming more diffused and are rapidly appearing in the arsenals of a growing number of countries. The dangers this poses are coming to be recognized not only by American military leaders, but also by those in the other major arms-supplying countries. Military professionals have warned that some of the arms supplied could boomerang and one day be aimed against the military forces of the very country that originally provided the weapons.[2]

Not only are conventional arms becoming much more widely available through the international arms market, but a number of additional countries are creating or augmenting their indigenous arms industries. More than forty-five countries already produce their own armaments. Most are seeking to export weapons to reduce their unit production costs and obtain other economic benefits. Increasingly, countries are purchasing the capability to manufacture armaments. China, for example, purchased a missile-producing factory from Russia and in turn reportedly has sold missile-producing technology to Iran and Pakistan. In its study, *Global Arms Trade: Commerce in Advanced Military Technology and Weapons,* the U.S. Office of Technology Assessment (OTA) determined that the United States and European states have in the past decade routinely transferred a great deal of advanced defense technology to less developed nations, often through complex co-production arrangements. As early as 1988, for example, India, Egypt, Indonesia, the Republic of Korea (South Korea), Taiwan and Brazil were producing forty-three major weapons under international licensing arrangements.[3] Several of these states have attained a significant military industrial capacity and are active arms exporters today.

Fifth, the weapons trade is now truly becoming an arms bazaar. More and more, advanced weapons are available to any country that can pay. They are enticingly displayed at international arms fairs from Abu Dhabi to Paris. One reason is the common, overarching problem the defense industries of all the traditional arms manufacturing countries are facing— overcapacity. In the interval since the end of the Cold War, defense budgets have decreased, and substantially less money is available for equipping the armed forces. Employment is going down, and valuable, experienced design teams are being disbanded. Defense conversion looks fine in theory but has proven impossibly difficult in practice. Labor mobility is low in most nations. Consequently, the export of arms is seen as *the* solution to the industry's problem and is being pursued fiercely. In the past some of the principal suppliers did not export their best equipment, but now they are increasingly willing to sell their top-of-the-line weapons. Moreover, they are also willing to sweeten deals by engaging, if necessary, in technology transfer agreements, offset arrangements, barters and bribes.

Sixth, the acquisition of excessive levels of conventional arms can destabilize a region. Witness the Persian Gulf: it is unlikely Iraq would have attacked Kuwait without the confidence provided—even if falsely placed— by its weapons stockpile. During the decade prior to the war, Baghdad purchased $77.5 billion (in constant 1993 dollars) worth of non-nuclear arms.[4] This massive accumulation, totally out of scale with Iraq's true security needs, encouraged Saddam Hussein in his aggression. It is noteworthy that practically all of Iraq's arms, and the technology used to manufacture weapons domestically, came from abroad.

Seventh, the purchase of arms continues to drain resources needed to ameliorate unacceptable social and economic conditions. Within the developing world, military expenditures are substantially greater than the combined public expenditures on health and education. There is a growing belief that development assistance should be conditioned upon limits to the military spending of recipient countries. Among some developed nations, the financial burden of acquiring advanced arms has increased public debt, contributing in a number of countries to a fiscal imbalance.

Eighth, in countries ruled by authoritarian regimes, there is a significant correlation between the acquisition of arms and human rights abuses. In many such countries the armed forces are used for internal policing that often results in the repression of ethnic and religious minorities. Weapons received from abroad have frequently been used to suppress dissident domestic groups and movements.

Ninth, there is a fast-growing, widespread use of small arms and light weapons in conflicts in the developing world. Weapons such as pistols, rifles, machine guns, grenade launchers, light mortar and landmines can be obtained very easily. They are the weapons used, often exclusively and in large quantities, in most ethnic conflicts and internal disputes. According to a United Nations (U.N.) study, these arms have been responsible for 90 percent of the casualties and deaths in recent wars.[5] Small weapons, which are often illicitly trafficked through black markets, present an especially difficult challenge with regard to their control or limitation.

Tenth, with the growth of terrorism, national and international, there are compelling reasons to be concerned about the easy availability of weapons in illicit world markets. To take one example, airplanes can be shot down by surface-to-air missiles that can be launched from the shoulder or from a vehicle or boat. The French Mistral, with a range of 5 miles and an altitude of 2.5 miles, is in the inventory of some thirty countries, and the equivalent Russian SA series missile is now readily found in the black market. Many of the American-made, shoulder-held Stinger missiles, which were widely dispensed by the United States to guerrillas fighting Russia in the war in Afghanistan, are now unaccounted for. U.S. Director of Intelligence John Deutch clearly viewed terrorism as a major worldwide threat.[6]

Eleventh, arming a region where there are pre-existing tensions and potential conflict, whether through the transfer of arms from outside powers or the local manufacture of weapons, may seriously exacerbate tensions and spur an arms race. Many believe that this pattern has existed in the Middle East and Persian Gulf for the past four decades. They now fear the consequences of the significant buildup in arms in parts of Asia in the 1990s. They argue that local arms races create or enhance regional instabilities, make any war that ensues more violent and destructive, and warn about the risks of the major arms-supplying nations being drawn into the conflict.

On the other side is the argument that arms transfers may deter aggression, restore a regional balance and thereby reinforce regional stability and peace. These differences in perceptions can only be resolved case-by-case. They are often at the heart of the debates about specific proposed arms sales. Nevertheless, the weight of history clearly argues in favor of arms restraint rather than weapons proliferation.

Twelfth, arms sales have the potential of creating serious political conflict. Economic competitiveness for arms sales among producing countries is to be expected and in most cases is manageable without leading to major political strains. But if disagreements about the desirability of particular arms transfers based on critical international security considerations do come into play, they can be serious. Chinese missile sales to the Democratic People's Republic of Korea (North Korea) and Pakistan have been a source of tension between Washington and Beijing. Indiscriminate Russian sales to rogue states whose international behavior or military potential are of particular concern—Iran, Iraq, Libya, North Korea and Syria—could be a still greater problem for the United States and the West. Given Russia's economic condition and the dire straits of its military-industrial complex industry—not to mention America's preponderance in the arms market—it is not at all clear that Moscow will agree to limit its sales because of arms control and non-proliferation considerations.

Thirteenth, and not least, if some type of multilateral arms restraint is to be achieved, it must be taken seriously and sought sooner rather than later. The United States, as the world's largest producer and exporter of arms, must take the initiative. It is still the case that only a very small number of countries account for an overwhelmingly large proportion of global arms sales. Six—United States, Russia/Soviet Union, France, United Kingdom, Germany and China—made 86 percent of the $190.8 billion (in constant 1995 U.S. dollars) in arms deliveries to the world in 1990–95.[7] In one of those years, 1993, the United States made 72.6 percent of the arms transfer agreements with the developing world.[8]

Creating multilateral controls or restraints on international arms sales is a world-class challenge of monumental complexity. It involves all the elements of world politics—national ambitions and insecurities, intra-state conflict resulting from ethnic and religious confrontations, regional balances, and international economic competitiveness.

Moreover, the trend line of the inherent difficulties is pointing upwards. The technology of weaponry is spreading across national frontiers because of cross-border mergers and acquisitions, cooperation in research and development, joint ventures and other business arrangements that are resulting in the globalization of defense industries. In some countries, governments are less and less able to control or strongly influence their nation's arms-producing companies. The manufacture of weapons is increasingly being undertaken through co-production or overseas co-assembly arrangements, which reduce the prime supplier's ability to control the end product. A number of developing countries—India, South Korea, and Taiwan, to give just a few examples—are upgrading their own military industries so as to make them less dependent on outside suppliers. The black market for illegal arms is growing and has characteristics similar to those of the global drug trade. What might be thought of as full-scope national controls are, therefore, becoming more illusive. More broadly, in the post-Cold War world, the economic benefits of arms sales are being given greater priority and the possible political case for avoiding them is receiving less.

For all these reasons, the passage of time does not necessarily favor the creation of effective arms transfer restraints. The need is urgent, and this international conundrum should be addressed now. Governments must be persuaded to take the policy aim of international controls over arms transfers much more seriously. This will only occur, however, if there develops a greater public understanding of its importance.

An Overview

This book is organized into five parts. Part One, containing chapters by Ian Anthony of the Stockholm International Peace Research Institute and Michael Klare of the Five College Program in Peace and World Security Studies, serves as an entry into the discussion of the international arms trade, treating the subject at a broad level. Anthony provides an overview of the weapons trade of the 1990s, underscoring the significant developments and trends of recent years. He evaluates the impact of the end of the Cold War, the Persian Gulf War and subsequent developments on the flow of arms. Current trends such as the increasing number of suppliers, the globalization of industries and the growing importance of dual-use technologies are transforming the arms trade and complicating efforts toward achieving conventional arms restraint.

While Anthony addresses the overt weapons trade, Klare's chapter covers the shadowy side of the commerce in arms, an aspect usually neglected in standard works on the subject. Klare argues that arms obtained through black market sales, secret procurement from foreign governments and the concealed deals that supply ethnic and insurgent warfare constitute a significant share of the global weapons traffic. Consequently, he believes that this subterranean trade has to be better understood to have an accurate picture of the conventional arms trade. Klare describes the character of this covert traffic and its evolution in the 1990s. He also makes specific suggestions for curbing this illicit trade.

Part Two covers the changing economics of arms production and sales. Ethan Kapstein of the University of Minnesota discusses the economic trends affecting arms supplier nations in the advanced industrialized countries, while Andrew Ross of the Naval War College addresses the emergence of arms industries in the developing world. Kapstein analyzes the economic forces creating pressures on national arms industries in the United States and Western Europe, such as rising costs and reduced domestic demand. He emphasizes the dominant position of the United States in the arms trade, following the dissolution of the Soviet Union, and argues that this market dominance will give the United States a high degree of autonomy. Ross explores the nature of arms production in the developing world and its connection to the quest of many countries for military autonomy. He discusses the economic incentives that have encouraged the growth of arms industries in some developing countries and assesses the prospects for the health of these enterprises in light of their recent commercial difficulties.

Following the macro perspective of the first two parts, the book turns in Part Three to specific examinations of the arms sales policies and practices of the four major arms-supplying nations or regions. In her treatment of the United States, Janne Nolan of the Brookings Institution traces the long tradition of "ambivalence" in its policies toward the arms trade. She finds, running through a number of past administrations, including the Clinton Administration, an enduring co-existence between policies that actively promote American exports and periodic initiatives to place restraints on the transfer of conventional weapons. Nolan discusses the lessons to be learned from the Soviet-American Conventional Arms Transfer talks of the Carter era and the later development of the Missile Technology Control Regime.

She reviews the varying domestic constituencies involved in the current American debate and proposes that limitation efforts be focused upon advanced technologies which are particularly dangerous and not yet widely dispersed.

Lawrence Freedman and Martin Navias of Kings College, London, discuss the impact of changed post-Cold War economic forces on the national arms industries of Western Europe. The authors foresee an erosion of the independence of national arms industries as stagnating domestic markets and rising production costs compel European companies to downsize and collaborate across national boundaries. Western Europe also faces, in acute form, the dilemma confronting all the major suppliers: the tension between increasing pressures to export arms and the desirability of restraining arms transfers. Decreased domestic demand motivates arms producers to search for export markets, while Western European governments' concerns over weapons proliferation have been raised as the result of their experience with Iraq. Underscoring the difficulties in creating an international arms control regime, Freedman and Navias suggest that export controls, whether global or within the European Union, threaten a domain of national sovereignty that European countries have traditionally protected.

Julian Cooper of the University of Birmingham writes on the arms sales practices of the former Soviet Union. Russia, inheritor of the lion's share of the Soviet weapons production capability, entered a period of uncertainty in the formulation of its arms sales policy following the precipitous decline in Russian arms exports in the early 1990s. Cooper underlines the potential economic importance of foreign arms sales to Russia's economy and points out the difficulties in introducing commercial practices into the government agencies overseeing the export of weapons due to the political motives that governed arms transfers in the past. Addressing the concern over the decentralization of authority in the Russian arms industry, Cooper notes that any fear of unrestrained Russian arms sales should be tempered by Russia's failure to make strong or sustainable inroads into the overseas arms market. Cooper also examines trends in the arms-exporting practices of Ukraine and the other Soviet successor states.

The arms sales practices of China have caused much anxiety because of Beijing's willingness to sell advanced weaponry to such "rogue" states as Iran, Pakistan and North Korea. Gerald Segal of the International Institute for Strategic Studies traces the Chinese government's shifting political and economic motives for arms exports and assesses the influence of the notorious Red Princes over the Chinese weapons trade. Segal largely attributes China's seemingly unrestrained transfers to its desire to use its commerce in arms as a political lever in the international community, as well as to its distrust of multilateral arms control agreements. In closing, he offers insights into dealing with Chinese sales, focusing in particular on curbing the regional arms races to which China has contributed.

Leaving the major suppliers, Part Four turns to the three largest regional markets for conventional arms—the Middle East and Persian Gulf, the Asia-Pacific area and South Asia (arms imports in Latin America and sub-Saharan Africa are comparatively small). Gerald Steinberg of the Bar-Ilan University in Israel and Abdel Monem Said Aly of the Center for Political and Strategic Studies in Cairo present contrasting Israeli and Arab perspectives on the flow of arms to the Middle East and the Persian Gulf, as well as the prospects for regional arms control. But their views differ less than might be expected. Steinberg examines the nature of demand and supply in the world's largest regional arms markets. He discusses the interconnectedness of the two sub-regional arms races, one stemming from the Arab-Israeli conflict and the other from the tensions among the oil-producing states of the Persian Gulf. Steinberg addresses the mix of political, security and economic motives that have traditionally encouraged the major arms suppliers to sell so freely and the recipients to purchase so willingly. While he harbors no illusions about the ease of establishing viable regional arms control restraints, he does point to a series of developments that suggest opportunities for such measures. These include the end of U.S.-Soviet competition in the Middle East, the progress made in the Arab-Israeli peace process, and the growing domestic economic pressures against continued high levels of national arms procurement. Steinberg also explores the possibility of an arms freeze in the Middle East and measures to limit the introduction of ballistic missiles into the region.

Said Aly emphasizes the contradictory pressures shaping conventional weapons proliferation in the Middle East. The progress made in the Arab-Israeli peace process offers some hope for arms control in the region. However, the Persian Gulf War gave rise to increased weapons procurement. While the war focused international attention on the dangers of unrestrained arms sales, it also prompted many states in the Middle East to seek the new advanced weapons systems that were estimated to have performed so successfully in the conflict. Said Aly also discusses the structural factors that have fueled the Arab-Israeli arms race for so long. He notes that the Arab quantitative advantage in military forces and the Israeli qualitative edge in weapons technology have resulted in an unstable, anxiety-ridden military balance, with each side fixating on its deficiencies. It is in the context of this complicated picture that he explores the prospects for arms control in the Middle East.

Next, Andrew Mack of the Australian National University discusses the arms trade and the opportunities for arms control in the Asia-Pacific region. In recent years, Asia-Pacific is the region where the demand for imported weapons has most accelerated. Mack describes the arms trading relationship of the Asian-Pacific nations with the three major suppliers to the region, the United States, Russia and China, and the emergence of a buyer's market, which is facilitating the acquisition of advanced military technologies. He examines the strategic implications for regional security of the type of advanced weapon systems the states of the area are purchasing. Mack concludes with an assessment of the relative potential of supply- and demand-side arms transfer restraints to curtail the flow of arms to the region.

Rodney Jones, of Policy Architects International and formerly of the U.S. Arms Control and Disarmament Agency, discusses the tense arms race in South Asia. He analyzes the trends in defense expenditures and arms imports there, paying particular attention to India and Pakistan. Both have the contesting pressures of a desire to modernize their military in the face of hostile neighbors while also wishing to reduce their military expenditures so as to allow resources to be spent on badly needed social and economic development. Jones emphasizes the potential danger of the South Asian race in conventional arms becoming linked with the development of nuclear weapons. In conclusion, he suggests various restraints on arms transfers in the hope of slowing the weapons trade in South Asia.

The book is brought to a close in Part Five with the chapters by Nicole Ball of the Overseas Development Council and Andrew J. Pierre, formerly of the Carnegie Endowment for International Peace. These chapters offer comprehensive suggestions on what might be done to manage and restrain the international flow of arms.

Ball explores conditionality and other economic instruments that the international financial community could employ to limit the demand for further arms purchases. She examines how the World Bank, the International Monetary Fund and concerned governments providing economic assistance could implement various means of leverage against arms-importing nations to encourage a reduction in their military expenditures. However, while specific conditionality—a direct linking of economic aid with reduced military spending—has drawn much attention, Ball notes that its use is problematic and is only one of the possible methods of leverage.

In the concluding chapter Pierre addresses the question of how to create an international regime for restraining conventional arms sales. He begins by reviewing past initiatives, such as the P-5 talks among the major suppliers, and evaluates current developments, such as the new Wassenaar Arrangement and the United Nations Arms Register. He notes the lack of a norm in the making of decisions on conventional weapons sales, as exists when dealing with nuclear non-proliferation. What needs to be done, he argues, is to *manage* the process of arms sales so as to avoid dangerous, destabilizing transfers to conflicted regions, as well as their other possible undesirable consequences, such as excessive spending on arms by poor countries or the placing of weapons in the hands of governments that abuse human rights. Pierre proposes a wide variety of restraint measures, including more intrusive national control regulations and legislative oversight, technology-based controls, regional arms control, and the strengthening and broadening of the Wassenaar Arrangement so as to make it an effective international regime. Beyond this, and given the increased economic pressures behind arms sales due to the post-Cold War overcapacity in arms production, he suggests a cooperative build-down of defense industries.

Notes

1. Ruth Leger Sivard, *World Military and Social Expenditures 1996* (Washington, D.C., 1996), 18–19.

2. See, for example, "Global Military Threats to the United States and Its Interests Abroad," prepared statement of Lieutenant General Patrick M. Hughes, U.S. Army, director of the Defense Intelligence Agency, before the U.S. Senate Select Committee on Intelligence, February 22, 1996, and the observations of Rear Admiral Edward Schaeffer, U.S. Navy, director of the Office of Naval Intelligence, as reported in *Arms Sales Monitor* (26) (30 July 1994), 3.

3. U.S. Congress, Office of Technology Assessment (OTA), *Global Arms Trade: Commerce in Advanced Military Technology and Weapons* (Washington, D.C., June 1991), 8.

4. Calculated from U.S. Arms Control and Disarmament Agency (ACDA), *World Military Expenditures and Arms Transfers, 1993–94* (Washington, D.C., 1995), 115; U.S. Arms Control and Disarmament Agency (ACDA), *World Military Expenditures and Arms Transfers, 1991–92* (Washington, D.C., 1993), 109. The data for the years 1980–82 were converted to constant 1993 dollars.

5. Swadesh Rana, *Small Arms and Intra State Conflict,* Research Paper No. 34, United Nations Institute for Disarmament Research (Geneva, 1995), v.

6. "Worldwide Threat Assessment Brief," statement of John Deutch, Director of Central Intelligence, to the Senate Select Committee on Intelligence (mimeo) (22 February 1996).

7. Richard F. Grimmett, *Conventional Arms Transfers to Developing Nations, 1988–1995,* Congressional Research Service, U.S. Library of Congress (Washington, D.C., 1996), 82.

8. Richard F. Grimmett, *Conventional Arms Transfers to the Third World, 1986–1993,* Congressional Research Service, U.S. Library of Congress (Washington, D.C., 1994), 19.

Part One

PATTERNS AND TRENDS IN ARMS TRANSFERS AND PROSPECTIVE DEVELOPMENTS

TWO

The Conventional Arms Trade

Ian Anthony

AS WITH almost all other aspects of international security, the new political environment created by the end of the Cold War has profoundly changed the trade in conventional arms. From the late 1940s to 1990, the global arms trade had been dominated by the two superpowers, the United States and the USSR, which used arms transfers to support wider foreign and security policies in the framework of the Cold War. To advance their positions vis-a-vis one another, both were prepared to subsidize exports of arms and to overlook their political differences with recipients, as those differences were considered to be subordinate to the Cold War competition. The United States and the Soviet Union were the two largest arms producers and exporters. As the leaders of alliances whose members possessed significant defense industries, the superpowers were also able to influence the arms transfer policies of other important suppliers.

It is difficult to exaggerate the significance for the arms trade of the collapse of the former Soviet Union and the resultant change in U.S.-Russian relations. This political change has removed the competitive ideological dimension from superpower-Third World relations. It has also reduced the willingness of either the United States or Russia to offer military assistance in the form of grants to arms clients. The resulting difficulty of financing imports of arms is likely to become a more and more significant restraint in the global arms trade.

The most important consequences of these changes for the arms trade are:

—The reduction (in many cases termination) of arms transfers and technology transfer programs between former members of the Warsaw Treaty Organization (WTO).

—The weakening of some bilateral relationships that have accounted for a major percentage of the total arms trade. Included are the relationships between the former Soviet Union and Afghanistan, Angola, Cuba, Iraq,

Libya, Nicaragua, the Democratic People's Republic of Korea (North Korea), Syria and Vietnam.

—The opening or reopening of arms transfers relations between countries, previously impossible for ideological reasons. The most important case is the re-establishment of military cooperation between Russia and China. The revision of the control lists and country lists applied by the Co-ordinating Committee on Multilateral Export Controls (COCOM) may also allow linkages to develop between the defense industries in countries that observe the COCOM embargo and former members of the WTO. Cooperation between Czechoslovakia and Israel on the development of a new jet trainer aircraft is a good example of this phenomenon.

Other elements of the new security environment will also have consequences for the international trade in arms. The most notable developments are:

—The reduction in expenditures for the procurement of domestic arms in the major arms producers—the United States, the former Soviet Union and Europe. They are also the primary exporters of arms.

—The large-scale transfers of second-hand military equipment within the members of the North Atlantic Treaty Organization (NATO) associated with the 1990 Conventional Forces in Europe (CFE) Treaty.

—Superpower sponsorship of efforts to resolve regional conflicts.

—The proportionately greater attention paid to industrial competitiveness and technology development in the United States, Europe and Asia. This factor has emerged as their perception of an overriding need to cooperate against a Soviet military threat has diminished.

These developments have made the global arms market increasingly complex, dynamic and difficult to forecast.

Recent Trends Among Major Arms Exporters

The Stockholm International Peace Research Institute (SIPRI) estimated the global value of the trade in major conventional weapons in 1995 at almost $23 billion in 1990 U.S. dollars. This figure, which refers to deliveries, is approximately the same value recorded for 1994 and suggests that the downward trend in the volume of arms transfers recorded after 1987 has stopped.[1]

The previous fall in the value of the global trade in major conventional weapons, which predated the most dramatic changes in East-West relations, can be attributed at least in part to economic factors. In some cases the factor has been a lack of foreign exchange combined with growing requirements for debt service. In some countries whose economies were highly dependent on returns from a single commodity or a small number of commodities, a fall in prices reduced the foreign exchange available for arms purchases.

Economic factors notwithstanding, the significant decline in the value of global arms transfers was largely a function of the dramatic political developments noted above. Therefore, while arms manufacturers in North America, Europe and the Commonwealth of Independent States (CIS) would certainly like to increase exports to compensate for the decline in domestic procurements, the shrinking market makes this strategy a questionable one except for a minority of producers. Current evidence suggests that U.S. producers are the best equipped to defend their position in overseas markets.

Given the decline in the value and volume of the arms trade, it is worth remembering why arms trade is important. With the exception of the United States, Russia, France, the United Kingdom and China, every country in the world depends on imported weapons to equip its armed forces. Even this characterization may be insufficiently sweeping. All countries import some weapon systems, subsystems and components, a loss of access to which would be disruptive. While it might be inaccurate to say that France and the United Kingdom depend on imported subsystems, they have not been immune from the growing interdependence of NATO and the intra-European nature of defense industrial production. As one commentator has observed:

> Technology has its own logic which is in the process of subverting the old model of the purely national, arsenal-type defence industry. The injection of market mechanisms, not only in competition for export markets but also on our home ground, has become a necessity and a possibility. Despite the political and strategic constraints which weigh on the European countries . . . the notion of a European defence industry market is no longer a fantasy.[2]

The United States

In 1995 the United States accounted for 42 percent of the total deliveries of major conventional weapons, compared with 28 percent in 1989. This dominant position resulted from the dramatic decline in the volume of arms exports from the former Soviet Union rather than from dramatic increases in U.S. sales. In 1995 Russia accounted for 18 percent of total deliveries, compared with 39 percent for the former Soviet Union in 1989.[3]

This concentration of market share with one supplier reversed the trend of the previous fifteen years, when a growing number of countries became active suppliers. The U.S. Arms Control and Disarmament Agency (ACDA) observed in 1990 that "since the late 1970s, almost all major arms have been offered by several exporters, and over two dozen countries offer munitions and simple support equipment."[4] Moreover, the extent of U.S. dominance may be greater than the statistics suggest. Major weapon systems per se are not the central factor in the military balance among states. Rather, it is the systems in combination with the required training, logistics and support. While there are several suppliers of advanced military hardware, by 1992 only the United States was able to deliver the full package, and only that package could give one side a decisive military advantage in an interstate war. Although European countries could achieve such a capability through collaboration, it would take time for the requisite degree of policy coordination to develop, if it did so at all.[5]

The argument U.S. administrations, both Democratic and Republican, have made most persistently in favor of arms sales has a military rationale: they contribute to the self-defense of a country facing an external military threat. However, the growth in the U.S. market share may strengthen the temptation to use arms transfers as a political rather than a military instrument—although the boundary between these two motivations has never been distinct, and both have always been considered during any specific case.

Where a recipient country has no realistic alternative supplier, a decision on a specific transfer may depend less on military need for the system (derived from assessing the capabilities of potential adversaries) and more on factors such as non-proliferation concerns, democratization or the treatment of minorities. This situation is perhaps most likely to arise in the Middle East, where important recipients of U.S. grant assistance are located, where Russia has severed its links with most regional countries, and where arms transfer policy clearly could be coupled to the peace process. There is also, however, evidence that U.S. policy in Pakistan is moving in this direction. While Pakistan can make a strong case for a military requirement, it is insufficient to overcome other barriers to the transfer of major weapons, especially high performance aircraft. The counterargument is that selling arms has become a commercial imperative to offset the decreased U.S. expenditure for research, development and procurement that has overtaken the military and political rationales for U.S. exports of arms.

According to official U.S. government data, the value of U.S. arms exports has increased since the end of the Cold War. The value of foreign military sales deliveries has risen from just over $7 billion in 1989 to $12 billion in 1995. The value of foreign military sales agreements, on the other hand, grew sharply in the period 1990 through 1993 but has since declined. In 1993 the Administration of President Bill Clinton reported $33 billion worth of foreign military sales agreements in fiscal year 1993, although by 1995 the total was down to $9.1 billion. Of these, slightly more than $15 billion (or 48 percent) were to the Near East and South Asia and $11.5 half billion (36 percent) to Saudi Arabia alone.[6] The extent to which a high market level is resumed will depend largely on the outcome of the debate in the United States about future procurement for domestic armed forces. Some politicians, notably in the U.S. Congress, have suggested continuing production of current generation equipment, improved through incremental changes to electronic, propulsion and weapons systems. If this becomes established policy, then logically U.S. dominance of certain important sectors of international production would be expected to increase, particularly aircraft and aerospace products, where the United States has many advanced systems at a mature stage of production.

Given that the producing companies are so far along the production curve for the F-14, F-15, F-16 and F/A-18 fighter aircraft, all of which have been exported, they could offer these aircraft at a comparatively low cost to new customers. These aircraft are more than sufficient to allow most air forces in the world to manage their allocated tasks. Moreover, international interest in an upgraded "F-15 plus," F-16 or F/A-18 Super Hornet would be very strong, if they offered new capabilities without requiring completely new logistical and maintenance systems to support them. This choice of arms would be an extremely cost-effective procurement for most foreign armed forces. The U.S. Air Force and Navy, however, are much less interested in these upgraded versions; they would prefer an entirely new generation of technology. The new generations of equipment will incorporate technologies subject to strict export control or, in some cases, prohibition. In addition, the unit procurement costs, maintenance costs, manpower and training requirements, and changes in force structure needed to operate such aircraft effectively would be a major obstacle to foreign sales. It would be many years before an F-22 fighter became available for export, if a buyer could even be found.

Developments of this kind have already occurred in the warship market: almost all the ships operated by the U.S. Navy are "unexportable," and the Navy has only reluctantly accepted more simple ships, notably the FFG-7 class, that have been widely exported.

Western Europe

Members of the European Community (EC) accounted for roughly 25 percent of the deliveries of major conventional weapons recorded in 1995—an increase from the roughly 20 percent recorded five years earlier. Within the EC in 1993–95, Germany increased its share of arms exports, while the shares of France and the United Kingdom declined. These three countries accounted for almost 85 percent of total EC exports of major conventional weapons.[7]

West European producers must deal with several problems. In general, there is both overproduction and market saturation. They tend to offer the same types of product in a shrinking market. Their patterns of exports are similar. A major share of their income comes from arms exports to the same customers. For example, Saudi Arabia and India together accounted for 55 percent of British and 21 percent of French exports of major conventional weapons from 1989 to 1993.[8] They dominate specific production segments, such as lightweight fighter aircraft, jet trainer aircraft, short-range surface-to-air missiles, light armored vehicles, diesel-powered submarines, frigates, corvettes, fast attack craft and mine warfare systems. Finally, only a relatively low level of equipment is being procured for domestic armed forces.

For all these reasons, rationalization of European production of arms appears inevitable. One logical step is to create a Europe-wide open market. Robert Verrue, director of the Directorate General responsible for the Internal Market and Industrial Affairs at the Commission of the European Communities, has accurately noted that "no single Western European government can afford to support companies producing a broad range of costly and sophisticated defence equipment."[9] Most independent observers have supported such a restructuring. Efforts at rationalization have, however, stumbled over the hurdle of governmental unwillingness to relinquish any vestige of national sovereignty.[10] The gridlock in European decision-making has meant that the restructuring taking place is not driven by a political or strategic agenda, but is being conducted by companies according to commercial and industrial logic.

A further issue is that West European companies, more than U.S. firms, will find their markets and product sectors challenged. Czechoslovak jet trainers, Russian diesel-powered submarines and Romanian wheeled armored vehicles will all compete directly with British, French, German and Italian products. U.S. companies, on the other hand, have largely vacated these production sectors. This decision in the United States to discontinue several kinds of arms production has made the U.S. market of central

Table 2-1. *Agreed Sales of Fighter Aircraft, 1987–92*

Country	1987	1988	1989	1990	1991	*1992 (Jan.–June)*	*1987–91*
United States	78	116	89	66	204	138	553
United Kingdom	n.s.	68	n.s.	30	n.s.	n.s.	98
France	n.s.	n.s.	n.s.	n.s.	n.s.	n.s.	0

Note: n.s. means no sales.

Source: SIPRI arms trade database.

importance to many West European companies and, in the future, even to some developing countries. The single largest overseas contract available to non-U.S. aircraft manufacturers in the 1990s will be for trainer aircraft for the U.S. Air Force and Navy. Few would argue that the Joint Primary Aircraft Training System (JPATS) requirement has major political or strategic implications. In addition, its value is estimated at $4 billion, of considerable economic significance to the companies that will bid for it.[11] Bidders will include FMA of Argentina, Embraer of Brazil and ENAER of Chile. Success or failure in winning this contract will go a long way to determining whether these companies have any future as producers of military aircraft.

West European companies are not competing successfully with the United States in many markets. The most severe competition for U.S. companies is usually other U.S. companies. In recent years, for example, McDonnell Douglas and General Dynamics have contested several fighter aircraft contracts. In Japan, Kuwait and the Republic of Korea (South Korea), General Dynamics' F-16 and McDonnell Douglas' F/A-18 were the final two aircraft in competition. The F-16 was successful in Japan and South Korea, the F/A-18 in Kuwait. In 1977 McDonnell Douglas' F-15 was selected over the F-16 as the basic air superiority fighter for the Japanese Air Self-Defense Force. The same two aircraft will shortly make up the bulk of the Israeli Air Force.

As table 2-1 shows, Western European aircraft have not fared well in competition with U.S. fighters. The French Mirage-2000 and the Swedish JAS-39 lost the competition in Switzerland and Finland (in both cases to the F/A-18), and South Korea and Japan rejected the Anglo-German-Italian Tornado fighter-bomber. The value of new U.K. fighter agreements in 1988 was attributable largely to the sale of forty-eight Tornadoes to Saudi Arabia. British Aerospace obtained the 1988 Memorandum of Understanding only because the U.S. Senate Foreign Relations Committee and House Foreign

Table 2-2. *Agreed Sales of Helicopters, 1987–92*

Country	1987	1988	1989	1990	1991	1992 (Jan.–June)	1987–91
United States	106	232	193	199	103	26	833
Europe							498
France	83	72	45	2	n.s.	n.s.	202
United Kingdom	n.s.	62	7	n.s.	35	n.s.	104
Italy	16	50	19	3	n.s.	n.s.	88
Germany, F.R.	80	16	2	6	n.s.	n.s.	104

Note: n.s. means no sales.

Source: Stockholm International Peace Research Institute (SIPRI) arms trade database.

Affairs Committee blocked the sale of F-15 fighter aircraft (British Aerospace did not secure a firm contract until 1992). Similar Congressional objections had permitted the United Kingdom to secure sales of seventy-two Tornadoes in 1985–86.[12]

Prior to the Gulf War, European companies were somewhat successful in selling lightweight and medium utility helicopters (table 2-2)—the great majority of these helicopter sales have involved utility helicopters. From 1987 to 1992, European companies sold about 40 percent of what U.S. companies sold (these figures, however, exclude the 175 helicopters sold overseas by Bell Helicopter Textron Canada, which is wholly owned by the U.S. corporation, Textron.)

In light of the success claimed for U.S. attack helicopters in the war against Iraq, many countries have investigated the possibility of purchasing the AH-64 Apache or the AH-1 Cobra. Thus, the U.S. companies are likely to dominate the specialized attack helicopters product sector. The number of these units sold is likely to be restricted by their cost and complexity. Western Europe currently produces only one specialized attack helicopter—the Italian A-129 Mangusta. The Franco-German company, Eurocopter, is producing another, the PAH-2 Tiger, but it will not be in production until the late 1990s at the earliest.

Former Soviet Union and Eastern Europe

The other major center of arms production and export at present is the former Soviet Union, in particular Russia, and Eastern Europe. Collectively this region accounted for over 22 percent of the deliveries of major conventional weapons estimated for 1995.[13] Of the Eastern European countries, the most important was the Slovakia. However, Poland and Romania also have the capacity to export arms produced in significant quantities. The same had been true of the former Yugoslavia. Bulgaria and Hungary have more limited arms industries.

No event could have had a more dramatic impact on the arms trade than the dissolution of the Soviet Union. For most of the 1980s, the USSR accounted for roughly 40 percent of the global trade in major conventional weapons, according to data from SIPRI. In 1995, Russia's share of 17 percent of global deliveries compares with 39 percent for the former Soviet Union in 1989.[14] It is probable the decline in other forms of Soviet arms trade—in particular transfers of subassemblies and components within the former WTO—has also been significant.

The Soviet Union had become a constructive participant in the conventional arms control process, including controls over arms transfers. The attitudes of the Russian government and industry are less clear. Statements for international consumption suggest the Russian government will be active in arms control discussions. Statements for domestic audiences by President Boris Yeltsin and by his principal advisor on the defense industry, Mikhail Malei, have underlined the need to maximize the earnings from arms exports.[15]

Exports by Russia's large arms industry are a long-term concern, even though Russia began taking steps to ensure central control over these sales in 1992. In the near term, however, it is the massive amount of equipment owned by the armed forces of the former Soviet Union that is the primary concern, given the lack of effective controls over inventory within the armed forces. Some equipment (the precise amount is not clear) is being sold without regard for its end use—including sales to criminals. This equipment is likely to cause security problems out of proportion to the financial value of the sales, which are relatively unimportant from an economic perspective.

Of the arms-producing centers in Eastern Europe, the newly independent countries of Slovakia and Ukraine are the most important. By 1987, Czechoslovakia employed more than 73,000 workers in arms production, and sales were valued at 29.3 billion crowns.[16] Seventy percent of this production was exported, 80 percent of that share to other members of the former WTO. However, after 1987 the value of arms production in Czechoslovakia declined from 29 billion crowns to 7.6 billion in 1991. This steep drop was largely the result of the collapse of the market for arms within the WTO. The value of exports from Czechoslovakia to non-WTO countries fell by less than 40 percent in the period 1987–91, compared with a reduction of 93 percent in sales to allies.

The fact that trade within the WTO was carried out on a soft currency basis has created a paradox—Czech exports are more profitable now than they were before 1989. The chairman of Aero, the manufacturer of the L-39 Albatross series of jet trainer aircraft, has said, "with half of the previous production, profit levels will be six to nine times higher."[17] However, this accomplishment depends on finding new customers. Czechoslovakia delivered more than 200 aircraft every year to its WTO allies, whereas current overseas orders stand at no more than 130, with deliveries spread over at least five years.

The former Czechoslovak government did not intend to stop all exports of arms. However, it did state that it aimed to reduce sales of specific categories of weapon—namely, tanks, infantry fighting vehicles and artillery. These systems were made under production agreements with the former Soviet Union and sold in line with foreign policy directives largely defined in Moscow. Prior to its dissolution Czechoslovakia had intended to concentrate on developing its aircraft industry (including as a subcontractor to Western prime contractors) and on selling radars, sensors and other passive reconnaissance systems. Whereas the aircraft industry is located in the Czech Republic, most of the heavy engineering plants, which would have been closed down under the previous program, are in Slovakia. The plans to reduce the production of land systems were made in Prague and are likely to be reversed by the new government in Slovakia.

Most of the non-Russian arms production capacity of the former Soviet Union was located in Ukraine, whose economy depends heavily on its machine-building and metalworking industries. However, these industries primarily manufacture subassemblies for shipment to Russia and do not have an independent capacity for system integration.[18] Not only has Ukraine lost much of its traditional market, but the nature of its industrial activity complicates the formation of new relationships. While bilateral agreements within the CIS should in theory have allowed continuity of interrepublic trade, the breakdown of the administrative system of the former Soviet Union has meant that few agreements have been implemented. Nevertheless, Ukrainian industrial and government representatives were active in 1992 in important Russian arms markets such as India. After several rounds of discussions, in October 1992 India and Ukraine concluded a trade deal, including military equipment.[19]

Polish arms producers face a similar problem with the collapse of their traditional markets—Warsaw Pact allies procured 63 percent of their exports in the 1980s. However, the size of the arms industry and its importance as an element of the Polish economy were less than in Czecho-slovakia.[20] Moreover, Poland has introduced perhaps the most restrictive export regulations of any former WTO country, embargoing sales to most of its traditional customers.[21] The Polish approach to dealing with this has been to appeal for potential arms suppliers to examine joint ventures in arms production with Poland. In this way, Poland is seeking to meet the needs of the Polish armed forces and the need for exports (within Poland's guidelines on exports).

Hungary is relatively a much less important arms producer than is either the Czech Republic and Slovakia or Poland. It has, however, taken a similar approach to dealing with the problems of overcapacity. The Hungarian government has stressed that economic reform has made foreign direct investment in the country an attractive proposition and that defense enter-prises are particularly worthy partners because of their skilled labor and experience in making high quality products. Hungary has an easier time making this case because it had, along with the former German Democratic Republic, supplied telecommunications and other electronic systems to the WTO.

Arms-Importing Countries

The dissolution of the Soviet Union also had an impact among the major importers of arms. The countries whose imports have declined the most—Afghanistan, Angola, Cuba, Czechoslovakia, Iraq, North Korea and Syria—were all major clients of the former Soviet Union through the 1980s. Only a small group of countries had both a high demand for major weapon systems and the resources to procure them. Ten countries accounted for more than 50 percent of the major conventional weapons delivered in 1995: China, Egypt, India, Indonesia, Kuwait, Malaysia, Saudi Arabia, South Korea, Taiwan and Turkey. Three of these countries—China, South Korea and Taiwan—are located in Northeast Asia, two—Kuwait and Saudi Arabia—on the Arabian Peninsula, and two—Indonesia and Malaysia—in Southeast Asia. Several of the countries can be grouped into what Barry Buzan, a British political scientist, would label "subregional security complexes": the Balkan region (Greece and Turkey); the Mid-dle East (Egypt, Israel and Saudi Arabia); Southeast Asia (Australia and Thailand); and East Asia (Taiwan and Japan). These security complexes contain many more countries than are listed here. However, grouping a significant number of major arms-importing countries in this way under-

Figure 2-1. *Changing Regional Percentage Shares of Total Deliveries of Major Conventional Weapons, 1984–95*

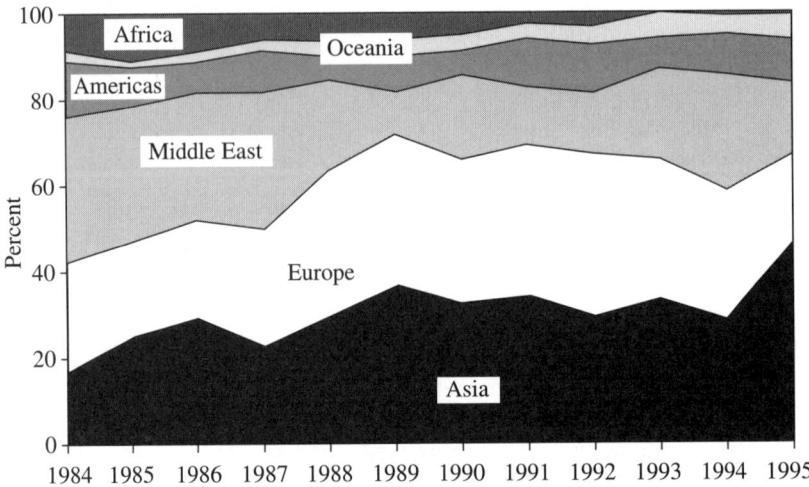

Source: Stockholm International Peace Research Institute (SIPRI) arms trade database.

lines that subregional approaches may be helpful in understanding the dynamics of arms acquisition.

Changing Regional Patterns

Figure 2-1 illustrates the changes in the regional shares for total deliveries of major conventional weapons in the decade 1984–93. The three primary markets for international arms transfers were Europe, the Middle East and Asia. Imports by African and South American countries dropped progressively over the decade, but neither of these regions, or Central America and Oceania, were more than minor importers throughout the period. The reductions in Central American and Sub-Saharan African imports are also tied to the changes in the former Soviet Union, in that the largest importers in these regions—such as Angola, Cuba and Nicaragua—tended to be Soviet clients. These countries would emerge as somewhat more important if data for transfers of small arms—man-portable systems such as assault rifles, machine guns, small caliber mortars and artillery and rocket-propelled grenades—were available, since this kind of equipment largely makes up the inventories of both the national and subnational armed forces in these regions.

North America is not a major arms-importing region. The United States meets almost all its equipment needs from local production, and the armed forces of Canada and Mexico are of limited size.

The Middle East accounted for 33 percent of imports in 1984 versus 23 percent in 1995. Asia accounted for 17 percent of deliveries in 1984 and 44 percent in 1995. Europe accounted for 24 percent in 1984 and 20 percent in 1995. In regional terms Asia and Europe replaced the Middle East as the primary market for major conventional weapons in 1988. The primary reason was the end of the 1980–88 Iraq-Iran war and subsequent reductions in imports, particularly by Iraq. The Iraq-Iran war was the most powerful force driving the trade in conventional arms in the 1980s, and the bulk of Iraqi acquisition of weapons took place in the first half of the decade, a time when the main concern of exporting countries about the region was the containment of Iran. Preventing the emergence of "another Iraq" was the primary objective of the initiatives to control transfers of arms in recent years. It is important to remember, however, that Iraq built up its military capacity in the early 1980s largely with the support of the Soviet Union, France and China, which were pursuing what they considered to be valid policy goals.

Regional governments in the Middle East continue to regard the preparedness of their own armed forces as the most important component of their national security. While this consideration is likely to lead to new orders for military equipment being placed with foreign suppliers, it is unlikely that Middle East sales will return quickly to the levels of the 1970s and 1980s. In general, the latter part of the 1980s were economically difficult throughout the region. Consequently, "Arab officers found themselves under increasing pressure to curtail their arms spending to free up funds for debt service and social investment."[22] In the post-Cold War political environment, it will be difficult for supplier states to argue the case for subsidized arms transfers, especially to oil-producing countries. Meanwhile, the Middle Eastern countries will find it more difficult to defend increased military expenditures now that Iraq is unlikely to re-emerge as a major arms importer. Much attention is now focusing on Saudi Arabian arms imports. While Saudi Arabia has long been a major arms-importing country, demography places a limit on how much equipment its armed forces can absorb.

While aggregate regional imports of arms are unlikely to grow over the medium term, spending by individual countries might. Iran has received a great deal of attention, but its future import pattern is difficult to predict. It is clear that Iran is investigating the possibility of modernizing its armed forces with foreign—probably Russian, perhaps European—weapons. However, public information about Iranian procurement is confused. What can be said with certainty is that shortly after the August 1988 ceasefire in the Iraq-Iran war, Iran began discussing a major arms transfer program with the former Soviet Union. During a visit of the Speaker of the Iranian Parliament, Ali Akbar Hashemi Rafsanjani, now President of Iran, to the Soviet Union in 1989, the two countries reached an agreement that led to the transfer of major platforms in 1990–91, including two, perhaps three, squadrons of MiG-29 fighter aircraft and several hundred T-72 tanks. Poland and Czechoslovakia supplied armored personnel carriers and other armored vehicles and artillery to Iran. This equipment was of an older generation but presumably represented a "package" of land systems intended to operate together. The Soviet Union also agreed to transfer up to three Kilo class submarines, one of which was delivered in 1992, along with training for Iranian crews and the construction of facilities in Iran to house and maintain the ships. Since 1990, reports about the extent of Iran's modernization have conflicted widely. Several reports suggested that in July 1991 Iran and the Soviet Union signed a new agreement that was said to include a large number of fighter aircraft—as many as 100 MiG-29s, 48 MiG-31s (a MiG-25 variant), A-90 Mainstay airborne warning and control system (AWACS) aircraft, SA-5 surface-to-air missile batteries and additional T-72s. Moreover, the MiG-29s and T-72s were to be produced in Iran.[23] If true, this agreement provided for the first-ever export of MiG-31 and A-90 aircraft. The reports were apparently based on three sources: a presentation of the Director of Central Intelligence, Robert Gates, before the House Armed Services Committee in March 1992; unnamed officials in Russia; and unnamed officials in Saudi Arabia. Other reports have been considerably more restrained. Some based on discussions with Iranian Air Force officers (including General Mansur Sattari, commander-in-chief of the Iranian Air Force) suggest that the number of systems transferred was no greater than that revealed in 1989.[24] What Russia also agreed to was sales of spare parts and the support needed to keep in service around 100 aircraft of Soviet origin that Iraq flew into Iran during the 1990–91 Gulf War. Iran has asked Russia for reassurance that, given the dramatic developments inside the former Soviet Union, it is still in a position to meet its commitments. The head of the Mikoyan agency, the maker of most of the systems that the Soviet Union apparently agreed in 1991 to provide, was interviewed during the first Air Show in Moscow, which had been organized to promote exports of Russian arms. He stated that no new orders had been placed for MiG fighters, including by the CIS, during 1991.[25] Meanwhile, testimony before Congress about Iranian intentions and military capabilities during a period when the United States and Saudi Arabia are discussing new arms agreements cannot be accepted entirely at face value. In short, it is too soon to say which version of Iranian procurement policy is correct.

The Asian region recorded consistent and sustained growth in arms imports between 1980 and 1989. Despite a reduction in Asian imports in 1990–91, this region may sustain the volume of arms transfers over the medium term. Several Asian countries have embarked on significant modernization programs—notably Japan, South Korea and Taiwan. China has begun to re-establish significant arms industrial links with Russia, and the modernization of the People's Liberation Army (PLA) represents a vast potential market. In Southeast Asia, Malaysia and Thailand seem on the verge of placing significant contracts for major conventional systems, perhaps including the establishment of submarine fleets for the first time. If the economic growth in the countries of East and Southeast Asia continues, these modernization programs may be sustainable, although the likelihood is not entirely clear. Establishing a submarine fleet from scratch is technically highly demanding and very expensive.

The Impact of the Gulf War: The Pressure for Greater Transparency

The full impact of the Persian Gulf War fought against Iraq between January 17 and February 28, 1991, has yet to be determined. The most immediate tangible impact was the removal of Iraq—previously the largest importer of major weapons—from the global arms market by the 1990 trade embargo. At some point, the United Nations will lift the full-scale sanctions imposed on Iraq. However, U.N. Security Council Resolution 715 of October 11, 1991, mandates that the international community should develop a mechanism for long-term monitoring of any future sales by other countries to Iraq. Iraq accepted this resolution "in principle" in February 1992. It is unlikely the Security Council would accept a lifting of the mandatory embargo against arms sales to Iraq as long as Saddam Hussein remains in power.

More intangibly—but in the long term perhaps most importantly of all—the Gulf War has changed government attitudes toward the arms trade. Suppliers have come to accept that the almost total secrecy surrounding the arms trade is counterproductive, since it assists the kinds of clandestine programs of weapons development that the inspectors of the United Nations Special Commission (UNSCOM) uncovered in Iraq. As discussed below, the impact of this development on the thinking in arms-importing countries is not yet clear.

After the Gulf War, Middle Eastern countries evaluated future equipment needs in the light of the new regional military and political environment the war produced. Initially, this evaluation led to some large contracts, but the war did not lead to any immediate, massive increase in the delivery of arms to the region. In fact, the value of the major weapons delivered to the Middle East in 1995 was around 38 percent of that recorded ten years earlier in 1986. This shift was largely a reflection of the termination of deliveries to Iraq. In the aftermath of Iraq's invasion of Kuwait, the United States in particular announced some significant arms sales agreements with several Middle Eastern states. Notable were large aircraft sales: Saudi Arabia, ninety-six F-15s; Egypt, forty-six F-16s; Israel, twenty-five F-15s and sixty F-16s; and Kuwait, forty F/A-18s. Other large transfers included Cobra attack helicopters to Israel and Bahrain and Apache attack helicopters to Egypt and the United Arab Emirates. The Patriot air defense missile system and Abrams main battle tank were each sold to several countries in the region. Whether there will be a further major increase in the flow of arms into the Middle East depends on four factors, none of which can be predicted with confidence: (1) the development of the regional security system; (2) the nature of the commitments made by extraregional powers, particularly the United States, to regional countries; (3) the outcome of the regional peace process initiated in 1991 in Madrid; and (4) the outcome of discussions about the control of arms transfers among major suppliers.

In the medium and long term, the 1990–91 Gulf War might act as a catalyst, forcing larger developing countries to rethink their attitudes toward the utility of military forces. The following evaluation of the underlying causes of the allied success in the 1991 Gulf War are widely accepted:

> The fundamental reason the ground campaign was so successful was that we knew where the Iraqi forces were, before and during the operation. The transparent nature of the opponent in this war stemmed from two things. One was the technical advantage of the United States, not only because of the reconnaissance assets, but because of the computer and communications capacities used to convert that information to real-time understanding of what existed, where and why. The other source of transparency was Iraq's having built its military from weapons and systems purchased from other nations. And those nations, for the

most part, were willing to share the specifications of hardware they had sold to the Iraqis. So Iraq was, in effect, naked. . . .[26]

Both these factors are certain to be closely studied by military establishments around the world. Logically, the first factor might be expected to boost the demand for dual-use goods, in particular, computers and telecommunications equipment. However, it is also likely to focus attention on the importance of the command and control of armed forces and their organization for military action, subjects beyond the scope of this chapter.

Confidentiality has been a condition of all arms transfer agreements, and recipients are expected to go to great lengths to prevent information about the performance of weapon systems from spreading without the approval of the supplier. Before the war against Iraq, suppliers—for fully understandable reasons—disregarded their commitments in this regard. In addition to revealing the technical characteristics of weapon systems, former WTO countries assisted with training exercises against the specific aircraft types used by Iraq. This sharing of information has already attracted comment in arms-importing countries. With reference to the current inventory of the Indian armed forces, an Indian commentator has noted that "with the dismantling of the Soviet Union much of the secrecy surrounding this equipment may have been jeopardized. During the Gulf War a large quantity of Soviet-built equipment was seized by the Allied forces, of which Pakistan was a part."[27] Israel has used arguments of this type as one reason to sustain a domestic defense industry regardless of its profitability.[28]

Different aspects of the question of confidentiality have also come up in the discussion of advance notification of arms agreements with countries in the Middle East, among the five permanent members of the Security Council, and in the discussion of the U.N. Register of Conventional Arms. In agreeing to the U.S. proposal for advance notification among the five permanent members of the Security Council, France, Russia and the United Kingdom all agreed, albeit reluctantly, to renegotiate agreements already signed in order to permit notification. China refused to agree to the principle of advance notification.

The information provided in the U.N. Register of Conventional Arms is public, unlike the P-5 process, which would have involved government-to-government exchanges. In the definition phase of the U.N. Register, the United States and the United Kingdom initially both objected to the public disclosure of the information in the Register and insisted on a broad aggregation of data, especially with respect to missiles.[29] In spite of its large defense industry, the United Kingdom in particular imports many kinds of missiles and wished to prevent the U.N. Register from being used over time to gauge the size of missile stockpiles and the pace at which they were being acquired. Paradoxically, the purpose of the Register is precisely to permit the monitoring of "excessive" stockpiles. The fact that Iraq—located in a region considered of strategic importance and by no means an international pariah before August 1990—could sustain a broad-based clandestine military program for more than 10 years focused attention on how little is known about the international arms trade.

In spite of some fine sounding statements about the value of transparency in the period during which the U.N. Register of Conventional Arms was established, few major exporting countries provide useful public data on arms exports in a manner that permits serious analysis.

Publicity about the volume of French sales to Iraq during the 1980s led the French Minister for Defense to provide the Parliamentary Commission for National Defense and the Armed Forces with an unprecedented fifty-five page report documenting French arms exports in 1991. However, the document was not made available to the public, and the Ministry of Defense did not commit itself to make such information available to the Parliament on a regular basis. In the United Kingdom not even members of the Parliament have been permitted access to detailed information about arms exports. Countries that disclose information about arms exports remain the exception and not the rule. From a national perspective, in a majority of democracies the Parliament does not exercise effective oversight over this aspect of foreign and security policy.

Several countries—Australia, Canada, Italy, Germany, Sweden, Switzerland and the United States—have produced country-by-country breakdowns of arms exports that show exactly which countries have received arms and military equipment. Only in this way is it possible for a domestic parliament to hold the government accountable for its export policy.[30] Of these countries, only Canada, Italy, Sweden, Switzerland and the United States produce reports annually. Canada's report omits sales to its largest customer, the United States, because export licenses are not required. Australia and Germany released information after a specific request by a member of Parliament (in both cases the member represented the Green Party).

Not all countries that produce export statistics explain the data fully, and in cases where explanations are given, they underline the difficulty of using the data for analysis. Some countries aggregate the figures for exports of arms and dual-use equipment; others release only an arms export figure.

Some countries release data on the value of the items delivered, others on the value of the items approved for export. Even the EC has a long way to go to achieve anything resembling transparency in the area of arms transfers.

Information on imports of arms is more available because a larger number of countries (although by no means a majority) publish annual or regular reports on the structure of their armed forces and often include the types of equipment deployed. Almost all the West European countries, Australia, Canada, India, Japan, New Zealand, South Korea and the United States produce such reports. While useful, few of these reports reveal the number of systems held or confirm their point of origin.

If the growing domestic and international pressure for greater transparency and disclosure of information on all aspects of arms trade succeeds in generating more information about the global arms market, some countries will use their continued commitment to confidentiality as a marketing device. Another effect is that countries with established defense industries will seek to preserve them, even where economic arguments might favor foreign purchases. Countries that continue to seek confidentiality may be aided by changes in the structure of the arms trade away from the transfer of major conventional weapons and toward different forms to trade.

The Changing Structure of the Arms Trade

Deliveries of major conventional weapons are of interest because they could, under certain circumstances, confer offensive capabilities to armed forces. Moreover, for practical reasons these systems are the easiest to monitor. As such, European conventional arms control has focused exclusively on major items. However, changes in the process of acquiring weapons in a geographically and politically diverse group of countries, together with the widening spectrum of technologies considered important for the effective application of military force, are likely to change the character of the arms trade. The most important of these changes are: the transnationalization of the arms industry; the emergence of new suppliers of certain kinds of products; and the growing importance of dual-use goods—especially in the field of telecommunications—to military establishments around the world.

The Transnationalization of the Arms Industry

Whereas arms production used to be a predominantly national activity, the recent trend has been toward transnational industrial organization. In the past, before transnationalization became prominent, there were multi-national arms procurement projects, carried out by teams of companies, each of which retained control over its own management, research and production activities and each of which was answerable to its own government.

Currently, the stagnating spending on arms procurement in the industrialized countries and the declining value of the international market for major conventional weapons are producing a restructuring of the arms industry toward transnationalization. Company strategies for dealing with the overall reduction in demand include (1) diversification through the development of new products or the purchase of companies or divisions whose activities are not defense-related and (2) shrinkage through the closure of production facilities and the laying off of workers. However, many companies prefer an alternative strategy of strengthening their position in the defense market at the expense of rivals, an approach that will lead to a concentration in the arms industry.

Where industry is privately owned, the main methods of concentration are mergers or the acquisition of one company by another, or the establishment of overseas subsidiaries. While cross-border mergers have been a feature in the civil sector for many years, mergers and acquisitions in the arms-producing sector are a recent development. Previously, manufacturers of weapon systems confined their international activities to cooperation, without significant foreign investment. In the past few years, in contrast, there have been important mergers and even some hostile takeovers involving large arms-producing companies. More companies have established significant shareholding in other firms, perhaps as a first move toward some more formal integration.

Arms-producing companies are also developing global supplier networks, often as a consequence of offset and counter-trade agreements with past customers. Governments in recipient countries have increasingly insisted on the establishment of close ties between local industry and companies located in the seller country as a condition of arms agreements. Even systems whose final assembly takes place in the United States or Western Europe are likely to have a growing percentage of foreign components. Moreover, company-to-company ties are not likely to impose any fixed barrier between commercial and military linkages. As an offset for the purchase of a military aircraft, for example, a company in the buyer country may be awarded a contract to supply components or major subassemblies such as wings for commercial aircraft assembled in the seller country. Countries in East and Southeast Asia in particular have sought to use their arms procurement policy in this way as an instrument of industrial policy. South Korea and Taiwan have also pursued this strategy but in addition have political and strategic motives underpinning their domestic arms production. Countries such as Indonesia, Malaysia, Singapore and Thailand cannot realistically aspire to an independent arms production capacity.

A continuing expansion of contractor-subcontractor relationships both within the industrialized world and between industrial and industrializing countries will create an extremely intricate transnational network of company-to-company ties of various kinds between arms producers. This in turn stimulates the creation of new capacities for the production of the components and subassemblies that make up major weapon systems and will increase the movement of such items.

As these kinds of ties develop, parallel mechanisms for technology transfer are likely to evolve. Government-sponsored, centrally controlled programs financed by supplier countries in support of political and strategic goals are likely to become less important. Perhaps more likely, technology transfer will occur through company-to-company contacts, perhaps even as an internal process between different elements of a multinational corporation. As yet, transnational corporations—some government-owned and some wholly in the private sector—do not dominate a global arms production system. There is, however, considerable evidence that the arms industry is restructuring broadly along the lines suggested here, although the speed of the process should not be exaggerated. The trend toward trade in semi-finished goods and technology packages—which occurred in other product sectors many years ago—will create a complex system that will be increasingly difficult to monitor and describe accurately. Equally, it presents a challenge to effective control over the proliferation of military technology.

The Emergence of New Centers of Production

In a different category from those countries that use arms procurement to stimulate company-to-company ties with industrialized countries for commercial reasons is a group of countries that continue to pursue the development of local arms industries for security-related reasons. This group includes China, Egypt, India, Iran, Israel, North Korea, South Africa, South Korea and Taiwan. These countries have moved the focus of their arms procurement policies one stage back in the cycle of military technology development. These countries import not only military equipment but also the machine tools and equipment needed to manufacture such items, and they spend time learning about the production process. While not yet able to develop advanced military systems from basic scientific research, most have a grasp of technology that permits significant arms production, including some innovation through the modification of equipment of foreign origin. Modification is a particular feature of Israeli, South African and Taiwanese arms production. Israel and South Africa have even been able to sell some modifications back to the country where the systems originated.

For the most part these countries have pursued independent arms production capacities either as part of an effort to achieve the status of a medium power or because of an overwhelming sense of insecurity. With the exception of North and South Korea, these countries share the characteristic that at some point in their independent history they experienced a sudden and unilateral cut-off of aid from a primary arms supplier. China, India, Iran, Israel, South Africa and Taiwan all suffered from an arms embargo imposed by one or more arms supplier. Egypt ended a major arms relationship itself when it expelled Soviet technicians.

Many of these countries have encountered major obstacles in their individual programs. This situation is not surprising—many of the technical problems involved in major defense programs have proved to be insoluble for almost all advanced industrial countries. Few advanced weapons will enter full-scale production in developing countries. Even if the technological problems are overcome, the absence of sustained investment through the procurement budget of domestic defense ministries at a level comparable with that made by the United States or Western European producers will be a major barrier.

The experience of a third category of new arms producers—Argentina, Brazil and Chile—illustrates that a significant level of domestic demand is a condition of arms production. These countries have to export arms to support their domestic arms industries, although they have failed to do so. Nevertheless, there is no evidence these countries are prepared to abandon their independent research and development (R&D). The larger developing countries in particular are likely to continue funding "technology demonstrator" programs whether these systems ever reach full-scale production. India, for example, has increased the funding for defense R&D from 2 percent of the total defense budget in 1982 to 4 percent of a larger budget in 1987. In contrast, the United States, France and the United Kingdom spent 12–13 percent of much larger defense budgets on military R&D. It is not difficult to see why the military technology gap between developed and developing countries is growing rapidly.[31] The situation among Arab countries is comparatively worse than that in India. There are no satisfactory data measuring the Arab defense industrial capability. Nevertheless, after surveying their scientific and industrial base to establish the potential arms production base, one Jordanian researcher has concluded that "the Arab countries are not particularly well-equipped for large-scale military industrialization. From one Arab state to another, the level of capability probably varies from low to medium."[32]

In addition to seeking access to technologies developed in North America and Europe, several of these countries are also discussing lateral cooperation among themselves. In 1956, the then-Prime Minister of Pakistan dismissed the idea of Muslim defense industrial cooperation on the basis that "zero plus zero equals zero." Today, however, the developing countries have many more possibilities for cooperation, with Israel perhaps the most sought-after partner. In addition to long-established contacts with South Africa and Taiwan, Israel has discussed cooperation with both China and India.

Dual-use Technologies

An increasing amount of technology is dual-use. That is, it has civilian utility but is also usable for military purposes. Preventing misuse by the purchaser of technologies sold for civilian applications is an arms control challenge that probably defies technical solution. While this problem is not confined to Iraq, information published by the Special Commission established by the United Nations under Resolution 687 drew attention to the role of foreign suppliers in the development of Iraqi military capabilities.

The obstacles to technical controls over such goods are many. Assuming that it were possible to devise a list of goods that should be controlled, problems of implementation and verification would remain. Effective implementation would have to be ensured without paralyzing international trade, restricting economic development by recipients or creating unfair competition among suppliers. The nature and volume of international trade disallow comprehensive monitoring of the movement of goods, and any system would have to depend on the participation of industry. Producers of tanks or fighter aircraft could be pressured into participating in a control regime, since they depend on government procurement for their existence. However, few companies engaged in producing dual-use items regard themselves as arms producers, and for many sales to defense ministries account for only a very small percentage of total sales. Most executives consider existing regulations to be an unnecessary burden. The cost and infrastructure required to carry out intrusive inspections or monitoring of the end-use of dual-use goods in the recipient country would be daunting. More importantly, most recipient countries object to such practices on political grounds.

While the United States and most West European countries prohibited sales of arms to Iraq during the 1980–88 Iraq-Iran War, no country imposed a trade embargo before the mandatory Security Council Resolution 662. Countries differed widely in their interpretations of which goods were subject to embargo. Officials at the British Department of Trade and Industry (DTI) made this observation and added that even after the Iraqi experience, the DTI had no plans to widen its definition of military equipment requiring special licensing procedures.[33] With hindsight, even the more restrictive U.S. export control list has been criticized as too permissive. According to one Commerce Department official, the renewal of U.S.-Iraqi diplomatic relations in November 1984 (suspended since 1967) provided "a strong impetus to the US business community to explore expanded trade relations with Iraq."[34] A review of the export licenses issued for sales to Iraq made by the Department of Commerce between 1985 and August 1990 showed that all licenses approved were in accord with regulations then in force.

Authorities in the United Kingdom and Germany continued to license exports of dual-use goods to Iraq until late in 1990. The United Kingdom approved its final license for the export of listed goods to Iraq August 5, 1990, three days after the Iraqi invasion of Kuwait.[35] In Germany, according to the Ministry of Economics, Trade and Industry, export licenses worth Deutschemark (DM) 136 million were granted for Iraq in 1989 and DM 30 million in 1990 under the Foreign Trade Act (Außenwirtschaftsgesetz, or AWG).[36] In the British case, the licensed items included armored vehicles, spare parts and combat support equipment.

These examples indicate how difficult it will be to develop a consensus on the need for foreign policy controls over dual-use goods, even among groups of countries with long experience with export regulation, such as COCOM. On the contrary, currently there is much greater pressure to deregulate international trade. Through the General Agreement on Trade and Tariffs, governments have stressed the need to stimulate trade in goods and services and remove the barriers to the free movement of goods. The introduction of open and general export licenses is a further symptom of the trend toward the deregulation of trade in dual-use goods.

Conclusions

Considering these developments, it is immediately clear that the global arms market is becoming increasingly complex, dynamic and difficult to forecast. The transfer of military technology cuts across global and regional security issues, international legal issues, industrial development and trade issues, and the ability of states to conduct territorial defense and power projection.

A full understanding of the global arms market and the dynamics that shape it requires multiple studies conducted from a wide range of different geographical and cultural perspectives by individuals with diverse expertise. It is now common to read that arms proliferation represents the primary security problem of the 1990s. If so, perhaps the necessary range of research skills can be mobilized and applied to the many issues and problems that fall under that very general heading.

Notes

1. Stockholm International Peace Research Institute (SIPRI), *SIPRI Yearbook 1996: Armaments, Disarmament and International Security* (New York, 1996), 465.

2. Francois Heisbourg, "Public Policy and the European Arms Market," in Pauline Creasey and Simon May (eds.), *The European Armaments Market and Procurement Cooperation* (London, 1988), 86.

3. SIPRI, *SIPRI Yearbook 1996,* 480.

4. U.S. Arms Control and Disarmament Agency (ACDA), "Diversification of Arms Sources by Third World Nations," *World Military Expenditures and Arms Transfers 1989* (Washington, D.C., 1990), 25. Before the end of the Cold War not all researchers into the trade in arms accepted that newly emerging suppliers had a significant role in the global arms market. At least one researcher argued that "despite the rising number of weapon suppliers, the dependency of less-industrialized states upon the superpowers appears to be increasing." Moreover, "through a delicate system of tacit rewards for restraint and punishment for infractions, both superpowers have been able to staunch the flow and regulate the level of armed hostilities in various regions of the world." Stephanie G. Neuman, "Arms and Superpower Influence: Lessons from Recent Wars," *Orbis,* XXX (1987), 711, 727. See also Stephanie G. Neuman, *Military Assistance in Recent Wars: The Dominance of the Superpowers,* Washington Paper 122, Center for Strategic and International Studies (Washington, 1986).

5. Edward A. Kolodziej, "Europe as a Global Power: Implications of Making and Marketing Arms in France," *Journal of International Affairs,* XLI (1988), 385–420.

6. "Foreign Military Sales, Foreign Military Construction Sales and Military Assistance Facts as of 30 September 1995," Defense Security Assistance Agency, U.S. Department of Defense (Washington, D.C., 1996), 1–15.

7. SIPRI, *SIPRI Yearbook 1996*, 511.

8. SIPRI data. European producers are now trying to establish themselves as major suppliers in East and Southeast Asia.

9. Robert Verrue, "The EC Commission as a Player in the Restructuring of the European Defence Industry," in *The Future of the European Defence Industry*, proceedings of a Business Policy Seminar held at the Centre for European Policy Studies (Brussels, 5 June 1992).

10. Michael Brzoska and Peter Lock (eds.), *Restructuring of Arms Production in Western Europe* (Oxford, 1992).

11. *Defense News* (9 December 1991). It should be noted that this contract is still relatively small in comparison with domestic expenditures. The development and procurement of the F-22 fighter are expected to cost $70–80 billion.

12. U.S. Congressional objections are described in chapter 4 of William D. Bajusz and David J. Louscher, *Arms Sales and the U.S. Economy* (Boulder, 1988), 63–70.

13. SIPRI, *SIPRI Yearbook 1996*, 465.

14. SIPRI, *SIPRI Yearbook 1994: World Armaments and Disarmament* (Oxford, 1995), 484; SIPRI, *SIPRI Yearbook 1996*, 463.

15. For two examples, see "Military Industrial Complex Keen to Export Military Hardware," *Moscow News* (12) (1992), and Yeltsin's statement to workers at the tank factory in Nizhnyi Tagil in *RFE/RL Research Report* XXV (1992).

16. All data on Czechoslovakia are from "Defence Conversion and Armament Production in the Czech and Slovak Federal Republic," a background paper from the Federal Ministry of Economy, Prague, presented to the NATO-C&EE Defence Conversion Seminar (Brussels, 20–22 May 1992).

17. *Interavia Aerospace Review*, XLVII (1992), 41.

18. *Quarterly Economic Review of the EBRD* (30 September 1992), 70.

19. *Asia-Pacific Defence Reporter* (June-July 1992), 25; *RFE/RL Research Report* I (30 October 1992), 61. The deal is to be financed in part through the barter of Indian consumer goods and in part in hard currency.

20. K. Zukrowska, "The Dilemmas of Polish Arms Industries," a background paper presented to the NATO-C&EE Defence Conversion Seminar (Brussels, 20–22 May 1992).

21. The Polish Ministry of Foreign Affairs has prepared a "negative list" of embargoed countries, which include Afghanistan, El Salvador, Iran, Iraq, Israel, Libya, Mozambique, Myanmar, Somalia, South Africa, Sudan, Syria, Taiwan, and Yugoslavia.

22. Yahya Sadowski, "Sandstorm with a Silver Lining?" *Brookings Review* X (1992), 8.

23. R. Mackenzie, "Iran Resurgent," *Air Force Magazine*, LXXV (1992), 78–81; Kenneth R. Timmerman, "Weapons of Mass Destruction: The Cases of Iran, Syria and Libya," paper presented at Simon Wiesenthal Center (Israel, 1992).

24. *Military Technology* (March 1992), 88; *Middle East International* (3 April 1992).

25. Rostislav Belyakov, Mikoyan Design Bureau, in *The Independent* (12 August 1992), 8.

26. James R. Blaker, "Now What, Navy?" *Proceedings of the US Naval Institute* (May 1992), 60.

27. M. Satish, "Defence Spending: Why It Must Be Increased," *The Economic Times* (New Delhi) (11 April 1992).

28. Aharon Klieman and Reuven Pedatzur, *Rearming Israel: Defense Procurement in the 1990s,* Jaffe Centre for Strategic Studies Study, XVII (Jerusalem, 1991), 23–50.

29. The process of defining the U.N. Register of Conventional Arms is described in Appendix 10F of SIPRI, *SIPRI Yearbook 1993,* 533–544.

30. Ideally, a complete list of export licenses should be provided, stating the recipient and the item licensed, as well as its value.

31. Ian Anthony, *The Arms Trade and Medium Powers: Case Studies of India and Pakistan 1947–90* (Hemel Hempstead, 1992), 122–123.

32. Yezid Sayigh, *Arab Military Industry: Capability, Performance and Impact* (Cairo, 1992), 196. Sayigh uses a model developed by Herbert Wulf in Nicole Ball and Milton Leitenberg (eds.), *The Structure of the Defense Industry* (London, 1983).

33. *The Independent* (3 April 1991).

34. U.S. export control practice vis-à-vis Iraq in the 1980s is described in the statement of Dennis Kloske, then under secretary of commerce for export administration, before the Subcommittee on International Economic Policy and Trade, Committee on Foreign Affairs, U.S. House of Representatives, 9 April 1991.

35. A list of all products licensed for export to Iraq from 1 January 1987 to 5 August 1990 is contained in Annex E of *Exports to Iraq,* Memoranda of Evidence for the Trade and Industry Committee, House of Commons (London, 17 July 1991).

36. Bundesministerium für Wirtschaft, "Erläuterungen zur statistischen Übersicht über die erteilten Ausfuhrgenehmigung," document no. BMWi V B 4–06 31 02, mimeograph. See also Ian Anthony (ed.), *Arms Export Regulations* (Oxford, 1991).

THREE

The Subterranean Arms Trade: Black-Market Sales, Covert Operations and Ethnic Warfare

Michael T. Klare

BY ALL ACCOUNTS, the legal, state-sanctioned trade in conventional arms has declined significantly since the late 1980s. Virtually all assessments, however, exclude illicit, black-market transfers of arms, which appear to be flourishing in many areas of the world.[1] Even though the black-market sales tend to be small *in dollar terms* compared with the legal sales, they play a critical role in sustaining the ethnic, separatist and insurgent conflicts that have proliferated in the post-Cold War era.

Individually speaking, most black-market enterprises are small operations composed of a few traffickers who deal in relatively small amounts of guns and ammunition. Collectively, however, these individual dealers constitute a global network of suppliers capable of providing belligerents with vast quantities of basic combat hardware. The imprint of this network can be found in such places as Cyprus, Mogadishu, Peshawar and Zagreb, where weapons of virtually every type and nationality are available for purchase at the right price.[2] Typically, these weapons have traveled a long and circuitous route to their destination—often having been used in a number of conflicts previously. While a certain percentage of these arms may be captured or destroyed along the way, there appear to be sufficient stocks of black-market weapons to satisfy even the most acquisitive buyers.

The term "black-market arms trade" actually encompasses a wide range of illicit and semi-legal activities. In its most pristine form, this trade entails the clandestine sale of weapons by private dealers who operate secretly and in violation of a country's export laws and regulations; if caught, they are likely to face stiff fines and/or prison sentences. The black-market trade also encompasses a variety of transactions that fall into a gray zone between fully legal and illegal sales. Such gray-market transfers, as they are called, can include clandestine deliveries by government agents to insurgent groups in another country or to "pariah" regimes that are subject to an international arms embargo. Also falling in the gray category are sales of military technology and dual-use equipment to "front" companies set up by the procurement officials of the pariah countries to acquire vital materials for secret projects to develop weapons of mass destruction and other advanced arms. Even when conducted in full compliance with the home country's export rules, such sales violate the intent of the various nonproliferation regimes.

This chapter looks at both black- and gray-market transactions but uses the generic term black-market arms trade to describe both. (The distinction between "black" and "gray" sales is discussed further below.) Because these transactions are illicit and/or hidden, black-market sales do not normally figure in the statistics published by ACDA, the Congressional Research Service and SIPRI, so that it is impossible to derive a total dollar figure on the subterranean arms traffic. Further complicating the task of monitoring black-market trade is that many transactions involve small arms, ammunition and spare parts—items that are relatively easy to conceal in comparison with tanks and jet planes, and are rather difficult to categorize and quantify. Nevertheless, it is possible to construct a rough picture of the underground arms trade from news reports on intercepted arms shipments and from judicial proceedings against apprehended arms smugglers.[3]

There has long been a black market of some sort in military hardware, but it was only in the 1980s that these sales accounted for a significant segment of the global arms traffic. As a result of the U.S.-imposed arms embargo on Iran (Operation Staunch) and the incessant sectarian warfare in Lebanon, black-market traffickers began to conduct multi-million dollar sales of arms and ammunition. To circumvent Operation Staunch and acquire spare parts for their U.S.-supplied weapons, the Iranians established a global network of underground arms smugglers—some of whom had direct access to U.S. military stockpiles.[4] The government of South Africa also came to rely on black-market sources for arms and equipment not available through legal channels because of a United Nations (U.N.) arms embargo imposed in 1977. During this period, moreover, the Reagan Administration tapped into black-market channels to transfer millions of dollars' worth of weapons to the Iranian government (as part of efforts to secure the release of American hostages held by Iranian-backed factions in Lebanon), and to aid anti-communist insurgents in Afghanistan, Angola, Cambodia and Nicaragua.[5]

The black-market trade received a major boost in the late 1980s when a number of emerging Third World powers—notably Egypt, Iran, Iraq, Libya, Pakistan and Syria—sought access to advanced Western technologies for use in the development of nuclear weapons, chemical munitions and ballistic missiles. Denied access to such technologies through state-sanctioned channels, officials in these countries have relied on corrupt businesses and black-market traffickers for the desired plans and materials.[6] Such methods were used with particular success by Iraq, which established an elaborate network of "front" companies in Europe and secured key technologies and components for its weapons programs from well-known firms in Germany, Britain, Switzerland and the United States.[7]

More recently, black-market traffickers have benefitted enormously from the outbreak of separatist and factional fighting in the states carved out of Yugoslavia and the former Soviet Union. Unable to obtain arms through legal channels, the insurgent and separatist armies in these countries have raided government arsenals and scoured the world for black-market weapons.[8] An impression of this underground trade is provided by news reports of illegal arms shipments to Bosnia, Croatia and Serbia in violation of a U.N. arms embargo imposed on the former Yugoslavia in September 1991.[9] In November 1991, for instance, a plane carrying eleven tons of rifles, mortars and bazookas from military arsenals in Chile was forced down in Budapest after failing to land in Croatia.[10] Similarly, in September 1992, an Iranian aircraft carrying 4,000 guns and 7 million rounds of ammunition—presumably intended for government forces in Bosnia—was intercepted at the Zagreb airport.[11] In January 1993, the North Atlantic Treaty Organization (NATO) vessels enforcing the U.N. arms embargo intercepted a cargo ship in the Adriatic carrying mortars, anti-tank rockets, rifle ammunition and Toyota jeeps believed destined for Bosnian government forces.[12]

Black-market channels of this type have also been a major source of arms for the belligerents in Somalia and Rwanda. After Somali dictator Mohammed Siad Barre fled Mogadishu in January 1991, the various clan-based militias that had cooperated in driving him from power began to fight among themselves over control of the country—killing thousands of civilians in the process and severely curtailing the production of foodstuffs. Although U.S. and Pakistani peacekeeping forces attempted in 1992–95 to stem the fighting and disarm the militias, these factions continued their internecine warfare—and launched occasional attacks on U.N. peace-keepers—using weapons procured from black-market suppliers in Africa and the Middle East.[13] In Rwanda, the Hutu-dominated government was able to procure arms on the open market until its defeat in 1994, while the rebel Rwandan Patriotic Front (RPF) was largely dependent on black-market sources for its weaponry.[14] Later, when the RPF assumed control of the country, remnants of the former Rwandan Army in Zaire themselves turned to black-market sources as they regrouped in preparation for re-newed fighting inside Rwanda.[15]

A similar pattern of illicit and underground sales can be detected in such contemporary conflict zones as Afghanistan, Angola, Cambodia, Colombia, Ethiopia, Kashmir, Kurdistan, Lebanon, Liberia, Myanmar, Peru, Sri Lanka and Sudan. Although the belligerents in these sectarian conflicts rarely possess tanks and other heavy equipment, they have succeeded in acquiring large stockpiles of small arms and light artillery, all of which have gen-erated high levels of death and destruction. With the growing frequency and ferocity of ethnic conflicts, black-market transactions are becoming an increasingly significant factor in the global arms traffic.

An alarming new development is emerging in these transactions. In-creasingly, black-market channels are being used to transfer heavier and more sophisticated weapons, including guided missiles. Of particular con-cern is the diffusion through black-market channels of shoulder-fired surface-to-air missiles (SAMs) such as the U.S. Stinger, the British Blow-pipe and the Soviet SA-7. These weapons, which the two superpowers supplied to their allies and clients (including friendly insurgent forces) during the 1980s, are now showing up in increasing numbers on the global black market.[16] The growing diffusion of such weapons has increased the intensity of ethnic and insurgent conflicts and complicated the task of international peacekeeping—a trend painfully evident in Bosnia, where shoulder-fired SAMs of these types have been launched at aircraft par-ticipating in U.N. relief operations.[17]

Black-market trafficking has also become a critical factor in the pro-liferation of military technology. As the traditional powers of the North intensify their efforts to stem the flow of sensitive nuclear, chemical and missile technologies to emerging powers in the South, nations seeking to circumvent such restrictions will increasingly rely on black-market sources.[18] Black- and gray-market channels are also a major source of technology for the development of tightly controlled conventional weapons, such as cluster bombs and cruise missiles, and of advanced production systems used in the manufacture of such systems. Increasingly, therefore, non-proliferation endeavors will have to focus on efforts to locate, identify and suppress clandestine sources of sensitive technology.

Definitions and Parameters

The growing importance of black- and gray-market transactions in global arms traffic makes it imperative that more be learned about this subterranean trade. However, a striking dearth of statistics and scholarly materials makes investigation very difficult: in comparison with the many well-researched studies on the licit arms trade, there are very few scholarly sources on the illegal arms traffic and no reliable statistics whatsoever. Researchers must rely on many fragmentary materials—court records, reports of thefts and intercepted arms shipments and anecdotal evidence.[19]

Black Versus Gray Sales

The term "black-market arms trade" is normally used to describe a wide variety of activities, ranging from patently illegal transfers of arms and ammunition to outlawed forces and regimes to more ambiguous transactions of varying degrees of legality, involving transfers of such items as missile and aircraft components, specialized tools and equipment, nuclear and chemical materials, and weapons plans and blueprints. As noted, transfers that are wholly illegal—that is, shipments conducted in knowing violation of a nation's statutes that are likely to result in fines and/or imprisonment for anyone caught—are usually described as "black" sales, while sales that violate national and international norms and policies, if not laws, are usually termed "gray" sales.[20]

Gray sales can encompass a wide variety of semi-legal and questionable activities. Typically, such transactions involve the knowing sale of dual-use items—that is, equipment that can used for both military or civilian purposes—to front companies operated by or for countries that are barred from receiving military (but not civilian) equipment. The sale of dual-use items to such countries for strictly civilian purposes normally is not illegal, as long as the supplier is confident that the items in question will not be diverted to military use. When the supplier suspects or has reason to believe that a particular shipment will be diverted, and goes ahead with the deal, the delivery is considered gray market. Much evidence of such trafficking has been uncovered in Iraq, where the U.N. Special Commission established by Security Council Resolution 687 has been investigating the links between Western companies and the Iraqi government's nuclear, chemical and missile programs.[21]

Gray sales can also entail arms deliveries by governments in violation of their own or international rules. Transactions of this type usually entail efforts by governments to arm or assist a pariah state that is subject to an international embargo—as with covert Israeli military sales to South Africa in the 1980s[22]—or to provide covert assistance to insurgent forces in another country—as when the Reagan Administration organized covert aid to the Nicaraguan contras. There is also evidence of hybrid cases wherein private firms sell dual-use equipment to pariah states with the expectation that government officials in their own country will overlook the risk of military diversion in accordance with certain foreign policy objectives, as, for example, when U.S. firms sold military hardware to South Africa during the Nixon era in the belief that the State Department tacitly approved of such deliveries.[23]

Scope and Magnitude

As noted, there are no statistics or reference works on the scope and magnitude of black-market arms trafficking. The only official documents that provide any concrete data are a report on illegal U.S. exports of firearms prepared by the Bureau of Alcohol, Tobacco and Firearms (BATF) of the U.S. Treasury Department,[24] and a listing of all major cases involving violations of U.S. export laws that have been prosecuted by the U.S. Department of Justice (DOJ) during the fifteen years ending in May 1995.[25] The DOJ document identifies 319 cases, of which 197 involve sales of arms or military technology to countries prohibited from receiving such items (the other 122 entail transfers of high-technology gear to the USSR and Eastern Europe, plus the sale of non-military products to Cuba, Libya and other countries subject to trade embargoes).

The DOJ document does not provide a dollar value for the arms and equipment involved in these cases, nor does it indicate what percentage of this hardware was actually delivered to the intended recipient. It does, however, provide a sweeping portrait of America's black-market arms business. Here, for example, are five characteristic entries from 1990:

—*U.S. v. Rudy Y. Tsai, et al.:* Unlicensed export of advanced Sidewinder missile components to Taiwan.

—*U.S. v. Jeremy Griffin, et al.:* Conspiracy and export of military aircraft and helicopter replacement parts to Iran.

—*U.S. v. Reginald Van Rossum, et al.:* Conspiracy to export a large quantity of components for missile guidance systems to South Africa.

—*U.S. v. Luis Arcila Giraldo, et al.:* Conspiracy to purchase and export 120 Stinger missiles to drug cartel in Colombia for use against government helicopters.

—*U.S. v. Ali A. Daghir, et al.:* Conspiracy and export of missile warhead detonation capacitors to Iraq.[26]

These and other such entries provide a rough indication of the types of commodities that move through black-market channels and of typical recipients. Of the 197 cases involving arms and military equipment, the largest number, equal to approximately 50 percent, entailed sales (or attempted sales) of missile and aircraft components to Iran, Iraq, Libya and South Africa. The next largest group, equal to approximately 25 percent, entailed sales of small arms and ammunition to drug traffickers and insurgents in Latin America, to the Irish Republican Army (IRA) in Northern Ireland, and to insurgent and separatist forces elsewhere. Other recipients included China, the Democratic People's Republic of Korea (North Korea), Hong Kong (presumably a transit point for illicit sales to China and other forbidden destinations) and Chile.[27]

The BATF report covers the diversion of firearms purchased at commercial gun shops in the United States to illicit foreign destinations. Although concerned solely with exports of small arms, the report indicates that such arms are being smuggled out of the United States in large numbers and that many are winding up in the hands of Latin American drug traffickers and other felons. The report provides extensive data on the techniques used by gun smugglers to transport the contraband out of the United States.[28]

The most common method of gun smuggling from the United States, according to BATF investigators, is through a "straw purchase" from a licensed arms dealer, usually in California, Texas or Florida (whose gun ownership laws are among the country's most lenient). "With this method," the BATF noted, "the smuggler or international arms trafficker pays a relative or associate [with valid U.S. identification] to purchase the firearm legally from a licensed dealer and then deliver the firearm for a usually small fee to the trafficker," who then arranges for its transport abroad.[29] Such activities have resulted in the illegal delivery of thousands of firearms from the United States to drug traffickers and insurgents in Mexico, Colombia and other Latin American countries.[30] One such gun, traced to a gun store in the Miami area, was used in the 1990 assassination of Bernardo Jaramillo-Ossa, a presidential candidate in Colombia;[31] another, traced to a gun store in California, was used in the 1994 assassination of Luis Donaldo Colosio, the leading presidential candidate in Mexico.[32]

The BATF and DOJ documents do not say anything about black-market smuggling from other countries, nor do they provide any information on gray- or black-market sales by agencies of the U.S. government. It is, however, possible to extrapolate from these documents and other fragmentary data to construct a rough picture of the underground trade in arms and technology. It appears that the black-market commerce in arms consists of three tiers, or export channels: first, small arms and weapons components transferred by individuals and small dealers operating wholly outside of the law; second, sales of arms and technology by officials of legitimate defense and industrial firms operating in the gray zone between legal and illegal exports; and third, clandestine arms deliveries by government agencies that seek to funnel arms and military equipment to pariah regimes or insurgent groups abroad. Although there is much overlap among the three tiers, nevertheless they constitute separate distribution systems for black-market armaments.

The Justice Department document and other published sources also provide information on the intended recipients of black-market arms and technology. It appears that the principal recipients are the pariah states of Iran, Iraq, Libya, North Korea and South Africa (while under apartheid), along with separatist and insurgent forces in such areas as Afghanistan, Angola, Bosnia, Cambodia, Colombia, Kashmir, Kurdistan, Lebanon, Liberia, Myanmar, Northern Ireland, Rwanda, Somalia, Sri Lanka, Sudan and the southern republics of the former Soviet Union. In addition, there is an underground flow of sensitive technologies and specialized materials— including materials for the production of nuclear, chemical and missile systems—to the pariah states and to such emerging powers as China, Egypt, Israel, Pakistan, Syria and Taiwan.

For obvious reasons, it is impossible to put a total dollar value on these transactions. Small arms and spare parts—the principal commodities carried through illicit arms channels—are far less costly than the tanks, planes and warships carried through licit channels. Not surprisingly, then, the typical value of black-market transactions (assuming such a figure were available) would certainly be much lower than that of most government-to-government sales. However, the roster of intercepted black-market deals includes many in the tens of millions of dollars, and some even higher.

Covert U.S. aid to the Afghan *mujahideen,* for instance, approached $1 billion a year in the late 1980s, and *The Economist* estimated in 1994 that the factions in Bosnia had procured $2 billion worth of arms and ammunition through black-market channels in the preceding year.[33]

Sources of Supply

Arms and technology are funnelled into black-market channels from a variety of sources. Many items are simply stolen from government arsenals, usually with the connivance of security guards and stock clerks—notoriously underpaid people—who have been bribed to look the other way or to falsify inventory records at the behest of black-market traffickers. In the mid-1980s, for instance, agents for Iran's clandestine arms procurement effort were able to obtain critical F-14 aircraft parts by bribing well-placed non-commissioned officers and civilian clerks at the U.S. Navy supply depot in San Diego.[34] More recently, the U.S. General Accounting Office (GAO) disclosed that undercover agents of the Federal Bureau of Investigation (FBI) of the U.S. Department of Justice had purchased some 7,422 items of military equipment, worth an estimated $13.8 million, from security guards and other personnel at U.S. military installations in Utah. Included in this contraband were weapons parts, night-vision devices, chemical warfare gear and three F-16 jet engines.[35] The GAO report blamed inadequate security measures for these thefts and called for a systematic strengthening of inventory controls at U.S. military facilities.[36]

Theft of government property is also a major source of black-market weaponry in Eastern Europe and the former Soviet Union. The international press is full of reports of military items being offered for sale by former members of the Soviet and Warsaw Pact forces.[37] In February 1992, for instance, arms dealers in Czechoslovakia reported that departing Russian military officers had offered to sell them everything from AK-47 rifles to tanks and combat helicopters.[38] Similarly, the armed militias fighting for control of South Ossetia in Georgia were reportedly using weapons purchased from the Russian troops that had been sent into the region as peacekeeping forces.[39] In 1991, moreover, U.S. authorities arrested a California businessman and six high-ranking Polish officials, including the former deputy chief of staff of the Polish Army, in a plot to sell an estimated $100 million worth of Soviet and Warsaw Pact munitions to Iraq.[40]

Another major source of black-market arms are legitimate defense contractors that knowingly or unknowingly sell military hardware to front companies or dummy firms in friendly countries that have been set up by corrupt merchants or the pariah states themselves to circumvent arms embargoes and non-proliferation measures. In one notable case from the mid-1980s, Hughes Helicopters sold eighty-seven Model-500 helicopters to a trading company in the Federal Republic of Germany, Delta-Avia Fluggerate, which in turn transferred the helicopters to North Korea. Because officials at Hughes were not aware of Delta-Avia's Korean connection, they were not charged with any wrongdoing. However, the two American owners of Delta-Avia, Ronald and Barry Semler, were subsequently found guilty of masterminding the whole operation.[41] More recently, officials of a Miami-based concern, AeroSystems, Inc., were indicted for smuggling large quantities of missile and aircraft parts to Iran via dummy firms in Singapore.[42]

Newly privatized or deregulated arms firms in Eastern Europe and the former Soviet Union have also contributed to the global supply of black-market armaments. Desperate for hard currency, many of these firms are selling surplus equipment to shady dealers without their governments' approval or, perhaps, with their tacit consent. In April 1992, for instance, customs officials in Estonia and Finland intercepted a shipment of 40,000 Makarov automatic pistols, the standard sidearm of Soviet forces, from a factory in Russia to a front company in England, Kennford Enterprises, which presumably planned to resell the pistols to black-market buyers elsewhere.[43]

War booty is another major source of black-market arms. Israel, for instance, contributed to the contra supply effort in Nicaragua by providing a boatload of Soviet weapons captured from Syria and the Palestine Liberation Organization (PLO) in the 1982 war in Lebanon.[44] Similarly, the United States and Saudi Arabia provided the Afghan mujahideen with millions of dollars' worth of Iraqi weapons captured during the 1991 Persian Gulf conflict.[45] Ironically, the contras and the mujahideen have themselves been accused of diverting some of their weapons to black-market dealers—the contras to drug traffickers in Colombia,[46] the mujahideen to Islamic insurgents in Kashmir.[47]

Dealers, Channels and Methods

As stated, the underground arms trade consists of three tiers, or channels, of delivery: individuals and small firms that knowingly violate export regulations by selling arms to illicit buyers; officials of legitimate defense firms that try to conceal illicit sales among their licit operations; and government agencies that supply arms to pariah regimes and insurgent organizations.

By far the largest number of individuals cited in the Justice Department's roster of significant arms export cases fall into the first category. Typically, such individuals seek to smuggle modest quantities of firearms or weapons parts to confederates or buyers in other countries where those products are in high demand. Such contraband is usually disguised as non-military cargo and shipped by boat, truck or cargo plane or, in the case of electronic devices and small missile and aircraft parts, carried as hand luggage by passengers flying on regular commercial flights.[48] Three characteristic examples of these operations from 1990 are:

—*U.S. v. Antonio Paredes, et al.:* Conspiracy to export 350 handguns and a large quantity of ammunition to the Philippines.

—*U.S. v. Francisco Estreda, et al.:* Large-scale arms smuggling operation involving shipment of arms to Ecuador in 1989–90, with sixty shotguns seized in Miami.

—*U.S. v. Saul Lieberman, et al.:* Attempted illegal export of a $500,000 portable Elint (electronic intelligence) system designed to detect and track military aircraft.[49]

Normally, the perpetrators of such endeavors are professional smugglers or unscrupulous businessmen, and the motive is profit. However, a significant group in this category consists of immigrants, or descendants of immigrants, from another country who seek to assist government, or anti-government, forces in their native country for essentially altruistic reasons. Members of this group include, for instance, Irish-Americans who support the IRA in Northern Ireland, Iranian-Americans who sought to assist their brethren in Iran during the Iran-Iraq war, and Yugoslav-Americans who have attempted to aid Croatian, Slovenian or Serbian forces in their homeland. The DOJ document lists many such cases, of which the following are illustrative:

—*U.S. v. Kevin McKinley, et al.:* Conspiracy to purchase a Stinger missile, 50-caliber machine guns, explosives and other military equipment for IRA associates in Northern Ireland in 1989–90.

—*U.S. v. Ivan Kapetanovic, et al.:* Conspiracy to purchase and export a large quantity of anti-tank missiles, mortars, machine guns, M-16 rifles, ammunition and night-vision scopes to Yugoslavia in 1991.[50]

Representatives of both groups—profit-minded smugglers and immigrants who support armed groups in their native countries—also figure prominently in black-market operations outside the United States, although information on such trafficking is relatively scarce. There are periodic reports of illegal arms operations in the European press, and similar reports occasionally surface in other areas. In the early 1990s, there was a spate of such reports from Europe and the Middle East, as shady dealers attempted to benefit from the burgeoning demand for black-market arms in the embattled remnants of Yugoslavia. In November 1991, for instance, the Italian police arrested eight people in Venice on charges of smuggling $5 million worth of firearms, mortars and ammunition to Croatia by boat;[51] two months later, Italian investigators smashed another smuggling ring that had been shipping arms to Croatia.[52] German customs officials have intercepted a number of illicit arms shipments destined for Slovenia and Croatia, several of which involved ex-Warsaw Pact weapons that had been bought up by Western European dealers and sold to agents for the various forces in the former Yugoslavia.[53]

The second category of supplier—corrupt officials of legitimate defense firms—is less numerous than the first (small-time dealers), described above, but typically operates on a much larger scale. These dealers are adept at hiding illicit transactions among a much larger number of legal sales, a technique that can fool customs inspectors, who have no reason to suspect that the *nth* transaction is any different from its perfectly legitimate predecessors. This approach is particularly suited to illicit transfers of spare parts and electronic systems, which normally receive less attention from customs inspectors than do sales of finished weapons systems. Dealers of this type are also skilled at using subsidiaries or front companies in third countries to evade restrictions on the direct sale of weaponry to proscribed recipients, such as Iraq or Libya. In these ways, such dealers often succeed in exporting large quantities of parts and equipment to illicit destinations over an extended period of time before they are caught—*if* they are caught.

A noted example of a major black-market dealer is James H. Guerin, the former owner of International Signal and Control Group (ISC), a Pennsylvania-based electronics firm. Long active in the military field, ISC was reported to have smuggled some $30 million worth of munitions and electronic equipment into South Africa between 1978 and 1989. Among the items the Justice Department cited in its 1991 indictment of Guerin and his associates are night-vision devices, missile components, navigation systems, missile tracking systems and grenade technology. At least some of this technology, moreover, was later transferred to Iraq and used in weapons employed by the Iraqi forces during the Persian Gulf War.[54]

Another major dealer in this category is Carlos Cardoen of Chile, a metallurgist with a doctorate from the University of Utah. During the Pinochet era, Cardoen built up a large industrial empire in Chile, with subsidiaries in Spain, South Africa, Greece and the United States. His parent firm, Industrias Cardoen S.A., sold thousands of cluster bombs to Iraq in the 1980s and is suspected of selling U.S. weapons technology to Iraq and South Africa—possibly in collusion with ISC.[55] In 1992, Cardoen was charged by U.S. officials with shipping 100 tons of weapons-grade zirconium to Chile for use in the manufacture of cluster bombs sold to Iraq.[56]

The third tier of suppliers—governments—includes a number of Western states, including France, Israel and the United States, that secretly provided military assistance to the apartheid regime in South Africa in violation of the U.N. arms embargo. During the Nixon Administration, for instance, it was the explicit but undeclared policy of the United States to provide South Africa with military technology and dual-use items that could be disguised as civilian products.[57] Many other examples of such trafficking occurred during the Iran-Iraq war, when officials of a number of European countries, including Austria, France and Sweden, authorized clandestine arms sales to Iran in violation of Operation Staunch.[58]

Equally common is the covert delivery of arms by government agencies to friendly insurgent forces in hostile countries, often by using black-market dealers as intermediaries (to disguise the origins of the weapons involved). The United States and the Soviet Union regularly employed such tactics during the Cold War in an effort to assist in the overthrow of Third World governments allied to their respective superpower adversary. In Angola, for instance, the Soviet Union aided the MPLA in its independence drive against Portugal; later, when the MPLA came to power with Cuban and Soviet assistance, the United States aided the guerrilla forces of UNITA in its drive to oust the MPLA. To support such operations, both superpowers developed a global network of underground supply channels to funnel arms and equipment to insurgent organizations.[59]

For the United States, covert activities of this sort achieved their greatest elaboration during President Ronald Reagan's second term, when aid to anti-Communist insurgents in the Third World became a primary focus of U.S. foreign policy. Sensing the time was ripe to reverse the "correlation of forces" that had earlier favored Soviet efforts in the Third World, Reagan and his associates launched a massive program of covert aid to underground forces in Afghanistan, Angola, Cambodia, Chad, Libya and Nicaragua—an effort known at the time as the Reagan Doctrine.[60] To aid the mujahideen in Afghanistan, for instance, U.S. and Pakistani agents established a secret supply line to guerrilla camps in Pakistan's remote border region, where rebel forces divided up several billion dollars' worth of arms and equipment.[61] Similarly, to aid the Nicaraguan contras, U.S. agents—led by Lieutenant Colonel Oliver North of the U.S. National Security Council (NSC) staff—created a clandestine supply network in Central America and secretly procured arms from China, El Salvador, Guatemala, Israel and elsewhere, using funds generated by the covert sale of missiles to Iran.[62]

By the end of 1990, most of the underground arms-supply systems set up by Moscow and Washington to aid insurgent forces in the Third World had been shut down, although the mujahideen were reported to be receiving arms from U.S.-backed sources as late as the summer of 1991.[63] Many other countries, however, are reportedly engaged in comparable operations of their own, for roughly similar reasons. The Pakistani government, for instance, is reportedly aiding pro-Pakistani insurgents in Indian-controlled Kashmir,[64] while Iran has been charged with aiding Kurdish guerrillas in Turkey,[65] and the African state of Burkina Faso has been accused of supplying arms to the rebel forces of Charles Taylor in Liberia.[66] Undoubtedly, many other such operations will come to light in the years ahead.

Beyond the Cold War: Black-Market Sales in a Changing International System

The dynamics of international arms trafficking are closely tied to changes in the international system as a whole, and this point is no less true of black-market trafficking than of state-sanctioned sales. It is not surprising, then, that the end of the Cold War has been accompanied by significant developments in the black-market arms trade. The breakup of the Soviet Union and Yugoslavia, for instance, has been accompanied by an upsurge in clandestine arms sales to separatist groups in both areas. Similarly, attempts by rising Third World powers to enhance their global power through the acquisition of nuclear, chemical and ballistic missile systems has produced a significant increase in illicit transfers of sensitive military technologies. It would be useful, therefore, to consider how ongoing changes in the international system may affect the black-market arms traffic in the years ahead.

Most analysts agree that the international system is experiencing a profound transformation as a result of the Cold War's demise and other recent developments. Some of these changes are quite visible, such as the emergence of new states in the territory of the former Soviet Union, while others are less so. While it is evident the bipolar system of the Cold War era has broken down, it is not yet apparent what sort of system will replace it. Nevertheless, some features of the new era are becoming evident. These include: an increased role in world affairs for non-state actors, including separatist groups, multinational corporations, international financial institutions and non-governmental organizations; the increasing integration of states and societies into the global market economy; the growing worldwide diffusion of Western consumer culture; the increased assertion of ethnic, religious, national and tribal identities; the international spread of advanced technologies; and, as a result of these other trends, a relative decline in the power and authority of nation-states. Together, these add up to a world in which power is diffuse and boundaries porous, and in which non-state actors will increasingly compete with nation-states for control over vital prerogatives and resources.[67]

Exactly how these various trends will unfold in the years ahead cannot be foreseen at this point. Several aspects of this emerging international environment, however, have particular significance for black-market arms trafficking:

(1) *The growing assertion of ethnic, religious, national and tribal identities.* Everywhere in the world today, multitudes of people are struggling for greater political recognition as a distinctive ethnic or national group and for the powers, prerogatives and resources that accrue to such recognition. As noted by Professor Myron Weiner of the Massachusetts Institute of Technology, "Peoples—however they define themselves by race, religion, language, tribe, or shared history—want new political institutions or new relationships within existing institutions." Where these demands are not being satisfied, "these groups may resist or flee across regional and international boundaries"—often provoking violence in the process.[68] Such violence often develops into insurgent or separatist struggles, as seen, for example, in the struggles of the Karen and Shan peoples of Myanmar, the Kashmiris in Indian-controlled Kashmir, the Kurds of Iraq and Turkey, the Ossetians of Georgia and the Tamils of Sri Lanka.[69]

Because all such groups are, by definition, non-state actors, they are normally excluded from participation in the licit, state-sanctioned arms trade. As such, they must rely on black-market sources for the arms needed to carry on their struggles. This process was clearly evident in the former Yugoslavia, where all the aspirants for autonomy turned to the black market for supplies of arms and ammunition, and it is increasingly evident in the embattled regions of the former Soviet Union. As struggles of this type proliferate in the years ahead, a corresponding increase in black-market arms trafficking can be expected.[70]

(2) *The emergence of ambitious, highly militarized Third World powers.* Accompanying the demise of the Soviet Union has been a relative increase in the military power wielded by ambitious Third World states, particularly those with the wherewithal to invest in domestic arms projects. Because these countries often face other heavily armed countries in their own region or because they seek to eradicate the gap in military power that historically has put them at the mercy of the major industrial powers of the North, many seek to acquire weapons of mass destruction along with the ballistic missiles needed to deliver them. The result is the emergence of "regional superpowers" armed with small but potent arsenals of nuclear and/or chemical weapons.[71]

In response to this trend, and especially in response to the Iraqi invasion of Kuwait, the United States and the other major powers have sought to tighten controls on the flow of sensitive military technologies and to establish new international restraints on the trade in conventional weapons. As part of this effort, then-President George Bush announced a series of new non-proliferation measures in the aftermath of the Persian Gulf War of 1991, including a "Middle East Arms Control Initiative" designed to curb the spread of both conventional and unconventional arms to the Middle East.[72] President Bill Clinton, who took office in 1993 with a pledge to further tighten non-proliferation controls, has adopted similar measures of his own.[73] While these various efforts have not necessarily resulted in the adoption of new international restraints, they have constrained the flow of advanced weapons and technology to aspiring Third World powers.[74]

The further tightening of arms export and non-proliferation controls in the North will be welcomed by many who fear the spread of conventional and unconventional weapons to potential belligerents in the South. Some Third World countries will, however, see these restraints as a barrier to their assumption of a more powerful world position or to the development of adequate defenses against other aspiring powers. It can be assumed they will try to circumvent the restraints, and, because they depend on foreign sources of technology and materials, they will be forced to rely on black-market channels to obtain key imports for their domestic military programs.[75] Indeed, the CIA reported in 1993 that Iran had initiated a secret procurement effort—much like the one conducted by Iraq in the late 1980s—to procure Western technology for its nuclear weapons program.[76]

(3) *The privatization and decontrol of arms industries in the former Soviet Union, Eastern Europe and China.* As part of the transition to post-Communist societies in Eastern Europe and the former Soviet Union, many of the large state-owned arms industries in these countries have been placed in private hands or been granted considerable autonomy from central government control. At the same time, these firms have lost much of their domestic business and been forced to place most arms exports, even those within the region, on a 100-percent hard-currency basis.[77] As a result of these wrenching changes and the desperate need to preserve jobs, these firms have been under enormous pressure to export as many arms as possible—even if they have to sell to the pariah countries or deal with black-market traffickers.[78] Hence, there have been numerous reports of illicit Russian, Ukrainian and Eastern European arms sales to the belligerents in the former Yugoslavia[79] and to other black-market customers elsewhere.[80]

A somewhat similar pattern is evident in China, where the government remains under Communist Party control but where many state-owned arms industries have been placed in private or quasi-private hands. To obtain hard currency for its own arms purchases, the People's Liberation Army began in the 1980s to place its military factories under private management—in many cases, a management controlled by children or relatives of senior military officers—and to encourage these firms to sell their products on the international market.[81] As in the case of Russia and the Eastern European countries, this process has led to many questionable sales to the pariah countries of the Middle East—including, in China's case, sales to Pakistan and Iran of missile technology whose transfer is barred under the Missile Technology Control Regime (MTCR), to which Beijing has pledged compliance.[82]

(4) *The multinationalization of the arms industry.* As noted by many analysts of the arms trade, economic forces in the post-Cold War era, notably the decline in domestic arms spending by the countries of NATO and the former Warsaw Pact, are forcing military firms to diversify their output and combine resources with other firms in order to compete better in the international marketplace.[83] This has led to strategic alliances between military firms in two or more countries and to joint arms projects in which each participating firm produces or assembles a particular part or section of the final product—resulting in what Stephanie Neuman of Columbia University has termed the "world tank."[84]

The emergence of the "world tank" has obvious implications for the control of licit arms transfers, in that such endeavors distribute production over several countries and thus diminish the degree of control any single government can exercise over arms exports from its territory.[85] The multinationalization of the arms industry also has considerable significance for the black-market trade, in that illicit buyers will be able to turn to a much larger number of outlets when attempting clandestine purchases of key parts and technologies. This outcome was clearly evident during the Iran-Iraq war, when Iran was able to procure spare parts for its U.S.-supplied aircraft and missiles from a number of sources, including Japan,[86] and is likely to become more so in the years ahead, when arms production is even more widely distributed.

(5) *The growth of multinational criminal enterprises.* As noted by many economists, the increasing pace of international trade and the increasing ease of international travel are facilitating the growth of multinational criminal organizations. "With the globalization of trade and growing consumer demand for leisure products," Phil Williams of the University of Pittsburgh noted in 1994, "it is only natural that criminal organizations have become increasingly transnational in character."[87] These enterprises have become adept at moving contraband—whether it be narcotics, alcoholic beverages, diamonds and gems, rare animals or guns—from one country to another, making full use of cellular telephones, facsimile machines and off-shore banking systems. As the nation-state's capacity to control and regulate international trade has diminished, moreover, these organizations have become an increasingly significant factor in the global economy.[88]

By all accounts, the growth of multinational criminal organizations has greatly facilitated the efforts of black-market arms dealers to transport munitions across international boundaries. The BATF has reported, for instance, that the small planes used to smuggle illegal drugs into the United States from Mexico and Colombia are often used to carry arms *out* of the country.[89] Similarly, officials in Italy report that the Italian mafia has become active in the shipment of arms and ammunition from Eastern Europe and the former Soviet Union into the Balkans and the Middle East. "Coinciding with the tumultuous developments that have changed the face of Eastern Europe," Defense Minister Fabio Fabbri observed in 1993, "we have noted a strong increase in clandestine trafficking of arms and explosives . . . [and] the mafia has often played the role of go-between for these movements of arms."[90] A very similar pattern of linkages between drug smuggling, arms trafficking and international criminal activity is evident in South and Southeast Asia, according to military analysts in India.[91]

As a result of (1) and (2), continued demand for black-market weapons and technology in the 1990s can be expected. Similarly, as a result of (3), (4) and (5), it can be assumed that black-market traffickers will experience increased ease in acquiring and transferring such products. These trends will surely contribute to the intensity and duration of local conflicts and complicate peacekeeping operations. At the same time, they will frustrate efforts to tighten controls on the proliferation of sensitive nuclear, chemical and missile technologies. It is essential, therefore, that the world community take steps to curb the subterranean trade in arms and military technology.

Curbing the Black-Market Arms Trade: Toward a System of Control

Unfortunately, until recently the world community did not view illicit arms sales as a major international concern. The only reference to the problem in a major international setting appeared in a 1991 report prepared for the U.N. Secretary-General on enhanced transparency in the conventional arms trade.[92] While commenting that the dollar value of black-market transactions is usually quite small, the report notes:

> [T]he consequences of the illicit arms trade can often be disproportionately large, particularly for the internal security and socio-economic development of affected States, but sometimes also for regional or even international security. Even small arms where transferred, directly or indirectly, to terrorist groups, drug traffickers or underground organizations can pose a danger to regional or international security and certainly to the security and political stability of the countries affected. . . . The illicit arms trade can also undermine efforts to negotiate political solutions to internal or international conflict.[93]

In recognition of these dangers, the report called on U.N. member states to tighten their controls over arms exports and to work together to combat black-market arms trafficking.[94]

Subsequently, in December 1991, the General Assembly adopted Resolution GA/46/36H, which calls on member states to tighten their arms export regulations and to take other steps to curb the illicit trade in arms. Specifically, the resolution urges member states

> . . . to ensure that they have in place an adequate body of laws and administrative machinery for regulating and monitoring effectively their transfer of arms, to

strengthen or adopt strict measures for their enforcement, and to cooperate at the international, regional, and subregional levels to harmonize, where appropriate, relevant laws, regulations, and administrative procedures as well as their enforcement measures, with the goal of eradicating the illicit arms trade. . . .

The United Nations has not, however, adopted any binding language to this effect, and no international treaties or agreements oblige governments to tackle the problem of illicit arms sales per se. (In contrast, parties to the Non-Proliferation Treaty, the Australian Chemical Suppliers Group and the MTCR are responsible for suppressing any illicit sales from their territory of the materials covered by these agreements.) If progress is to be made in curbing the black-market arms trade, the world community will have to develop and impose new systems of control.

Given the diversity and complexity of the black-market arms traffic, no single measure or agreement is likely to result in a significant reduction in illegal trafficking. Rather, it will be necessary to adopt an array of control measures at both the national and international levels. Together, such measures would constitute a "black-market control regime" modeled after the web of national and international measures that constrain the trade in nuclear and chemical munitions.[95]

At the national level, such a regime would encompass measures to protect domestic arms stockpiles against theft and to tighten controls on military exports. In the United States, these measures would include:

—Tightened security at military storehouses and arsenals, with enhanced oversight of stock clerks and warehouse staff.

—Mandatory seminars for all domestic arms exporters on U.S. export regulations and on the procedures to follow in the case of suspicious transactions.

—Increased transparency of U.S. arms exports (including direct commercial sales) through reinstatement of Section 657 of the Foreign Assistance Act, a measure that mandated the release of an annual statistical survey of U.S. weapons sales until repealed during the Reagan era.

—An increase in the number of investigators assigned to suspicious arms export cases by the U.S. Customs Service and the BATF.

—Increased oversight of licensing operations at the State Department's Center for Defense Trade Control, with additional staff appointed to investigate suspicious transactions.

—Automatic cut-off of U.S. arms deliveries, whether through commercial or governmental channels, to nations that fail to establish suitable security procedures for the protection of U.S.-supplied weapons in their possession.

—Imposition of selective trade sanctions against nations that fail to curb illegal arms trafficking within their territory.

A similar array of measures should be adopted by other governments, particularly those with large military stockpiles or production facilities. In the case of countries that have lacked such controls in the past, the United States should be prepared to assist in the development of modern customs services and to provide the hardware (computers, inspection devices and so forth) where necessary. Such aid is particularly needed in the post-Communist states of Eastern Europe and the former Soviet Union, many of whose governments lack experience in such operations.

At the international level, a black-market control regime should include measures to enhance the identification, monitoring and apprehension of black-market arms traffickers. Specific measures should include:

—Development of uniform, counterfeit-resistant "end-user certificates" (the documents supplied by the intended recipient of a licit arms shipment attesting to the validity of the sale) by the United Nations for use by all arms-exporting countries.

—Establishment of a special research and intelligence bureau on black-market arms trafficking at INTERPOL or the United Nations, with data links to national police and intelligence services. This bureau would collect and disseminate data on suspected international arms smugglers and assist national law enforcement agencies in locating, monitoring and prosecuting such traffickers.

—Establishment of bilateral or multilateral monitoring and interdiction teams—modelled on those developed to combat the illegal drug trade—to curb the smuggling of arms across particular borders (for example, the U.S.-Mexican border) or within a particular region (such as the Caribbean and the Eastern Mediterranean).

—Development and use of special U.N. teams to detect and prevent clandestine arms deliveries that violate U.N. embargoes, such as those imposed on the states of the former Yugoslavia.

The measures described above, if taken singly, will not produce a significant reduction in black-market arms trafficking. If combined, however, they could constitute a global filtering system that would stop much, if not all, illegal trafficking. The United States and the other major weapons suppliers should take whatever steps are necessary to implement this program at the national level and to work together at the international level to effectuate the other initiatives.

The goal of these efforts should not be to stop *all* black-market arms trafficking—such a goal, like that of zero drug trafficking, is clearly impossible. Rather, the goal should be to reduce the illicit arms trade to such an extent that potential belligerents—whether states or non-state actors—lose confidence in their ability to sustain military operations or secret arms-development projects through black-market channels. If this goal can be accomplished, states and non-state actors will no longer believe they can overcome international arms embargoes and non-proliferation efforts in their aggressive pursuit of territorial or political advantage and will be compelled to seek negotiated solutions to their disputes with other states and actors.

Notes

1. The author first analyzed this underground traffic in "Secret Operatives, Clandestine Trades: The Thriving Black Market for Weapons," *Bulletin of the Atomic Scientists,* XLIV (1988), 16–24.

2. For a report on arms trafficking in Cyprus, see Chris Hedges, "Cyprus Shores Wash Dirty Money," *New York Times* (15 June 1995). For a description of the gun market in Mogadishu, see Keith B. Richburg, "Somalia: Guns and 'Rambo' Are the Rule," *Washington Post* (14 September 1992). On black-market sales in the Peshawar area, see Edward A. Gargan, "As Afghan Veterans Limp into Town, Pakistan's Gun Traders Are Buying," *New York Times* (24 February 1992); Molly Moore, "Gun-Ho in Pakistan," *Washington Post* (31 August 1992). On gun sales in Zagreb, see Blaine Harden, "Croatians Seek High-Tech Arms on World's Black Market," *Washington Post* (15 August 1991); Philip Sherwell, "Croats Shop for Weapons," *Washington Times* (1 November 1991).

3. For an earlier attempt at such an analysis, see Klare, "Secret Operatives." Edward J. Laurance provides another attempt in "Political Implications of Illegal Arms Exports from the United States," *Political Science Quarterly,* CVII (1992), 109–140. See also "The Covert Arms Trade," *Economist* (12 February 1994), 21–23; Aaron Karp, "The Rise of Black and Gray Markets," in Robert E. Harkavy and Stephanie G. Neuman (eds.), *The Arms Trade: Problems and Prospects in the Post-Cold War World* in *The Annals of the American Academy of Political and Social Science,* 535 (1994), 175–189.

4. See Glen F. Bunting and H. G. Reza, "Agents Tipped to Thefts of Parts for Iran 2 Years Ago," *Los Angeles Times* (21 August 1985); William C. Rempel and Larry Green, "London Center of Iran Arms Smuggling," *Los Angeles Times* (3 September 1985); Gaylord Shaw and William C. Rempel, "Billion Dollar Iran Arms Search Spans U.S., Globe," *Los Angeles Times* (4 August 1985).

5. For an overview of these efforts, see Michael T. Klare, "Subterranean Alliances: America's Global Proxy Network," *Journal of International Affairs,* XLIII (1989), 97–118. See also Raymond W. Copson, *The Reagan Doctrine: U.S. Assistance to Anti-Marxist Guerrillas,* Issue Brief 87005 (Washington, D.C., 11 March 1988).

6. In 1988, for instance, U.S. authorities arrested two officers attached to the Egyptian Embassy in Washington, D.C., and an employee of Aerojet Solid Propul-

sion Co. of California on charges of attempting to smuggle 432 pounds of carbon-carbon fiber—a material used on the nose cones of ballistic missiles—to military plants in Egypt. Documents seized in connection with the arrests indicated that high officials of the Egyptian military, including Field Marshal Abdul-Halim Abu Ghazala, the Egyptian defense minister, were directing the smuggling operation. See Philip Shenon, "U.S. Accuses 2 Egyptian Colonels in Plot to Smuggle Missile Material," *New York Times* (25 June 1988); Richard W. Stevenson, "Egyptian Official Is Accused in Role in Smuggling," *New York Times* (25 October 1988); Patrick Tyler, "High Link Seen in Cairo Spy Case," *Washington Post* (20 August 1988).

7. See Kenneth R. Timmerman, *The Death Lobby* (Boston, 1991).

8. See Daniel K. Nelson, "Ancient Enmities, Modern Guns," *Bulletin of the Atomic Scientists,* XLIX (1993), 21–27.

9. For sample reports of such activities, see Roger Cohen, "Arms Trafficking in Bosnia Goes on Despite Embargo," *New York Times* (5 November 1994); Steve Coll, "Despite U.N. Embargo Weapons Sneak into Balkans," *Washington Post* (14 February 1993); "The Covert Arms Trade"; Harden, "Croatians Seek High-Tech Arms"; Edith M. Lederer, "Arms Reach Factions [in Yugoslavia] Despite U.N. Embargo," *Philadelphia Inquirer* (11 August 1992); Sherwell, "Croats Shop for Weapons."

10. According to news reports from Chile and Hungary, as transcribed by the U.S. Department of State, Foreign Broadcast Information Service (FBIS). See FBIS-LAT-91–239 (12 December 1991), 40, and FBIS-LAT-91–242 (17 December 1991), 40. See also Nathaniel C. Nash, "Chilean Arms Shipment to Croatia Stirs Tensions," *New York Times* (11 January 1992).

11. Michael R. Gordon, "Iran Said to Send Arms to Bosnians," *New York Times* (10 September 1992).

12. Michael R. Gordon, "Weapons Shipment Intercepted on Way to Bosnia," *New York Times* (26 January 1993).

13. For background, see Alison Mitchell, "In an Armed Land, Somalis Live and Prosper by the Gun," *New York Times* (5 January 1993); Rakiya Omaar, "Somalia: At War with Itself," *Current History* (May 1992), 230–234; Keith B. Richburg, "The Guns of Mogadishu," *Washington Post* (6 December 1992).

14. See Human Rights Watch Arms Project, *Arming Rwanda* (New York, 1994). See also Stephen D. Goose and Frank Smyth, "Arming Genocide in Rwanda," *Foreign Affairs,* LXXIII (1994), 86–96.

15. See Human Rights Watch Arms Project, *Rwanda/Zaire: Rearming with Impunity* (New York, 1995). See also William Branigan, "Arms Flows to Rwandans Raise Alarms," *Washington Post* (19 July 1995).

16. See, for instance, Alan Malcher, "Author Goes on 'Shopping Spree' in Illicit Arms Market," *Armed Forces Journal* (August 1989), 40–41. See also Aaron Karp, "Arming Ethnic Conflict," *Arms Control Today,* XXIII (1993), 8–13.

17. "Stinger Caused Bosnia Crash," *Flight International* (29 September 1992), 23.

18. For a discussion, see Leonard Spector, *Nuclear Ambitions: The Spread of Nuclear Weapons 1989–1990* (Boulder, 1990), 29–49.

19. The author has assembled a large collection of such materials and drew on them heavily when preparing this chapter. Edward J. Laurance used a similar approach that relied on many of the same sorts of materials in "Political Implica-

tions." See also Laurance's earlier essay on this topic, "The New Gunrunning," *Orbis*, XXXIII (1989), 225–237.

20. The author first discussed the distinction between black and gray sales in "Secret Operatives," 18–20. Aaron Karp discusses this distinction in Stockholm International Peace Research Institute (SIPRI), *SIPRI Yearbook 1988: World Armaments and Disarmament* (Oxford, 1995), 190–192, and in "The Rise of Black and Gray Markets."

21. See Keith Bradsher, "Report Links German Companies and Scud Parts," *New York Times* (8 December 1991); Victor Mallet and Mark Nicholson, "UK Probes Companies on UN List of Iraq's Suppliers," *London Financial Times* (16 October 1991); R. Jeffrey Smith, "13 Firms Named as Sources of Nuclear Items for Iraq," *Washington Post* (12 December 1991); R. Jeffrey Smith and Marc Fisher, "German Firms Primed Iraq's War Machine," *Washington Post* (23 July 1992); Michael Wines, "Documents Said to Name Iraq Suppliers," *New York Times* (30 September 1991).

22. See Yossi Melman and Dan Raviv, "Israel and South Africa's 'Unnatural Alliance' Is Under Siege," *Washington Post National Weekly Edition* (9 March 1987), 23–24; David B. Ottaway and R. Jeffrey Smith, "U.S. Knew of Israel-S. Africa Missile Deal," *Washington Post* (27 October 1989); R. Jeffrey Smith, "Israel Said to Help S. Africa on Missile," *Washington Post* (26 October 1989); Russell Watson, "Israel's Deal with the Devil?" *Newsweek* (6 November 1989), 52.

23. When charged with illegal sales of firearms to South Africa in 1978, the Winchester Division of the Olin Corporation affirmed in a legal brief that

> . . . the Winchester employees principally responsible for dealing with the State Department on export license matters over the years developed the belief that the Department was "winking" at the representation [by the company] that arms sent to South Africa were [said to be] destined for other countries.

Jacobs, Jacobs and Grudberg, "Memorandum on Behalf of Olin Corporation," brief filed in U.S. District Court, New Haven, Conn. (20 March 1978).
For a discussion of this and other such cases, see Michael T. Klare, "Evading the Embargo: Illicit U.S. Arms Transfers to South Africa," *Journal of International Affairs*, XXXV (1981), 15–28.

24. U.S. Department of the Treasury, Bureau of Alcohol, Tobacco and Firearms (BATF), Office of Law Enforcement, Firearms Division, *International Traffic in Arms (ITAR)*, Report to Congress (Washington, D.C., no date).

25. U.S. Department of Justice (DOJ), Criminal Division, Internal Security Section, Export Control Enforcement Unit, "Significant Export Control Cases, January 1981 to May 31, 1995" (Washington, D.C., no date) (mimeo).

26. DOJ, "Significant Export Control Cases," 36–41.

27. Ibid., entire document. For a similar analysis covering cases stretching from 1981 to 1988, see Laurance, "The New Gunrunning," 226–229.

28. BATF, *International Traffic in Arms*, especially sections I and IV.

29. Ibid., 7.

30. See J. Michael Kennedy, "Easy Purchase of Guns Makes U.S. a Hub for Illegal Exports," *Los Angeles Times* (19 May 1991).

31. BATF, *International Traffic in Arms*, 28.

32. In this case, the gun was manufactured in Brazil and then exported legally to the United States for resale by retail gun dealers. See James Brooke, "Brazil a Leading Supplier of Handguns to U.S.," *New York Times* (21 August 1994).

33. "The Covert Arms Trade," 21.

34. See Bunting and Reza, "Agents Tipped to Thefts,"; Ralph Frammolino, "Theft of Navy Parts: Tale with a Disturbing Moral," *Los Angeles Times* (2 September 1985); Gaylord Shaw, "Navy Thefts Spark Fear for National Security," *Los Angeles Times* (17 July 1985).

35. U.S. General Accounting Office (GAO), *Internal Controls: Theft at Three Defense Facilities in Utah,* Report GAO/NSIAD-91–215 (Washington, D.C., 1992).

36. GAO, *Internal Controls,* 1–7.

37. See, for instance, "Russian Troops Hold Clearance Sale," *London Times* (4 August 1992), which describes overt sales of military equipment by departing Russian soldiers at the open air market in Legnica, Poland. See also reports on illegal arms sales by destitute Russian soldiers in the eastern part of Germany in *Welt Am Sonntag* (Hamburg) (7 October 1990), (as translated in FBIS-WEU-90–197 [11 October 1990], 8); "Soviet Military Offers Up Equipment in Garage Sale," *Washington Times* (12 March 1991).

38. "Soviet Weapons Sold on Czech Black Market," *Baltimore Sun* (14 February 1992).

39. Francis X. Clines, "Strife in Georgia Republic Reveals Danger in New Soviet Separatism," *New York Times* (24 March 1991).

40. Dean Baquet, "U.S. Arrests 7 in Plot to Sell Ex-Soviet Arms to Iraq," *New York Times* (1 April 1991). See also DOJ, "Significant Export Cases," 45.

41. See Stuart Auerbach, "Copter Shipments Detailed," *Washington Post* (27 February 1985); "Coast Exporters May Have Sent Helicopters to North Korea," *New York Times* (3 February 1985); DOJ, "Significant Export Control Cases," 22.

42. David E. Sanger, "Inquiry Has Japan Edgy: Was It an Arms Conduit?" *New York Times* (14 July 1991); Tracy Thompson, "2 Defense Firms Indicted in Jet-Part Deals with Iran," *Washington Post* (5 September 1991); DOJ, "Significant Export Control Cases," 44.

43. "Former Soviet Borders Prove Easy Target for International Gun Runners," *London Financial Times* (29 April 1992); Kjell Engelbrekt, "Bulgaria and the Arms Trade," *RFE/RL Research Report* (12 February 1993), 44–53.

44. See *United States of America vs. Oliver L. North,* docket no. CR.88–80, "Stipulation of Facts Acknowledged by the United States Government to Be True."

45. See Steve Coll, "Afghan Rebels Said to Use Iraqi Tanks," *Washington Post* (1 October 1991); Robin Wright and John M. Broder, "U.S. Will Send Iraqi Arms to Afghan Rebels," *Los Angeles Times* (19 May 1991).

46. Douglas Farah, "Traffickers Said to Buy Contras' Arms," *Washington Post* (18 November 1990).

47. Selig Harrison, "Sparks of War in Kashmir," *Washington Post National Weekly Edition* (30 April-6 May 1990).

48. For a discussion, see BATF, *International Traffic in Arms,* 122–125, 143–146, 157–158, 168–172.

49. DOJ, "Significant Export Cases," 37–38.

50. Ibid., "Significant Export Cases," 36, 42.

51. FBIS-WEU-91–218 (12 November 1992), 26.

52. FBIS-WEU-92–006 (9 January 1992), 17.

53. See, for instance, "Tank Seizure by Germany Spotlights Arms Trade by Czechoslovakia," *Washington Post* (2 February 1992). See also FBIS-EEU-91–244 (19 December 1991), 28; FBIS-WEU-92–148 (31 July 1992), 5.

54. Michael Isikoff, "Defense Firm, Officials Indicted in Arms Sales, Diversion to Iraq," *Washington Post* (1 November 1991). See also DOJ, "Significant Export Cases," 44; Alan Friedman and Richard Donkin, "The Man Who Used CIA Tools to Commit a $1.1 Billion Fraud," *London Financial Times* (12 June 1992). Guerin, who once worked with the U.S. Central Intelligence Agency (CIA), is now serving a fifteen-year prison sentence for his role in managing the diversion to South Africa.

55. Shirley Christian, "Chilean Arms Maker Plays Role in Iran-Iraq War," *New York Times* (22 July 1987); Andy Pasztor, "Investigators Say Chilean Dealer Smuggled U.S. Weapons to Iraq," *Wall Street Journal* (20 November 1991).

56. William Booth, "U.S. Accuses Chilean of Iraqi Arms Deal," *Washington Post* (7 April 1992). See also DOJ, "Significant Export Cases," 48.

57. See Klare, "Evading the Embargo," 19–21.

58. See Michael Brzoska, "Profiteering on the Iran-Iraq War," *Bulletin of the Atomic Scientists,* XLIII (1987), 42–45; Karen DeYoung, "Sweden Has an Irangate of Its Own," *Washington Post National Weekly Edition* (21 September 1987); Steven Greenhouse, "An Iran Affair Is Emerging in France," *New York Times* (2 November 1987); Pranay Gupte, "Rhetoric and Reality in the Iranian Arms Trade," *Forbes* (19 October 1987), 33–35; James M. Markham, "An Iran Arms Scandal Is Embroiling France," *New York Times* (8 November 1987).

59. For a discussion, see Klare, "Subterranean Alliances;" John Prados, *Presidents' Secret Wars: CIA and Pentagon Covert Operations Since World War II* (New York, 1986); Richard Shultz, "Soviet Use of Surrogates to Project Power into the Third World," *Parameters,* XVI (1986), 32–42.

60. For a discussion of the Reagan Doctrine and its implementation, see Copson, *The Reagan Doctrine;* Klare, "Subterranean Alliances," 101–116; Michael Ledeen, "Fighting Back," *Commentary,* LXXX (1985), 28–31; Prados, *Presidents' Secret Wars,* 355–401; Patrick E. Tyler and David B. Ottaway, "Reagan's Secret Little Wars," *Washington Post National Weekly Edition* (31 March 1986). Then-Secretary of State George Shultz discussed the "correlation of forces" theme in "New Realities and New Ways of Thinking," *Foreign Affairs,* LXIII (1985), 706–707.

61. See Steve Coll, "Anatomy of a Victory: CIA's Covert Afghan War," *Washington Post* (19 July 1992); Robert Pear, "Arming the Afghan Guerrillas: A Huge Effort Led by U.S.," *New York Times* (18 April 1988); Tim Weiner, "How the U.S. Used a Secret Global Network," *Philadelphia Inquirer* (29 February 1988).

62. For the full background on the NSC's contra operation, see U.S. Congress, House Select Committee to Investigate Covert Arms Transactions with Iran and Senate Select Committee on Secret Military Assistance to Iran and the Nicaraguan Opposition, *Report of the Congressional Committees Investigating the Iran-Contra Affair* (Washington, D.C., 1987). For details on the efforts made by North and his associates to procure arms for the contras from foreign governments, see *U.S. vs. North,* "Stipulation of Facts."

63. See Coll, "Afghan Rebels."

64. See John Anderson, "Pakistan Aiding Rebels in Kashmir," *Washington Post* (16 May 1994); Steve Coll, "India, Pakistan Wage Covert 'Proxy Wars,'" *Washington Post* (8 December 1990).

65. "Iran Is Reported to Aid Turkish Kurds in Iraq," *New York Times* (25 October 1992).

66. Barbara Crosette, "U.S. Recalls Envoy to Burkina Faso," *New York Times* (6 November 1992).

67. For a discussion, see Lawrence Freedman, "Order and Disorder in the World," *Foreign Affairs,* LXXI (1992), 20–37; Pierre Hassner, "Beyond Nationalism and Internationalism: Ethnicity and the New World Order," *Survival,* XXXV (1993), 49–65; Stanley Hoffmann, "A New World and Its Troubles," *Foreign Affairs,* LXIX (1990), 115–122; Robert Jervis, "The Future of World Politics," *International Security,* XVI (1991/92), 39–73; Miles Kahler, "The International Political Economy," *Foreign Affairs,* LXIX (1990), 139–151; William H. McNeill, "Winds of Change," *Foreign Affairs,* LXIX (1990), 152–175; Myron Weiner, "Peoples and States in a New Ethnic Order?" *Third World Quarterly,* XIII (1992), 317–333. See also essays in "What Future for the State?", *Daedalus,* CXXIV (1995).

68. Weiner, "Peoples and States," 318.

69. Ibid., 317–333.

70. For a discussion, see Weiner, "Peoples and States," 330–331.

71. For a discussion, see Michael T. Klare, *Rogue States and Nuclear Outlaws* (New York, 1995) 130–203; W. Thomas Wander and Eric H. Arnett (eds.), *The Proliferation of Advanced Weaponry: Technology, Motivations, and Responses* (Washington, D.C., 1992).

72. "Fact Sheet: Middle East Arms Control Initiative," *U.S. Department of State Dispatch* (3 June 1991), 393–394.

73. See "Fact Sheet: Nonproliferation and Export Control Policy," White House press statement (27 September 1993).

74. See "Progress in Middle East Arms Control," statement by Under Secretary of State Reginald Bartholomew before the Foreign Affairs Committee, U.S. House of Representatives, March 24, 1992, in *U.S. Department of State Dispatch* (30 March 1992), 241–243.

75. For a discussion, see Spector, *Nuclear Ambitions,* 29–49; Wander and Arnett, *The Proliferation of Advanced Weaponry,* especially the essays in Part III.

76. See "Testimony by Director of Central Intelligence Before Senate Governmental Affairs Committee, February 24, 1993," 8–9 (mimeo).

77. SIPRI, *SIPRI Yearbook 1992: World Armaments and Disarmament* (Oxford, 1993), 279–281, 380–390.

78. On the plight of Czechoslovakian arms manufacturers after the collapse of Communism and their continuing sales to Iran and Syria, see "Hard-Pressed Czechs Retain Arms Trade," *New York Times* (3 May 1991); "Question of Arms Points Up Czechoslovak Divisions," *Financial Times* (London) (11 March 1992); Mary Battiate, "Czechoslovakia Considers Selling Tanks to Syria, Iran," *Washington Post* (7 May 1991); John Tagliabue, "Czechoslovaks Find Profit and Pain in Arms Sales," *New York Times* (19 February 1992). On Russian sales to Iran, see "Iran Acknowledges Buying Soviet Warplanes," *Philadelphia Inquirer* (6 February 1992); Christopher Drew, "U.S. Says Iran Buying Soviet Subs," *Chicago Tribune*

(22 January 1992); Richard H.P. Sia and Mark Matthews, "Iran Buying Submarines to Control Gulf Entrance," *Baltimore Sun* (5 February 1992).

79. See, for instance, Baquet, "U.S. Arrests 7;" Lederer, "Arms Reach Factions;" Peter Maass, "East Bloc's Cold War Arsenals Are Arming Ethnics," *Washington Post* (8 July 1991).

80. See "Former Soviet Borders Prove Easy Target," and "Soviet Weapons Sold on Czech Black Market."

81. For background on Chinese arms export activities, see *SIPRI Yearbook 1992*, 370–376; John W. Lewis, Hua Di and Xue Litai, "Beijing's Defense Establishment," *International Security*, XV (1991), 87–109; William C. Triplett, II, "China's Weapons Mafia," *Washington Post* (27 October 1991); Eden Y. Woon, "Chinese Arms Sales and U.S.-China Military Relations," in Thomas C. Gill (ed.), *Essays on Strategy VII* (Washington, D.C., 1990), 123–148.

82. See Bill Gertz, "Pakistan-China Deal for Missiles Exposed," *Washington Times* (7 September 1994); Michael R. Gordon, "U.S. Worries That China May Again Sell Missiles," *New York Times* (9 November 1989); Nicholas D. Kristoff, "U.S. Feels Uneasy As Beijing Moves to Sell New Arms," *New York Times* (7 June 1991); Jim Mann, "U.S. Fears China Is Seeking Missile Deal with Iran," *Los Angeles Times* (3 April 1992); Elaine Sciolino, "CIA Report Says China Sent Arms Components," *New York Times* (22 June 1995).

83. See Ethan B. Kapstein (ed.), *Global Arms Production* (Lanham, Md., 1992).

84. Stephanie G. Neuman, "The Arms Market: Who's On Top?" *Orbis*, XXXIII (1989), 509.

85. For a discussion, see Neuman, "The Arms Market," 527–529; Janne E. Nolan, "The Global Arms Market After the Gulf War: Prospects for Control," *The Washington Quarterly*, XIV (1991), 128–130.

86. In 1991, Japan Aviation Electronics Inc., a subsidiary of Nippon Electric Co., was accused of selling parts for F-4 aircraft and Sidewinder missiles (which it produced under license from U.S. firms) to Iran in the mid-1980s, when Operation Staunch was in effect. See Peter Grier, "U.S. Raps Knuckles of Japanese Firm for Jet Parts Sale," *The Christian Science Monitor* (8 April 1992); Ronald J. Ostrow, "Japan Firm Fined for Iran Military Sales," *Los Angeles Times* (12 March 1992); T. R. Reid, "Missile Parts Said Smuggled by Tokyo Firm," *Washington Post* (6 July 1991).

87. Phil Williams, "Transnational Criminal Organizations and International Security," *Survival*, XXXVI (1994), 97.

88. Williams, "Transnational Criminal Organizations," 96–113. See also R. T. Naylor, "The Structure and Operation of the Modern Arms Black Market," in Jeffrey Boutwell, Michael T. Klare, and Laura W. Reed (eds.), *Lethal Commerce: The Global Trade in Small Arms and Light Weapons* (Cambridge, Mass., 1995), 45–57.

89. BATF, *International Traffic in Arms*, 15.

90. From an interview in *Panorama* (Milan) (19 September 1993), as translated in FBIS-WEU-93-186 (28 September 1993), 33.

91. See Tara Kartha, "Southern Asia: The Narcotics and Weapons Linkage," a paper prepared for delivery at the Pugwash Workshop on the Proliferation of Small Arms and Light Weapons (New Delhi, 21–23 October 1995).

92. U.N. General Assembly, "Study on Ways and Means of Promoting Transparency in International Transfers of Conventional Arms," Report of the Secretary-General, Report A/46/301 (New York, 9 September 1991).

93. U.N. General Assembly, "Study on Ways and Means," 46.

94. Ibid., 48–49.

95. The author first proposed a regime of this sort in "Deadly Convergence: The Perils of the Arms Trade," *World Policy Journal,* VI (1988–89), 163. This concept was further developed by the author in "Gaining Control: Building a Comprehensive Arms Restraint System," *Arms Control Today,* XXII (1991), 9–13.

Part Two

THE CHANGING ECONOMICS OF ARMS PRODUCTION AND SALES

FOUR

Advanced Industrialized Countries

Ethan B. Kapstein

ECONOMIC FORCES are pounding the world's defense industries. Rising costs and declining defense budgets are combining to transform the industry in ways that will leave it unrecognizable by the end of the 1990s. Specifically, this transformation will leave the United States with overwhelming domination of the marketplace, unless the European countries pool their resources to create a single procurement market, or the Asian countries, especially China, make a significant effort to expand their military capabilities.

The implications of this transformation for arms exports are stark. It is likely the United States will be virtually the sole supplier of advanced weaponry in many market segments, particularly aerospace. To the extent proliferation of such weapons is a problem, it will be the result mainly of American imprudence, although sales of old Russian hardware and the diffusion of defense engineers from the former Soviet Union and perhaps Western Europe to foreign countries will undoubtedly pose some dilemmas.

This chapter presents an economic analysis of the defense industry that largely eschews discussion of the U.S. government's declared policy with respect to arms sales. The argument advanced here is that economic forces will be the main policy driver when it comes to arms trade, irrespective of what the government says. As U.S. defense industries become more aggressive in world markets, foreign companies will find themselves at a tremendous competitive disadvantage. The only "hope" for the inefficient producers in Europe and elsewhere that are facing this onslaught will be an adverse change in the international security environment that causes states to increase their defense budgets and, in turn, shift scarce resources into

Table 4-1. *Defense Spending by NATO's Big Four Countries, 1986–94*
(Billions of constant 1989 dollars)

Year	United States	United Kingdom	France	Germany, F.R.
1986	311.7	36.4	33.8	33.9
1987	309.0	35.8	34.9	33.9
1988	305.1	34.0	34.9	33.6
1989	304.1	34.6	35.3	33.6
1990	295.4	33.9	35.2	33.5
1991	275.0	32.4	35.0	32.0
1992	266.0	32.0	34.8	31.7
1993	267.2	35.1	34.2	30.4
1994	251.4	34.8	35.9	29.0

Note: 1993 and 1994 figures are in current year dollars.

Source: Author's estimates, based on interviews; U.S. Arms Control and Disarmament Agency (ACDA), *World Military Expenditures and Arms Tranfers* (Washington, D.C., various issues), and, for 1993 and 1994, International Institute for Strategic Studies, *The Military Balance* (London, 1995).

defense industrialization. In the absence of such "environmental" changes, policy choices will only influence industrial reorganization and consolidation at the margins.[1]

The Macroeconomic Environment

The best single indicator of a state's security environment is its defense budget. Since 1986, defense budgets in most of the Western alliance countries have declined; France was a telling exception until 1989, but since then its defense budgets have fallen slightly in real terms. (See table 4-1 for data on defense spending in the United States, the United Kingdom (U.K.), France and the Federal Republic of Germany.)

The statistics in the table are important for several reasons. First, they demonstrate that sharp cuts in defense expenditures had already taken place by 1991, particularly in the United States, where defense spending fell over 13 percent since 1986, a Reagan-era high point.[2] Sharp cuts have continued: by 1994 the defense budget had dropped to just over $250 billion.

Second, they show that European defense cuts have been more modest than those in America. However, this result should not be surprising. During the Cold War, the United States devoted approximately 40 percent of its defense budget to security problems associated with the Soviet threat to Western Europe. By providing the bulk of the defense in the North Atlantic Treaty Organization (NATO) countries, Washington enabled its allies to "free ride," spending less than they would have otherwise.[3] Europe was thus able to focus its defense budgets on problems of particular concern to it. Since many of those problems remain (such as French interests in Africa, British defense of the Falklands or peacekeeping efforts in the former Yugoslavia), its defense budgets must continue to support such missions.

Nonetheless, sheer size is important when it comes to defense budgets, and the statistics reveal just how much bigger the U.S. defense market is than that of its European allies, despite the relative differences in budget cuts. According to a RAND Corporation report, "the combined defense expenditures for the thirteen European NATO allies in 1990 were $147 billion, about half the size of the U.S. defense budget."[4] The ratio in defense procurement is similar—European procurement budgets totaled some $46 billion in 1990, compared with over $80 billion for the United States.

Presumably, the decline in domestic defense spending could be offset by increases in arms exports. Here, too, the picture for the international arms industry is relatively bleak from the macroeconomic perspective. World arms imports peaked in 1984 at $64 billion; by the early 1990s they had fallen to $45 billion—despite a fillip in the aftermath of the Iraq war—and were expected to fall somewhat further.[5] In the past, competition in this market was sharp, and it may be expected to continue. It is argued here, however, that the market is now America's to lose, and if U.S. firms capture or retain a substantial share of the marketplace, it will offset in large measure the declines in domestic procurement.

In short, the macro data on defense suggest an industry that faces declining aggregate sales. As firms face this zero-sum environment, they will see increasing competition as they try to maintain their market share. In that competitive world, American firms will have a commanding advantage.

The Microeconomic Environment

While the macroeconomic environment for nearly all defense firms is gloomy, the ability of individual companies to survive and prosper varies greatly. As a starting point, this section briefly describes the defense industry structures found in the United States and Western Europe.

The United States

One important finding is that, despite the export orientation of European firms, American companies are well-placed to capture global markets. An examination of the prime contractors in the U.S. defense industry reveals the following characteristics:

—*Concentration.* Overall, the U.S. defense industry has not been particularly concentrated, especially compared with its European competitors. According to Jacques Gansler, in the 1980s the top 100 firms accounted for 75 percent of the annual turnover, comparable to that found in many commercial sectors.[6] However, some segments of the defense industry are more concentrated than others, and the trend toward greater concentration is inevitable. Only one firm builds aircraft carriers (Newport News), while only two build jet engines (General Electric and Pratt & Whitney). In contrast, whereas seven firms produced airframes in the 1980s, merger activity had reduced that number to four by the early 1990s. Downstream in the lower tiers of subcontractors, the industry becomes more diffuse, but even there shrinking budgets are forcing these firms to close.

In seeking to survive, some defense firms have developed "niche" strategies: McDonnell Douglas has focused on naval aircraft; General Dynamics had been the Air Force's supplier of choice (in 1992 General Dynamics sold its F-16 operations to Lockheed); Lockheed has focused on its "skunk works," the high-technology shop that produced, among other systems, the Stealth fighter; and, increasingly, Northrop has become a subcontractor for the other primes. This degree of specialization distinguishes American firms from their European counterparts.

Not surprisingly, the industry is now becoming increasingly concentrated. Merger and acquisition activity—such as the Lockheed-General Dynamics deal mentioned above—is leading to a rationalization of production lines and greater efficiency. As American firms become leaner and hungrier, they will become fierce competitors in the world markets.

—*Rising costs.* In most mature industries, products become less costly to produce as companies ride the learning curve toward optimal efficiency. This pattern has not occurred in defense production. The aerospace industry provides a telling example of stubbornly rising cost structures. In 1970, U.S. firms shipped a total of 3,500 military aircraft with a value of $4 billion. In 1975, the industry shipped 1,700 units with a value of $4 billion. In 1980, it shipped 1,000 military aircraft with a value of $6 billion. In 1985, although the number of aircraft dropped to 919, their value had trebled to $18 billion.[7] During this period, it should be noted, the number of airframe manufacturers dropped by only one firm.

These rising cost structures, which are found universally, can only be supported by large internal markets. Among the Western industrial nations, only the United States has a sufficiently large market to enable it to build new weapons systems on its own. A united Europe would certainly be of sufficient size to execute many modern programs; there are military and commercial precedents, such as the Airbus. The future of the European defense industry will therefore be a function of the continent's ability to create a united procurement market.

—*Research and development (R&D)-intensive.* In 1990 the United States devoted almost $40 billion to defense-related research, development, training and evaluation. In addition, the major U.S. defense contractors spent perhaps $1 billion of internally generated funds for R&D. This funding has brought the United States to the verge of a "military technology revolution," characterized by stealth technology, precision-guided munitions, laser technologies and new generations of radars. America's military technology capacity, demonstrated partly during Operation Desert Storm, is central to its future domination of the defense marketplace. In addition, as discussed below, the United States dwarfs Western Europe when it comes to total R&D spending.

—*Defense dependence.* Despite talk among some public officials and policy analysts of the need for defense "conversion," the prime contractors remain firmly in the defense business and have little inclination to shift their production to unrelated areas. Over 70 percent of McDonnell Douglas' sales in 1990 came from defense, while virtually all of General Dynamics' sales were defense-related. Over $6 billion of Raytheon's $9 billion in sales were for defense, and for Martin Marietta the figures were $5.6 out of $6 billion. United Technologies, parent of Pratt & Whitney, was among the most diversified of the prime contractors, relying on government work for only one-third of its sales.[8] These numbers indicate a fierce determination to retain defense work and, as will be argued, to seek export markets to offset the decline in domestic procurement.

Given these characteristics, how has the industry responded to the sharp cuts in defense procurement budgets? Three strategies have emerged. First is a trend toward mergers and acquisitions, which inevitably will create a more concentrated industry. Prominent examples include Loral's acquisitions of Ford Aerospace in 1989 and LTV's missile division in 1992, Lockheed's acquisition of Sanders Associates in 1986 and General Dynamics' F-16 line in 1992. Since then, Northrop and Grumman have merged, as have Lockheed and Martin Marietta.

While the defense electronics sector was the target of most of this activity in the 1980s, mergers have become prominent in the aerospace segment in the 1990s and will remain so for the rest of the decade. As the industry becomes more concentrated, the government will have to become increasingly concerned with anti-trust issues and other problems of competition.

Second, firms have engaged in teaming and co-production arrangements to develop next-generation weaponry. Such arrangements came into vogue in the United States as soon as the changes in the economic environment appeared on the horizon beginning in 1986. According to General Dynamics,

> . . . as a result of the increased financial commitments required for new weapons systems, the Company is developing teaming arrangements to compete for new programs. The Company is currently teamed with the Boeing Company and Lockheed Corporation to produce two prototypes of the Advanced Tactical Fighter. The Company, teamed with McDonnell Douglas Corporation, was awarded a development contract for the US Navy's Advanced Tactical Aircraft (A-12). Teaming arrangements with companies in other countries are in place for the M1 Tank, U.S. Army's Single Channel Ground and Airborne Radio System and for the FS-X Aircraft.[9]

Third, the industry has become increasingly export-oriented. According to a report by the U.S. Office of Technology Assessment (OTA), "General Dynamics projects overseas sales to increase from 17 percent in the mid-1980s to about 50 percent in the mid-1990s, while Martin Marietta plans to move from 8 percent in foreign sales in 1991 to about 20 percent in 1994."[10] These sales will play a significant role in offsetting the decline in domestic procurement.

Western Europe

The industry characteristics and responses noted above provide a baseline against which it is possible to compare the structures and strategies of European defense firms. Although some European firms are formidable competitors internationally, they have been hampered by having their bases in relatively small domestic markets. Should integration of defense procurement proceed on the continent, they could conceivably challenge American hegemony in weapons production. This shift is unlikely to happen, however, as the Europeans continue to protect national producers in one way or another. Further, European countries will continue to "buy American" in many cases, as suggested by the recent Swiss and Finnish purchases of U.S. aircraft. In terms of overall competitiveness in defense hardware, it is thus difficult to imagine the European allies—alone or jointly—reaching the technological level now enjoyed by the United States.

The European defense industries face a significantly different economic environment than their American counterparts do. The single most important difference is that neither European defense budgets nor defense markets are unified. They are relatively small and diverse, as countries continue to purchase most of their weapons from domestic suppliers. While collaborative production and procurement have become increasingly prominent on the continent, it has not been an unqualified success. In the autumn of 1992, for example, Germany withdrew from one of Europe's most important collaborative ventures, the Hermes space shuttle, and it pulled out of and then re-entered the European Fighter Aircraft program (EFA, which is to be built with Britain, Italy and Spain) after intensive political wrangling. While it is popular to speak about "Europe" as if it were already integrated, in defense production and procurement the continent is far from a united actor. Instead, it remains a collection of sovereign states.

Aside from market segmentation and relatively small defense budgets, French defense executive Phillipe Cothier has noted the following factors working on European defense industries: (1) the revolution in electronics, which has largely bypassed European industry; (2) new military strategies that rely on mobility and surprise; (3) the high costs of weapons development, which no single state can support; (4) an attempt to develop dual-use technology—technology that has both civilian and military applications; (5) increasing competition in the export markets; and (6) the rise of Japan as an industrial power and, perhaps, as a producer of weaponry and aerospace in the next century.[11]

The implications of these trends for European defense firms, Cothier says, are clear. They include consolidation, with both vertical and horizontal integration becoming prominent in the defense industry; privatization, with European states moving toward divestiture of defense firms, especially in the United Kingdom; and an increasing need for fresh capital and new financial linkages, with non-defense firms taking an equity share in defense-related industries.

Consolidation is perhaps the most important step in terms of the industry's survival. In fact, a veritable merger boom has taken place on the continent since the 1980s. Daimler-Benz's subsidiary, Deutsche Aerospace (DASA), acquired MBB and Dornier, creating a pole of defense-aerospace expertise in Germany. Similarly, the U.K.'s GEC acquired the electronics firms Marconi and (along with the German company Siemens) Plessy; the result is a defense electronics giant. France's Thomson combined with CSF to create the other major player in defense electronics, while the French government has promoted a merger between famed jet fighter manufacturer Dassault and Aerospatiale. Also during the 1980s, British Aerospace merged with Royal Ordnance, creating Europe's single largest defense enterprise. In short, Europe is shedding its excess capacity and creating a handful of major players in the defense markets.

The firms that have emerged are far larger than most Americans would imagine. The 1990 revenues of British Aerospace and Thomson would rank these companies among the top ten U.S. defense industries.[12] If European defense procurement eventually becomes consolidated as the European defense industry, the result would be a pole fully capable of challenging the United States in some market segments, including missiles and a variety of smart weapons.

Instead, European procurement policy is working at cross-purposes, leading to what has been called a "crisis" for the industry. In the United Kingdom, Tory governments have promoted more competition among defense industries and adopted a "value for money" policy in weapons procurement. It has welcomed bidders from foreign countries, notably the United States, and bidders have been seeking a role in new programs. In some cases, such as the airborne warning and control system (AWACS) radar program, American weapons have been chosen over British designs, albeit with substantial domestic co-production requirements. In the coming decades, the government may yet open the market increasingly to off-the-shelf purchases of foreign weaponry. According to a distinguished British defense economist, Keith Hartley, "pressures to reduce defense spending will lead governments to consider buying from abroad (importing) to obtain further cost savings in their equipment programs."[13]

An alternative model to the value-for-money policy is collaboration within the European defense industry. In some respects, collaboration—at least as practiced to date—represents the antithesis of defense production based on the principles of comparative advantage. The reason is that collaboration has proceeded on the basis of *juste retour,* or "to each according to his contribution." In the words of Harvard's Andrew Moravcsik,

> According to this principle, the share of work each participating nation receives, as well as the burden of financing it bears, is proportional to the percentage of the production it procures. Once this basic rule is established, the precise tasks allotted to each country are carefully negotiated, generally with efforts made to distribute the technologically challenging portions equitably.[14]

As can be seen, juste retour basically results in the cartelization of a weapons system. Production is not necessarily economically efficient, but instead reflects the political power of the largest purchaser. Because of these inefficiencies, collaborative programs have, despite some successes, been widely criticized, and many have been abandoned. Most recently, Germany pulled out of Europe's major collaborative venture of the 1990s, the EFA program, citing inflated costs. It re-entered the deal only after reaching an agreement to build a significantly less expensive weapon.

Traditionally, European defense firms have had three strengths relative to their American counterparts. First, they have been far more diversified in their corporate activities. The largest defense firm, British Aerospace, relies on defense for only 40 percent of corporate sales, while Thomson-CSF derives 65 percent of its revenues from defense. Matra, the French missile and space firm, has been the most defense-dependent, with military work accounting for 70 percent of sales. It is further notable that most European defense firms have managed to reduce their defense dependence since the 1980s, something American companies have failed to do.

Second, and perhaps because of their diversification, European companies have tended to spend a larger share of their revenues on R&D than their American counterparts have. Much of this R&D has been explicitly aimed at dual-use technology, although to what end is not yet entirely clear. The French company Thomson, for example, spends about $1 billion a year on R&D, a rate that compares favorably with the largest American defense firms. Overall, however, European R&D spending is still dwarfed by that of the United States.

The third strength has been an export orientation. During the 1980s, France exported about 30 percent in value terms of the armaments it produced, the United Kingdom 20 percent. In contrast, the rate for the United States was 10–15 percent. The export-driven nature of the European defense business has, along with substantial government political and financial assistance in selling arms, made these firms sharp, some Americans would say "unfair," competitors for their U.S. counterparts.

Nevertheless, systemic factors are overwhelming these traditional advantages. In the words of one of Europe's most perceptive observers on these matters, Francois Heisbourg, "Rapid and costly change, the contraction of traditional markets, the stagnation of European defense budgets, [and] the remarkable American R&D effort; such is the scene confronting Europe's defense industry."[15] In this environment, the European firms, which still must sell to small, national markets, are unlikely to prosper.

Clearly, the key to their survival is a commitment to Europeanization of the defense market, a shift that remains on the distant horizon and that may be unreachable. According to a senior British defense official, collaboration on arms procurement will only occur case-by-case; it is unlikely Europe will create a single defense procurement authority in the foreseeable future.[16] In this environment, American defense firms will be able to "divide and conquer," offering less expensive products than those produced on the continent.

In sum, Europe's defense industries are facing their most challenging period since the end of World War II. While they have consolidated in an effort to survive, weapons procurement has remained segmented and stubbornly nationalistic. At the same time, despite a number of pan-European high-technology programs, the firms have not been keeping up with the military-technology revolution gripping the United States. In the absence of a Europe united on weapons procurement, the future of its defense industry is bleak.

Export Markets

In an era when domestic defense budgets are shrinking, firms must necessarily look to the export markets to prosper. The power of economic forces is well-demonstrated by American actions—as opposed to policies—in this area. Despite concerns over the proliferation of advanced weaponry, and despite calls for limitations on arms sales, since World War II the United States has consistently sought to increase exports in the wake of domestic downturns.[17]

An analysis of the data supports this conclusion. Between 1970 and 1975, for example, U.S. defense spending fell with the withdrawal from Vietnam. The procurement budget in particular dropped by over 10 percent. In this same period, the value of agreements for foreign arms sales grew by nearly 65 percent, as the Nixon Doctrine led to large transfers of arms to Iran, Israel and other friends and allies. Conversely, arms agreements remained flat during the entire Reagan-era arms buildup—and it should be recalled that Reagan had no particular ideological opposition to arms sales, as did his predecessor Jimmy Carter. During that period, the domestic procurement budget rose by nearly 20 percent. Since 1986, arms agreements and deliveries have again risen sharply—between 1986 and 1990, the value of arms agreements increased by 21 percent.[18]

In 1992 alone, the United States signed agreements for arms sales worth $24.1 billion, up from $6.5 billion in 1987. According to *The New York Times,* "foreign orders should account for at least 25 percent of American arms production . . . up from about 15 percent [in 1992]. . . ."[19] In the rush to sell arms, even traditional pariahs, such as Taiwan, have been targets for business: in September 1992 President George Bush agreed to sell 150 F-16 fighters to Taiwan for nearly $6 billion, over the loud protests of China. President Bill Clinton, who as a candidate was critical of the arms trade, has followed suit as a salesman for American weapons.

To the extent that export sales are a zero-sum game, America's willingness to provide arms to friends and allies has meant a net loss of sales by European defense firms. The Taiwan deal is a case in point. Given America's unwillingness in recent years to provide advanced fighters to Taiwan, the French government had hoped to sell it 120 Mirage 2000 fighters from Dassault. To promote the sale, the French had offered a comprehensive economic package that even included the construction of a *très grande vitesse* (TGV) train on the island and a nuclear power plant. For its part, Dassault was desperate to export, having failed to sell a single plane overseas since 1986. As part of its campaign, the French government dispatched industry minister Dominique Strauss-Kahn to Taipei during the summer of 1992. Hopes for a major sale were dashed by President Bush's decision to offer Taiwan the F-16. Finally, the French won just half the total sale.[20]

The British, too, have felt the impact of the American assault. British arms exports fell from 2.4 billion pounds sterling in 1989 to 1.8 billion in 1991, largely because of a drop in Middle East sales. According to one British defense official, the outlook over the long term is "gloomy." American marketing in Saudi Arabia and Kuwait, especially during and after the Iraq war, has placed European weapons at a distinct disadvantage. During the presidential election, President Bush took a direct interest in selling American weapons such as had not been seen in many years. It was reported he placed significant pressure on the Kuwaiti government to purchase the American Abrams tank rather than the British challenger.[21]

The United States has not only offered off-the-shelf sales of advanced weaponry to its friends and allies. More troubling from both an economic and security perspective, it has also been willing to co-produce weapons with foreign purchasers. In these cases, the buyers are not purchasing weapons off-the-shelf; rather, they are purchasing "know-how" and production capability. Thus, another important issue for students of the arms trade is the manner in which foreign sales are executed.

America's sale of F-16s to Turkey in 1983 is a case in point. One U.S. government official described the deal as follows:

> Turkish F-16s are produced in part by a jointly-owned aircraft manufacturing plant built in Turkey by a Turkish aerospace firm and by General Dynamics. The aircraft or important subsystems of it are manufactured jointly in Turkey to help offset the cost of the buy by providing employment, technology transfer, and investment in new plants and equipment...The hundreds of trained engineers and thousands of skilled workers as well as the machinery and facilities involved in these projects represent a sizable investment and will allow Turkey to assemble other aircraft and join international consortia as a serious partner.[22]

The implications of these co-production deals are troubling, as noted, from both an economic and security standpoint. In terms of the economic perspective, co-production represents a form of protectionism, as governments intervene in the marketplace to promote the development of an indigenous defense industry capability. It is ironic the United States, which champions free trade and the international division of labor in the commercial sphere, supports the growth of inefficient defense industries in countries around the world.

In terms of security, co-production is no less troubling. As William Keller of the OTA has written,

> the proliferation of the ability to produce modern arms (emanating principally from the United States and Europe) has led directly and indirectly to the arming of our adversaries as well as our friends. U.S. companies have played a major role in the transfer of sophisticated defense technology to Europe, Japan, and elsewhere.[23]

In general, the United States has supported arms co-production arrangements in those cases where it was necessary to "win the sale." Defense firms have argued that their European counterparts would have been willing to make technology transfer arrangements even more generous than those advanced by the United States. Co-production has thus become a marketing tool.

Conclusions

As this period of American domination over the arms market begins, the question is whether such military and economic power could facilitate control over the future proliferation of conventional weapons. The United States is in a stronger position than ever to determine the rules of the game for the arms trade. Its actions will provide the baseline for acceptable practices.

The manner in which arms sales proceed will be an important issue for both the economic and security environments as the millennium approaches.

This chapter suggests that the driving force behind the arms market in the coming decade will be economics, not politics. That is to say, with the decline in domestic defense budgets, most firms will have no choice but to increase exports if they are to survive and prosper. In the international marketplace, the United States has a commanding advantage in many market segments, particularly those associated with aerospace.

The United States, however, may be creating problems for itself when it exports advanced weaponry, for these sales can create an upward spiral in a particular region. To offset the impact of a particular sale, surrounding countries will themselves try to buy more weaponry. In many cases, they, too, will seek to "buy American." This pattern has already developed in the Middle East, where the United States is selling top-of-the-line equipment to both Israel and Saudi Arabia. Ironically, the United States is now fueling an arms race on its own.

History supports this economically driven perspective of the arms trade. During previous downturns, the United States has always looked to increase exports, official rhetoric to the contrary. The converse also seems to hold true. When domestic procurement budgets rise, foreign sales appear to decrease, even when administrations are in place that ostensibly favor exports. In sum, the most powerful predictor of American arms sales is the level of the domestic procurement budget.[24]

For those who seek to control American arms exports through public policy, such conclusions must be disturbing. However, an economics-based approach also suggests the promise of arms control. If the analysis presented here is correct, the United States will exercise something like monopoly power over the high-technology end of the conventional arms trade. European firms lag behind financially and technologically, and Third World defense industries are falling into bankruptcy. This fact leads to the conclusion that, should weapons proliferation remain a significant policy problem in the 1990s and beyond, it will be largely the result of American imprudence.

Notes

1. I thank Professor Kenneth Mayer of the University of Wisconsin for highlighting the importance of environmental factors in defense-industrial transformation.

2. See the Clinton Administration's "Bottom-Up Review," U.S. Department of Defense (Washington, D.C., 1993); for a sample of the extensive press reporting,

see James Flanigan, "Cutting Budget Means Cutting Defense," *Los Angeles Times* (11 August 1993); "Where's the Bottom?" *New York Times* (3 September 1993).

3. For the classic argument, see Mancur Olson and Richard Zeckhauser, "An Economic Theory of Alliances," *Review of Economics and Statistics,* XLVIII (1965).

4. James B. Steinberg, *The Transformation of the European Defense Industry* (Santa Monica, 1992), 9.

5. U.S. Arms Control and Disarmament Agency (ACDA), *World Military Expenditures and Arms Transfers* (Washington, D.C., various years).

6. Jacques Gansler, *Affording Defense* (Cambridge, Mass., 1989), 245.

7. U.S. Department of Commerce, *U.S. Industrial Outlook: 1990* (Washington, D.C., 1990), 25–26.

8. See the various corporate annual reports, which break out defense, nondefense, and international sales.

9. General Dynamics, *Annual Report: 1989.*

10. U.S. Congress, Office of Technology Assessment (OTA), *Building Future Security: Strategies for Restructuring the Defense Technology and Industrial Base,* OTA-ISC-530 (Washington, D.C., 1992), 103. It should be noted that since the report was written, General Dynamics sold its combat aircraft division, manufacturer of the F-16, to Lockheed.

11. Phillipe Cothier, "The European Defense Industry," a paper presented at the workshop on International Arms Collaboration, Harvard University (Cambridge, Mass., 1989).

12. Steinberg, *The Transformation of the European Defense Industry,* 69.

13. Keith Hartley, et al., "The Economics of UK Defence Policy in the 1990s," *RUSI Journal,* CXXXV (1990), 49–53.

14. Andrew Moravcsik, "The European Armaments Industry at the Crossroads," *Survival,* XXXII (1990), 65–85.

15. Francois Heisbourg, "Public Policy and the Creation of a European Arms Market," in Pauline Creasey and Simon May (eds.), *The European Arms Market and Procurement Cooperation* (London, 1988), 68.

16. Ministry of Defence, interview with the author (London, 15 December 1992).

17. For a detailed study, see Tobey Susan Weintraub, "U.S. Arms Exports, Military Procurement and the Defense Industrial Base: Economic Motivation Behind National Security Policy," senior honors thesis, Harvard College (Cambridge, Mass., 1992).

18. Weintraub, "U.S. Arms Exports"

19. Eric Schmitt, "Arms Makers' Latest Tune: Over There, Over There," *New York Times* (4 October 1992).

20. Jean Leclerc du Sablon, "Chine-France: Le Casse-tête des Mirage," *Le Figaro* (3 August 1992).

21. Data and information from an interview with a source at the Ministry of Defence (London, 15 December 1992).

22. Daniel C. Grynaviski, "The International Arms Market and the Transfer of Military Production Capability," Executive Research Project S23, Industrial College of the Armed Forces (Washington, D.C., 1991).

23. William Keller, "Global Defense Business: A Policy Context for the 1990s," in Ethan B. Kapstein (ed.), *Global Arms Production: Policy Dilemmas for the 1990s* (Lanham, Md., 1992), 68.

24. For further development of this point, see Weintraub, "U.S. Arms Exports."

FIVE

Developing Countries

Andrew L. Ross

THE DIFFUSION of military power and capabilities has been an abiding concern of analysts and policymakers alike. This concern is evident in the three waves of research on the post-World War II spread of military resources from the advanced industrial countries of the North to the developing countries of the South. The first wave of research focused on the horizontal proliferation of nuclear weapons.[1] Despite its relatively slow pace, nuclear proliferation continues to command attention.[2]

The focus of the second wave of analysis was the proliferation of conventional, rather than nuclear, weapons. This spread of weaponry conventional armaments has provided the developing countries of Africa, Asia and Latin America with fungible military power.[3]

As became increasingly evident during the 1970s and 1980s, a growing number of developing countries were not only importing but also producing, and even exporting, modern, sophisticated weapons systems such as jet fighters, armored fighting vehicles, missiles, rocket systems and naval combat vessels. The result was an extensive body of literature on the emergence and growth of the developing world's indigenous defense industrial capabilities.[4]

This third wave of research contributed substantially to an understanding of the variety of political, military and economic factors underlying the initiation and expansion of defense industrial programs, the increasingly extensive range of weaponry in production, the processes of defense industrialization and the role of external inputs. It generated numerous, although largely non-cumulative, country case studies. It also provided assessments of the political, military and economic implications of defense industrialization in the developing world. Still, analytical understanding of the phenomenon of defense industrialization in the developing world remains incomplete.

This chapter provides a basis for assessing the prospects for the production and export of conventional weaponry by developing countries. The approach is to: (1) place defense industrialization in historical context; (2) delineate the larger processes that have enabled developing countries to initiate and develop defense industrialization programs; and (3) foster recognition of how defense industrialization could reduce the dependence and enhance the autonomy of developing countries. Defense industrialization in developing countries evolved in response to their post-colonial military dependence. It was facilitated by the adverse conditions confronting suppliers in the international arms market, by industrial development and by the central role of the state in the late industrializing countries. Changes in two of these factors—the dynamics of the international arms market and the developmental role of the state—will significantly shape the future of the developing world's arms production and export ventures.

Military Dependence

The multifaceted economic dimension of dependence has profoundly influenced inquiry into the nature of relations between the developing countries of the South and the advanced industrial countries of the North.[5] Still there is a military as well as an economic dimension to dependence.[6] Indeed, the military dependence of the South upon the North has been the defining characteristic of political-military relations between North and South following the world wars and decolonization.

Military dependence, along with economic dependence, are the historical legacies of colonialism. Under colonialism the local military forces of Latin America, Africa and Asia were integrated into the command structures of the imperial powers. The European colonial powers commanded, trained, financed and equipped the indigenous military forces. The virtual absence of a modern manufacturing capability limited the procurement of military equipment within the colonial territories to, at most, small arms and ammunition. Only manpower was consistently generated locally.

Consequently, developing countries were militarily as well as economically underdeveloped upon their entry as formally sovereign actors into the interstate system. They had few capable, experienced officers—colonial military training emphasized following orders and the development of administrative skills rather than command and initiative. In addition, state systems for collecting the revenue from society that was needed to finance the establishment and expansion of military forces were either unreliable or non-existent. Perhaps most important, there were no significant indigenous sources of military equipment.

Confronted with an array of external and internal security threats resulting from the poor correspondence between state and society, developing countries had little alternative but to import their arms, given their inadequately equipped military forces and the lack of a scientific, technological and industrial capability to manufacture vital military hardware locally. They became dependent upon foreign arms suppliers for a crucial security input.

With some exceptions, during the early post-colonial period developing countries tended to get their arms from their former colonizers. Long accustomed to being commanded, trained, financed and equipped by Northern militaries, most defense establishments in the developing world were reluctant to cut those close ties. The colonial military relationship was thereby maintained in the post-colonial world. For instance, former British colonies imported arms from the United Kingdom; the former French colonies turned to France.[7] Subsequently, as the United States and the Soviet Union struggled for global influence, they displaced the former European colonial powers as the dominant suppliers of arms to developing countries.

The vulnerabilities and limitations on policy and behavioral autonomy inherent in dependence upon foreign arms suppliers did not escape notice. The developing world's leaders repeatedly noted the adverse consequences of relying upon foreign sources of military equipment. In an address before the United Nations (U.N.) General Assembly, S. Rajaratnam, the foreign minister of Singapore, vividly depicted the dangers of dependence upon foreign arms suppliers:

> ... the most dangerous consequences are political. The flow of arms carries with it a measure of dependency on the part of the client on the seller of arms not unlike that prevailing under the old imperial system. ... The massive flow of arms to the third world confronts it with a new danger. It is, first of all, a drain on the economies of third world countries; but even more important is the fact that it creates a new form of dependence on the great Powers, which can exploit the third world's dependence on them to manipulate them, to engineer conflicts between them, and to use them as proxies in their competition for influence and dominance.[8]

As a result of their dependence on Northern arms suppliers, developing countries have been vulnerable to arms embargoes, interruptions in the flow of spare parts for imported equipment, restrictions on the end-use and resale of foreign weaponry, and attempts by suppliers to use supply relationships to exert undue influence on their foreign and domestic policies. They have also had to contend with the in-country presence of the often numerous Northern military technicians and advisors sent to train their military personnel in the operation, maintenance and repair of complex imported systems.

Military Import-Substitution

Escape from the dilemmas of military dependence has been sought through military import-substitution.[9] The substitution of domestically manufactured weapons for imported weapons is thought to offer the prospect, albeit long-term, of nationalizing arms acquisition and enhancing military self-reliance. By developing the capability to produce increasingly greater proportions of the military equipment they require, developing countries expect to reduce, and in the long run perhaps even eliminate, their debilitating dependence upon arms imports.

Defense Industrialization

Military import-substitution can be conceptualized as a long-term process that assumes different forms as it progresses through five distinct developmental stages. During the first three stages, as defense industrial capabilities develop, dependence upon imported military technology gradually displaces dependence on imported military hardware. Subsequently, during the more difficult fourth and fifth stages, indigenous technological capabilities are brought to bear to overcome technological dependence. Therefore, military import-substitution involves coping with both the import and technology forms of military dependence.

The first stage of defense industrialization involves simply the assembly of imported arms. Military equipment is still acquired from abroad but is imported in the form of prefabricated components for assembly in-country. The foreign supplier provides technical training and assists in initiating weapons assembly programs. Technical training includes not only instruction on assembly skills, but also instruction on the utilization of equipment employed in weapons inspection, evaluation and testing.

In the second stage, weapons components are manufactured under license agreements with foreign suppliers. In addition to the assembly of the weapon, an increasing proportion of the components are fabricated domestically.

In the third stage, military import-substitution results in the actual production of complete weapons, albeit foreign-designed, under license.

Technological skills and manufacturing capabilities developed in the earlier stages are utilized to modify, redesign or reproduce foreign weapons systems in the fourth stage. It marks the first appearance of indigenous technological capabilities—in the form of either system modification or reverse engineering.

The production of domestically designed arms occurs in the fifth stage. During this last stage, production typically progresses from reliance on domestic research and development (R&D) that still incorporates components designed or produced abroad to autonomous, indigenous R&D with minimal foreign input.[10]

Demonstrated Defense Industrial Capabilities

Large-scale, widespread military import-substitution in the developing world is a relatively recent phenomenon. In 1950, only Argentina, Brazil, Colombia and India were producing any of the four types of major conventional weapons—aircraft, armored vehicles, missiles and naval vessels.[11] The data presented in tables 5-1 and 5-2 indicate that a significant number of countries initiated military import-substitution programs since 1950.

Table 5-1 lists the developing countries engaged in the production of aircraft, armored vehicles, missiles and naval vessels in 1960, 1970 and 1980. The demonstrated production capabilities of each country within the four major defense industrial sectors have been coded according to the five developmental stages outlined above. Data for the entire range of production capabilities exhibited in each sector have been provided.

The number of developing countries with military import-substitution programs increased from only four in 1950 to fifteen in 1960, nineteen in 1970 and twenty-nine in 1980. In 1960 only two of the four types of conventional weapons were being produced: aircraft and naval vessels. In 1970 and 1980 all four types were being produced. The number of countries with aircraft and naval programs increased, respectively, from seven and fourteen in 1960 to sixteen and twenty-seven in 1980. The number of countries producing armored vehicles went from five in 1970 to eight in 1980, while the number of those manufacturing missiles rose from three to nine.[12] No country in 1960 and only one in 1970, India, was producing

Table 5-1. *The Developing World's Arms Producers: Demonstrated Capabilities in 1960, 1970 and 1980*

Country	Aircraft			Armored vehicles			Missiles			Naval vessels		
	1960	1970	1980	1960	1970	1980	1960	1970	1980	1960	1970	1980
Argentina	3,5	5	2,4,5		3	3,4			4,5	5	5	1,3,5
Bangladesh												5
Brazil	3,5	5	1,2,3,5		5	4,5			3,4,5	5	4,5	3,5
Chile	5		3,4							5	3,5	3
Colombia										5		5
Dominican Rep.										5		5
Egypt	3,4	3,5	5	4	4				3	4	5	4,5
Gabon											5	5
India	1,3,5	2,3,5	3,4,5	3	3		3	3		5	3,5	3,4,5
Indonesia	4	4	1,3,4							5	5	4,5
Israel	1,3	3,5	2,3,4,5			5		5	4,5		5	3,4,5
Ivory Coast											5	5
Korea, North			3							3,4,5	3,4,5	3,4,5
Korea, South			2			3,4			4		3,5	3,5
Madagascar												3
Malagasy Rep.												4
Malaysia												3,5
Mexico										5	5	3
Myanmar										3,5	5	
Nigeria			1									
Pakistan			1,3				3	3				
Peru			1							5	5	3,5
Philippines			2									5
Senegal												5
Singapore										3	3	3,5
South Africa		3	2,4		3	4,5			4,5			3,5
Sri Lanka												5
Taiwan		3,4	3,4			4			3,4		5	3,5
Thailand			3							5	4,5	5
Venezuela												3

Note: Stages of demonstrated manufacturing capabilities—1. licensed assembly; 2. licensed assembly/component production; 3. licensed system production; 4. system modification/redesign/reproduction; and 5. manufacture based on indigenous R&D. Blanks mean no demonstrated capability.

Source: Revised version of data that first appeared in Andrew L. Ross, *Arms Production in Developing Countries: The Continuing Proliferation of Conventional Weapons,* RAND Note N-1615-AF (Santa Monica, Calif., 1981).

systems in each of the four categories. By 1980, however, eight countries—Argentina, Brazil, Egypt, India, Israel, the Republic of Korea (South Korea), South Africa and Taiwan—had demonstrated across-the-board production capabilities.

Without exception, there were a larger number of countries producing aircraft, armored vehicles, missiles and naval vessels at each stage in 1980 than in 1960. The number of countries with stage five aircraft programs, for

Table 5-2. *Demonstrated Capabilities of the Developing World's Arms Producers, Mid-1980s*

ABC[a]	Developing ABC[b]	Lesser producers[c]	Minor producers[d]
Argentina	Chile	Algeria	Bolivia
Brazil	Colombia	Bangladesh	Burkina Faso
Egypt	Indonesia	Dominican Republic	Cameroon
India	Korea, North	Gabon	Congo
Israel	Mexico	Honduras	Cuba
Korea, South	Pakistan	Iran	Ecuador
South Africa	Philippines	Ivory Coast	Ethiopia
Taiwan	Thailand	Madagascar	Ghana
		Malaysia	Guatemala
		Morocco	Guinea
		Myanmar	Iraq
		Panama	Jordan
		Peru	Nepal
		Senegal	Nigeria
		Singapore	Saudi Arabia
		Sri Lanka	Sudan
		Trinidad & Tobago	Syria
		Uruguay	Tunisia
		Venezuela	

[a]ABC: Across-the-board capability—demonstrated ability to produce military aircraft, armored vehicles, missles, naval vessels, and small arms and ammunition.

[b]Developing ABC: Developing across-the-board capability—demonstrated ability to produce both two or three of the four major types of weapons systems and small arms and/or ammunition.

[c]Lesser producers: Demonstrated ability to produce one of the four major types of weapons systems; may also produce small arms and/or ammunition.

[d]Minor producers: Demonstrated ability to produce small arms and/or ammunition.

Source: Derived from data presented in Michael Brzoska and Thomas Ohlson, "Arms Production in the Third World: An Overview," 16, and Appendix 2, "Register of Indigenous and Licensed Production of Major Conventional Weapons in Third World Countries, 1950–84," 305-349, in Michael Brzoska and Thomas Ohlson, (eds.), *Arms Production in the Third World* (London, 1986), 49.

instance, increased from four in 1960 to five in 1980, while those with stage five naval programs increased from twelve to twenty-two. From 1970 to 1980 the number of countries with stage five armored vehicle programs went from one to three and from one to four in the case of stage five missile programs. The pattern is repeated for each of the other stages across all four categories of military import-substitution programs. Most important, by 1980 twenty-two of twenty-nine countries had a stage five production capability within at least one of the four defense industrial sectors.

According to data from the Stockholm International Peace Research Institute (SIPRI), the source for table 5-2, a total of fifty-three developing countries were manufacturing either one of the four types of major conventional weapons or small arms and/or ammunition by the mid-1980s. Thirty-five of the fifty-three were producing at least one of the four types of major weapons. Eight countries had demonstrated an across-the-board capability, and another eight were developing across-the-board capabilities. An additional nineteen had the ability to produce at least one of the major types of conventional weapons and small arms and/or ammunition. In eighteen countries military import-substitution was limited to small arms and/or ammunition.

The data in table 5-3, which come from a survey of the developing world's defense industrial capabilities prepared for the U.S. Department of Defense, again show an increase in military import-substitution ventures and indicate the continued growth in defense industrialization efforts in the 1980s. As before, it can be seen that although developing countries were producing only aircraft and naval vessels in 1960, they were producing all four types of major weapons systems in 1970, 1980 and 1990. In the latter year, the number of countries manufacturing the aircraft, armored vehicles, missiles and naval vessels for which data are provided in table 5-3 was either greater than or at least equal to the number of producers in 1980. The increase in the number of countries producing armored vehicles between 1980 and 1990 is particularly striking.

Despite the attention they have attracted, especially since the Gulf War, ballistic missiles are, as yet, a relatively limited addition to the developing world's expanding array of defense industrial capabilities. According to the data in table 5-3, only five developing countries were building surface-to-surface missiles in 1990. However, former U.S. Director of Central Intelligence William H. Webster, in testimony before the Committee on Foreign Relations of the U.S. Senate in 1989, forecast that "by the year 2000, at least fifteen developing countries will be producing their own ballistic missiles."[13] Eleven developing countries with deployed ballistic missiles and eight countries with a total of eighteen ballistic missile development programs were included among the sixteen developing countries identified by the U.S. Arms Control and Disarmament Agency (ACDA) as either possessing or developing ballistic missiles in 1989.[14]

According to one count, "Twenty-two countries in the Third World currently possess ballistic missiles or are actively attempting to acquire them. Thirteen of these countries have programs to design and build ballistic missiles, and at least 15 have operational missile forces."[15] Of the twenty developing countries identified by *Arms Control Today* as either having

Table 5-3. *Defense Industrialization in the Developing World: Number of Producers of Selected Equipment, 1960, 1970, 1980 and 1990*

Equipment	1960	1970	1980	1990
Aircraft				
Fighters	1	1	5	7
Trainers (jet)	2	4	3	4
Trainers (basic)	5	5	11	11
Maritime reconnaissance			2	4
Transports	1	3	7	7
Engines	1	2	5	5
Avionics			3	3
Armored vehicles				
Tanks		2	5	7
Armored personnel carriers			2	11
Armored cars		2	2	6
Reconnaissance vehicles			2	5
Missiles				
Surface-to-air			5	5
Air-to-ground			3	3
Air-to-air		1	5	5
Surface-to-surface		1	3	5
Anti-tank		1	7	7
Naval vessels				
Frigates	1	1	4	6
Corvettes	2	2	1	6
Patrol craft—Forward Air Control	8	11	21	21
Submarines			3	7
Amphibious craft	1	2	3	8
Support craft	6	4	7	11

Note: Blanks mean no producers.

Source: Science Applications International Corporation, *Diffusion of Military Technology and Its Implications for U.S. Defense Policy*, prepared for the Director, OSD Net Assessment, under MDA903-89-C-0163, 1 June 1989 through 30 September 1990, vol. 1, *Executive Summary*, 3 and vol. 2, *Technical Report*, 45.

acquired or been engaged in developing ballistic missiles by early 1992, four were attempting to develop them independently and seven were engaged in joint efforts.[16] While only one independent effort had resulted in an operational ballistic missile capability, seven countries possessed operational ballistic missiles developed under bilateral or multilateral projects. While it can still be said that operational ballistic missile capabilities in the developing world are primarily a result of imports rather than production programs, by early 1992 seven of the seventeen developing countries with operational ballistic missile capabilities had augmented their imported capabilities with either independent or joint development efforts. The Missile Technology Control Regime (MTCR) established in 1987 by Canada, the Federal Republic of Germany (West Germany), France, Italy, Japan, the United Kingdom and the United States has not entirely forestalled the spread of ballistic missile technology to the developing world.

An increase in the monetary value of the arms produced has accompanied the increases in the number of developing countries attempting to substitute indigenously produced weapons for imported weapons, the range of weaponry produced and the level of domestic inputs. According to SIPRI data, the value of aircraft, armored vehicles, missiles and naval vessels produced by developing countries in 1950 was a mere $2 million. In 1960, it was still only $11 million. In 1970, however, the value of major conventional weapons manufactured by developing countries reached $274 million, in 1980 $980 million and in 1984 $1.1 billion. From 1950 through 1984, developing countries produced $12.7 billion worth of major conventional weapons. Ranked according to value of production, the leading producers in that period were India, Israel, South Africa, Brazil, Taiwan, the Democratic People's Republic of Korea (North Korea), Argentina, South Korea and Egypt.[17]

Additional data on the value of the major conventional weapons manufactured by these leading producers—minus North Korea and plus Chile, Indonesia, Pakistan and Singapore—are presented in table 5-4. This group of countries, which includes all eight of the producers with across-the-board capability and three of the producers with developing across-the-board capabilities identified in table 5-2, produced some $87.4 billion worth of major conventional arms during the years 1965–90. Over $65.1 billion of total production, or 75 percent, consisted of what SIPRI, the source of these data, identified as indigenous production, roughly equivalent to stages four and five production. For eleven out of twelve of these countries (Taiwan was the exception), the value of production was greatest during either the 1980–84 or 1985–89 period, the two most recent of the five periods for which production data have been aggregated. The value of production peaked in 1980–84 for five countries and during 1985–89 for six. Total production by the twelve countries peaked during the most recent period, 1985–89. The five leading producers in this group of twelve were, in descending order, India, Israel, South Africa, Brazil and South Korea.

Table 5-4. *Estimated Value of Conventional Weapons Production in Selected Developing Countries, 1965–90*
(Millions of constant 1990 U.S. dollars)

Country	1965–69	1970–74	1975–79	1980–84	1985–89	1990	Total
Argentina							
A	64	73	455	1,362	884	157	2,995
B	3	13	64	810	567	148	1,605
C	67	86	519	2,172	1,451	305	4,600
Brazil							
A	22	471	1,609	2,194	1,559	478	6,333
B	—	117	35	293	270	25	1,060
C	22	588	1,964	2,487	1,829	503	7,393
Chile							
A	23	5	—	84	280	34	426
B	3	5	—	73	153	31	265
C	26	10	—	157	433	65	691
Egypt							
A	100	65	170	436	744	162	1,677
B	35	—	—	240	415	122	812
C	135	65	170	676	1,159	284	2,489
India							
A	1,500	3,108	3,559	4,855	6,262	1,198	20,482
B	839	1,706	1,559	1,762	4,428	1,078	11,372
C	2,339	4,814	5,118	6,617	10,690	2,276	31,854
Indonesia							
A	14	10	54	172	613	116	979
B	—	—	21	105	230	43	399
C	14	10	75	277	843	159	1,378
Israel							
A	144	2,308	3,141	4,205	2,799	609	13,206
B	—	—	25	50	27	—	102
C	144	2,308	3,166	4,255	2,826	609	13,308
Korea, South							
A	1	50	535	2,098	1,023	454	4,161
B	1	—	310	1,316	690	267	2,584
C	2	50	845	3,414	1,713	721	6,745
Pakistan							
A	—	—	18	20	25	9	72
B	—	—	5	5	29	6	45
C	—	—	23	25	54	15	117
Singapore							
A	6	129	391	1,276	719	93	2,614
B	4	102	189	232	99	—	626
C	10	231	580	1,508	818	93	3,240
South Africa							
A	395	876	1,526	2,006	5,262	730	10,795
B	119	432	796	432	629	37	2,445
C	514	1,308	2,322	2,438	5,891	767	13,240
Taiwan							
A	19	263	807	14	114	188	1,405
B	14	102	676	—	—	113	905
C	33	365	1,483	14	114	301	2,310

Note: A—indigenous production; B—licensed production; and C—total production. —means data not provided by SIPRI.

Source: Ian Anthony, "The 'Third Tier' Countries: Production of Major Weapons," in Herbert Wulf (ed.), *Arms Industry Limited* (Oxford, 1993), table 17-1, 370–373.

Arms Exports

To increase their economic viability, the arms industries of developing countries have also been producing arms for export. According to data from ACDA, the export performance of developing countries at times compared favorably with that of advanced industrial countries during the 1980s.[18] From 1979 to 1989, arms exports by developing countries grew by 6.8 percent, whereas exports by developed countries declined by 1 percent. During the second half of this period, 1985–89, arms exports by developed countries dropped 4.5 percent, while those by developing countries increased 0.5 percent. From 1989 to 1993, however, the export performance of developing and developed countries alike deteriorated significantly. Arms exports by developing countries fell 26.7 percent, while those by developed countries dropped 22.9 percent.[19]

Developing countries also experienced an increase in their share of world arms exports during the 1980s. In 1979, developing countries had a market share of only 4.2 percent. By 1989, it was 9.7 percent. ACDA counted six developing countries among the twenty leading arms exporters during the years 1985–85 and eight among the top twenty in 1989.[20] By 1993, however, the developing world's share of the arms market had declined to 7.9 percent, and only three developing countries were to be found among the fifteen leading arms exporters.[21] According to data on arms exports from SIPRI, which employs a more restrictive definition of "developing country" than does ACDA, developing countries exported $23.7 billion worth of major conventional arms during the years 1985–94.[22] Significantly, even though their arms exports declined to only $1.5 billion in 1994 after a peak of $4.7 billion in 1987, exports by developing countries grew at an average annual rate of 2.9 percent from 1985 to 1989 before declining 3.3 percent from 1990 to 1994. Arms exports by advanced industrial countries, in contrast, declined at an average annual rate of 0.5 percent from 1985 to 1989 and 10.5 percent from 1990 to 1994. In addition, the developing world's share of total arms exports increased from 6.2 percent in 1985 to 6.9 percent in 1994. Six developing countries—Brazil, China, Israel, North Korea, Pakistan and South Korea—made SIPRI's list of the twenty-five leading exporters during the years 1990–94.[23] Of the eleven leading suppliers of conventional arms to the developing world from 1987–94 identified by the Congressional Research Service (CRS), four—China, ranked fourth, North Korea, seventh, Israel, ninth, and Brazil tenth—were developing countries.[24]

The Market, Industrialization and The State

Efforts to develop domestic arms production have been motivated primarily by the explicitly political goal of reducing dependence on unreliable and unpredictable foreign suppliers.[25] However, such motivation is not sufficient to enable a country to initiate military import-substitution programs. Three distinct factors have allowed developing countries to establish these military programs. At the systemic level, adverse conditions for suppliers in the international arms market have enabled developing countries to acquire foreign military technology. At the national level, industrial development has provided the economic foundation for defense industrialization and enabled developing countries to absorb foreign defense technology. In addition, the central developmental role of the state in the late industrializers has facilitated defense industrialization.

The International Arms Market

The dynamics of the international "arms bazaar" have played a vital role in defense industrialization in the developing world. As noted, events in the international arms market made it possible for developing countries to acquire the military production technology crucial to the success of military import-substitution programs. Adverse market conditions in the advanced industrial suppliers resulted in a buyer's market that allowed importers in the developing world to require production technology as part of arms transfer packages.

An increasing number of countries became active as arms exporters during the 1960s, 1970s and 1980s. Both the number of arms exporters and of countries that consistently, rather than sporadically, exported arms increased. As the number of suppliers rose, market concentration declined, and commercialization and competition increased.[26] The supply side of the international arms market became progressively less oligopolistic and more diffuse as the market shares of first-tier suppliers, the United States and the former Soviet Union, declined and those of the second and third tier, for instance, Brazil, China, France, Germany, Israel, Italy, the United Kingdom and the former Czechoslovakia and Yugoslavia, increased. Economic and commercial incentives to export arms came to rival the political and military incentives. Those economic and commercial incentives included the need to earn foreign currency, pay for imports, improve trade and payments balances, maintain a viable defense industrial base, lower production costs and generate economies of scale by stretching production runs and spreading out R&D costs, and increase dependence on foreign sales, especially at the industrial sector and firm levels.

More suppliers in the market meant more competition. The commercialization of arms exports and the dependence of suppliers on exports meant intense competition. Suppliers exuberantly pursued buyers. Marketing and advertising became a more prominent feature of sales promotion. More countries and firms hawked their wares at the increasing number of military trade shows. Heightened competition prompted major suppliers to relax restrictions on arms exports, offer compensatory trade agreements and attractive long-term financing arrangements—easy credit at below market rates—that bordered on subsidies, export technologically sophisticated, front-line equipment—at times even before it became available to national defense forces—and develop products exclusively for the export market.

This array of supply-side developments—the increase in the number of suppliers, declining market concentration, increasing commercialization and dependence upon exports, and increasing competition—collided with stagnating demand. In the 1980s, demand for weapons in the developing world and globally dropped off considerably.[27] What were, from the perspective of suppliers, adverse market conditions transformed the international arms market from a seller's into a buyer's market. Market conditions led exporters to conclude agreements that would not have been considered in the supplier-dominated market of the 1950s and 1960s. Exporters found they had to accommodate importers by offering ever more attractive, and creative, sales-promoting inducements. These inducements often took the form of direct and indirect compensatory trade agreements (offsets) that are now familiar features of the international arms trade. These agreements involved co-production, licensed production, subcontractor production, technology transfer, countertrade and even the provision of investment capital.[28]

Developing countries attempting to substitute locally manufactured for imported weaponry became adept at exploiting the market. Market conditions enabled them to require that suppliers not only provide arms but also assist in their progressive local assembly and production. Technology and expertise as well as hardware had to be transferred: exporters were confronted with the choice of losing sales or providing buyers in the developing world with the arms *and* the technology needed to establish and develop military import-substitution programs. Unable to resist buyer demands in a buyer's market, exporters inevitably succumbed to market dictates.

Major suppliers such as France, Germany, the United Kingdom, the United States and even the former Soviet Union provided developing countries with military technology that played a crucial role in the establishment and development of their military import-substitution programs, particularly in the first three of the five stages of defense industrial development. Market developments therefore provided developing countries with a critical opportunity to acquire foreign technology. However, it was their prior industrialization that enabled the developing countries to capitalize on that opportunity.

Industrial Growth

Despite the economic setbacks of the 1980s, developing countries made remarkable economic and industrial progress during the post-World War II period. Significant—and often dramatic in the case of the newly industrializing countries (NICs)—economic and industrial growth in the developing world was sufficient to indicate that international economic power had become less concentrated and more dispersed. The economies of some developing countries now rival those of the advanced industrial countries. Northern direct investment and the transfer of industrial technology to the South helped produce Hobsonian offspring able to challenge the economic dominance of the industrial North. The economic and industrial progress that provided the basis for that challenge also enabled developing countries to engage in military import-substitution.

The economies of some developing countries, especially their manufacturing sectors, have grown substantially since 1945, at rates that compare favorably with those of the advanced industrial countries. A significant proportion have grown as fast as, or even faster than, the industrialized countries. Several, particularly East Asian NICs such as Singapore, South Korea and Taiwan, have achieved particularly noteworthy growth rates.

With manufacturing growing more rapidly than gross domestic product (GDP), the expansion of industry and manufacturing set the pace of economic growth. Generally, developing countries have achieved higher manufacturing growth rates than have the advanced industrial countries. Double-digit manufacturing growth rates, especially in countries such as Brazil, Indonesia, Singapore and South Korea, have not been uncommon. This growth in manufacturing has altered the structure of the economies of many developing countries. The proportion of GDP attributable to agriculture has declined while that attributable to manufacturing has risen. The proportion of GDP contributed by manufacturing in developing countries has come to rival that in the advanced industrial countries.

Although the economic and industrial growth of developing countries has often been viewed as dependent development,[29] or indebted industrialization,[30] it has provided the economic and industrial foundation for the initiation and development of military import-substitution programs. Development of an industrial infrastructure, iron and steel plants, and chemical, electronics, transportation and shipbuilding industries contributed directly to defense industrial potential. Industrialization provided a managerial elite, trained scientific and technical manpower and a corps of skilled workers. In short, economic and industrial growth enabled developing countries to absorb and build upon the military technology acquired from established defense producers in the advanced industrial countries.

The State

As Alexander Gerschenkron recognized in a seminal piece,[31] the state typically has played an extensive role in the economic development of the late industrializers. In European economic development, the state played the smallest role in the early industrializers, especially England, where private capital financed the growth of industrial activities. On the continent, *crédit mobilier* banks financed industrial growth in France, Germany, Austria, Italy, Belgium and other small industrial countries. In Russia, however, the last major country in Europe to begin industrializing, the state "assumed the role of the primary agent propelling the economic progress in the country."[32] In developing countries as in Russia, the state took upon itself "the role of the primary agent" in industrialization.

Industrial development in the late industrializers required a high level of state involvement. Protection against foreign competition alone necessitated extensive state manipulation of market forces. Since the first country went industrial, subsequent industrializers have been confronted with the pressures of foreign competition. Infant industries typically require protection from foreign competition to survive, and only the state can effectively provide protection. The example of successful industrialization provided by the early starters also created enormous expectations with which only the state could cope.

In the late industrializers, the state has been performing, to a greater or lesser extent, at least six developmental functions that have been transferred to the military import-substitution process.[33] Perhaps the most significant function is that of director. As director, the state attempts to guide the course of economic development, determining the direction and quality of economic and industrial growth. The state establishes priorities and attempts to set a timetable of economic growth. Frequently, state economic ministries draw up multi-year development plans designed to enable the state to meet its economic and industrial goals.

A second function of the state is that of protector. As the "gatekeeper" between intrasocietal and extrasocietal flows of action,"[34] the state can protect fledgling local industry from import competition by erecting a protective wall of tariff and non-tariff barriers to stem the inflow of foreign products. In this way it can foster import-substitution and provide secure domestic markets for national producers.

Third, the state provides an industrial infrastructure and the basic services without which modern industry cannot survive. Typically the state builds a transportation infrastructure (roads, railroads, airports and harbors) and communications networks (mail and telephone), and assists in the development of energy resources (oil, gas, coal, nuclear and hydroelectric).

The fourth function of the state is that of financier. The state often supplies the start-up capital for new and perhaps risky industrial ventures that, although necessary for industrial progress, commercial lending institutions might avoid. To support its third and fourth functions, the state has endeavored to institute more reliable and effective systems for extracting resources.

The state's final two functions are producer and consumer. The state itself may assume the role of producer. When it does so, it typically invests in large-scale and often financially risky industrial enterprises that private, domestic firms are perhaps incapable of establishing. State ownership of the means of production also serves to limit foreign penetration and enhance national control of key industrial sectors.

In its role as consumer of national products, the state contributes its purchasing power to domestic demand so as to enlarge the domestic market. The state procures products ranging from office supplies and equipment to motor vehicles and computers from domestic suppliers.

Historically, the role of the state in defense industrialization has paralleled its role in the larger processes of economic development.[35] The state has played a crucial role in the development of defense industrial capabilities and the acquisition and assimilation of foreign military technology ever since the military revolution of 1550–1660.[36] In the developing world, the state has been instrumental in the initiation and expansion of military import-substitution programs. The state has determined which sectors of the arms industry to emphasize, set the timetable for the development of military import-substitution programs and coordinated private and public sector activities. It has rather easily and effectively protected infant defense industries from foreign competition. The domestic military market, after all, is a state-dominated market; a monopolistic state decides what is to be purchased and from whom. It has even protected non-military products manufactured by defense firms from foreign competition.

The state has been a primary source of investment capital for private firms initiating or expanding production of military equipment. Since commercial lending institutions have often been reluctant to provide capital in the early stages of military import-substitution programs, private enterprises have had to turn to the state. In such cases, when the private sector has been either unwilling or incapable, the state has taken on the function of producing military equipment. In some developing countries, such as India, defense production, until quite recently, has been exclusively a public sector activity. To assure that arms production will be nationalized rather than internationalized, the state has restricted foreign penetration of the defense industrial sector through either public sector production, which excludes foreign investment altogether as in India, or limited foreign investment in private sector defense firms as in Brazil in the past. Whether defense production occurs primarily in the public sector, as in, for example, Argentina, India, Indonesia, Israel, Nigeria, Pakistan, Peru, Singapore and Taiwan, or is distributed across both private and public sectors as in Brazil, Chile, the Philippines, South Africa, South Korea and Thailand, the state typically absorbs the cost of R&D and plays a key role in acquiring the foreign defense technology necessary for defense industrial advancement.

The Impact of Import-Substitution on Dependence

It is apparent that a significant number of developing countries are attempting to substitute domestically manufactured for imported arms in an effort to reduce their dependence upon foreign suppliers. The continued role of foreign technology in defense industrialization, however, has led analysts to question the effectiveness of military import-substitution as a counterdependence strategy. Military import-substitution has relied heavily upon military technology acquired from the same sources that developing countries have been dependent upon for arms. The developing world's defense industrial products continue to include equipment assembled or produced under license. Even indigenously designed products generally incorporate imported components or components manufactured under license.

The central role of foreign technology in military import-substitution prompted one noted observer to write that "there is, for the time being, no short-term or even medium-term fulfillment of the desire of developing countries to reach a high degree of self-sufficiency in arms production."[37] Others have argued that "Domestic production creates other dependencies . . .,"[38] that military import-substitution results only in the replacement of dependence on arms imports with dependence on technology inputs[39] and even that military import-substitution increases dependence.[40] A number of analysts have advanced these arguments[41] and have asserted that developing countries have not, and will not, attain military self-reliance.[42]

The proposition that military import-substitution reduces military dependence and enhances autonomy cannot be summarily dismissed, however. The reduction and elimination of military dependence are long-term processes. Military import-substitution is a necessary part of that process and is more than a mere change in the form of dependence. Dependence is not absolute but relative; there are degrees of dependence. The transition from import dependence to technological dependence is a change not just in the form but also in the character of military dependence. Military import-substitution involves a transition from importing arms to importing and assimilating the technology required progressively to assemble, manufacture and develop arms.[43] This transition represents a reduction in the level of military dependence. Even the adaptation and integration of imported systems evident in the add-on and add-up engineering denigrated by Ball[44] amounts to technological progress that reduces dependence.

A crucial difference between import dependence and technological dependence must be acknowledged: the former is static while the latter is dynamic. Military dependence inevitably is static when only arms are imported. When technology instead of, or even in addition to, arms is imported and assimilated, a dynamic form of dependence that has an inherent potential to reduce dependence replaces a static form.[45] Suppliers cannot withdraw technology once acquired; it is not as amenable to controls by suppliers. Conditions in the arms market are such that a supplier will, at best, find it difficult to exploit the leverage derived from the transfer of technology. Most significantly, technology, unlike arms, can be built upon. Already, arms industries built upon imported technology are providing

increasing proportions of the developing world's arsenals and thereby reducing the need for imported hardware. One observer who empirically
examined the relationship between arms production and arms imports by
developing countries concluded that "Indigenous arms production in the
third-world has tended to reduce the importation of arms."[46] As development and production experience accumulates, dependence on foreign technology should decline as well.[47]

The data in table 5-5 show that domestic production can account for a
significant production of total arms acquisition in developing countries. It
accounted for 50 percent or more of total arms acquisition during the years
1965–90 in three of the twelve countries (table 5-5)—Brazil, Singapore and
South Africa. In two more, India and South Korea, domestic production was
the source of over 40 percent of total acquisition. Argentina and Israel met
34 percent of their acquisition through domestic production, and Taiwan 27
percent.

The share of domestic production in total arms acquisition in this group
of countries is increasing. In ten of the twelve countries, the level of
acquisition met through domestic production was greatest during the two
most recent of the five periods for which data are available (table 5-5). In
the case of the two exceptions, India and Taiwan, the data for 1990 indicate
they may have reversed that trend.

Significantly, indigenous production has accounted for a larger proportion of total acquisition than has licensed production. The proportion of
acquisition met by licensed production never exceeded that met by indigenous production in any of the twelve, either during the entire period
1965–90 or during any of the five five-year periods. In none of the twelve
countries did licensed production account for more than 17 percent of total
acquisition during the years 1965–90.

A number of analysts have argued that the developing world's arms
manufacturers cannot but fail to keep technologically apace with their
counterparts in the advanced industrial countries, that the defense technology gap between North and South will continue to expand, and that the
weapons produced in the developing world are technologically obsolete by
Northern standards.[48] However, to reduce dependence on imported arms
and technology, military import-substitution need not result in the production of military equipment equivalent to that produced by advanced industrial countries, nor narrow the defense technology gap between North
and South. It need only provide weaponry incorporating technology appropriate for the military environment of the developing world. That tech-

Table 5-5. *Arms Production as a Proportion of Total Acquisition, Selected Countries, 1965–90*

(Percent)

Country	1965–69	1970–74	1975–79	1980–84	1985–89	1990	1965–90
Argentina							
A	9	4	16	23	41	49	22
B	<1	1	2	14	26	46	12
C	10	5	19	36	67	95	34
Brazil							
A	4	20	34	70	48	76	43
B	—	5	7	9	8	4	7
C	4	25	41	80	57	80	50
Chile							
A	14	1	—	5	25	12	8
B	2	1	—	4	14	11	5
C	16	1	—	9	39	22	14
Egypt							
A	2	1	8	4	9	12	5
B	1	—	—	2	5	9	2
C	3	1	8	6	15	21	7
India							
A	30	37	36	35	23	42	30
B	17	20	16	13	16	38	17
C	47	58	52	48	38	79	47
Indonesia							
A	6	3	6	7	25	41	14
B	—	—	2	4	9	15	6
C	6	3	8	10	34	56	20
Israel							
A	1	23	35	43	45	73	34
B	—	—	<1	1	<1	—	<1
C	1	23	35	44	45	73	34
South Korea							
A	<1	4	16	43	27	50	27
B	<1	—	9	27	18	30	17
C	<1	4	26	70	45	80	43
Pakistan							
A	—	—	1	1	1	2	1
B	—	—	<1	<1	1	1	<1
C	—	—	1	1	2	3	1
Singapore							
A	16	12	31	79	75	98	52
B	11	10	15	14	10	—	12
C	27	22	46	94	85	98	64
South Africa							
A	24	31	40	54	57	66	48
B	7	15	21	12	7	3	11
C	32	46	60	65	67	69	59
Taiwan							
A	1	14	31	2	10	34	16
B	1	5	26	—	—	20	11
C	2	19	57	2	10	55	27

Note: — means data not provided by SIPRI. A—indigenous production/total acquisition; B—licensed production/total acquisition; and C—total production/total acquisition.

Source: Derived from data presented in Ian Anthony, "The 'Third Tier' Countries: Production of Major Weapons," in Herbert Wulf (ed.), *Arms Industry Limited* (Oxford, 1993), table 17-1, 370–373.

nology generally is of an intermediate or even, in some cases, low level. Comments by a Pakistani Minister of State for Defense, Ghulum Sarwar Cheema, make that point:

> Because of our experience in the past, we have come to the realization that the equipment of our defense forces should be as far as possible locally manufactured so that, in our hour of need, we don't have to take a bowl and go begging. So, our first priority is indigenisation, self-reliance, local production. The hardware might not be in step with the latest technology, but as long as you are making it yourself, whether in collaboration with others or as a solo effort, you can sleep well at night. It might not be as sophisticated as what the people next door have, but at least we have full control over it; we will not be running short of numbers, we understand how to use it and the technology involved.[49]

Developing countries with the most advanced military import-substitution programs have already reduced the level of their dependence upon foreign arms and, to a lesser extent, technology. If technological independence can be defined as

> the ability to decide what technologies to import from abroad; under what terms and conditions to accept them; how to adjust and adapt them to national requirements, assimilate and diffuse them, derive the maximum benefits in terms of national skill formation; and determine the balance between imported and nationally developed technologies . . .,[50]

there can be little doubt that progress has been made in the reduction not only of dependence on imported arms but also technology. In the long run, technological dependence in the defense sector, as in other industrial sectors, can be overcome.[51]

Brazil

Despite the continuing difficulties confronting Brazilian producers of aircraft, armored vehicles, and missiles and rocketry, Brazil provides an illuminating example of the potential of arms producers and exporters in the developing world. Brazil has pursued a concerted across-the-board military import-substitution strategy since the mid-1960s.[52] Its aircraft and armored vehicle industries have been at the forefront of military import-substitution efforts. Latin America's premier aircraft manufacturer, Empresa Brasileira da Aeronautica S.A. (Embraer), was founded by the Brazilian government in 1969. During the 1970s, Embraer produced the EMB-326 Xavante advanced trainer/strike aircraft, the first Brazilian-built jet aircraft, under license with Aeronautica Macchi of Italy. It then produced a series of commercial aircraft under license with Piper. Domestically designed aircraft produced by Embraer include military and commercial versions of the Bandeirante transport/maritime patrol aircraft, in production since 1970, the Xingu transport/trainer, in production since 1977, and the Brasilia transport/reconnaissance aircraft, available since 1985. Embraer's internationally acclaimed Tucano military trainer, which first flew in 1980 and went into service with the Brazilian Air Force in 1983, became a "sales leader in the military turboprop trainer field."[53] A dedicated strike aircraft, the AMX, was developed in collaboration with Italy's Aeritalia and Aermacchi. One analyst concluded an assessment of the Brazilian aviation industry by writing that "the overall picture created is that of a capable, efficient and diversified industry. It is one that is already producing quite substantial returns in terms of independence from foreign sources. . . ."[54]

Brazil's private sector armored vehicle producers kept pace with its aviation industry. Engesa began series production of Urutu armored personnel carriers and Cascavel armored fighting vehicles at its plant in Sao Jose dos Campos in 1974. Two more Engesa wheeled armored vehicles, the Sucuri tank destroyer and Jararaca armored scout car, had been developed by the early 1980s. Engesa also developed two tracked armored vehicles, the Osorio battle tank and the Ogum tracked weapon platform/general-purpose carrier. Another firm, Bernardini, has rebuilt and upgraded M-3 and M-41 tanks and developed the Tamoyo battle tank in cooperation with the Brazilian Army.

Brazil's missile and naval military import-substitution programs are also at advanced stages. Avibras Aerospacial has provided the military with a variety of Avibras-designed rocket systems and has developed air-to-air, air-to-surface and medium-range ballistic missiles. It has also furnished the space program with the Sonda series of sounding rockets. Another firm, Orbita, has developed prototypes of surface-to-air, air-to-air and anti-tank missiles. The Arsenal de Marinha do Rio de Janeiro has supplied the Brazilian Navy with British-designed Niteroi class frigates constructed under license during the 1970s, and Brazilian-designed Inhauma class frigates built during the 1980s. The Arsenal de Marinha in Rio de Janeiro and Verolme in Porto Allegre have built West German Type 209 submarines, and there are plans to build nuclear-powered submarines. Brazilian shipyards also have built locally designed patrol craft.

As a result of the success of its military import-substitution programs, Brazil met fully half of its arms acquisition needs through domestic production during the years 1965–90. The level was even greater during the

1980s—80 percent from 1980–84 and 57 percent from 1985–90. In 1990, domestic production again was the source of 80 percent of the military equipment acquired by Brazil's armed forces.[55] Military import-substitution has also enabled Brazil to export its defense products.[56] Brazil's arms exports were more than its imports during the years 1983–93.[57] Between 1977 and 1988, Brazil's defense industries exported to thirty-five countries.[58] While the claim by the U.S. Office of Technology Assessment (OTA) that Brazil's export success during the 1980s made it the world's sixth leading arms merchant[59] was overblown, according to SIPRI data, Brazil was the world's nineteenth leading arms exporter during the years 1990–94.[60]

Significantly, Brazil's import-substitution and export performance have been founded primarily upon indigenous R&D capabilities rather than licensed production. Of the nearly $7.4 billion worth of military equipment produced in Brazil from 1965 to 1990, 86 percent was based on indigenous R&D efforts, only 14 percent on licensed production.[61] Even though the content of its defense industrial products has not been exclusively domestic, there can be little doubt that Brazil's military import-substitution programs enhanced its policy and behavioral autonomy and reduced its dependence upon arms imports.

India

While the Brazilian case may be particularly dramatic, there are other examples of successful military import-substitution. India has also developed an across-the-board defense industrial capability.[62] Its nine public sector defense industrial enterprises have extensive experience in the assembly, licensed production and design of weapons systems. Hindustan Aeronautics Limited (HAL) has assembled and produced under license three British aircraft, the Vampire and Gnat fighters and the HS-748 military/commercial transport, French Aérospatiale Alouetta and Lama helicopters, Anglo-French Jaguar supersonic strike aircraft, West German Dornier 228 transports, and Soviet MiG-21s, MiG-23s, MiG-27s and MiG-29s. The Ajeet fighter, a modified and upgraded version of the Gnat, was produced until 1981. Aircraft of its own design produced by HAL include the Marut fighter, HT-2 and Kiran trainers, and the Pushpak and Krishak light planes. A light combat aircraft and an advanced light helicopter are under development. India's Defence Research and Development Organisation and HAL have also recently developed a radar-absorbent coating that reduces the radar cross-section of aircraft by up to 70 percent, providing India with a stealthy advantage over regional adversaries.[63]

Armored vehicles produced at the Avadi Heavy Vehicles Factory include the Vijayanta tank, a modified version of the British Chieftain, and the Soviet T-72 main battle tank. The Combat Vehicle Research and Development Establishment is developing the Arjun, an indigenous main battle tank. Bharat Dynamics Limited has assembled or produced Aérospatiale SS-11 anti-tank missiles, Milan anti-tank missiles and Soviet Atoll air-to-air missiles under license. Indigenously developed missiles include the Prithvi and Agni surface-to-surface missiles, the Nag anti-tank missile, and the Trishul and Akash surface-to-air missiles.

India's three public sector shipyards—Mazagon Dock Limited, Garden Reach Shipbuilders and Engineers Limited, and Goa Shipyard Limited—have constructed indigenously designed frigates, patrol craft, landing craft, minesweepers and support vessels and have built British Leander class frigates and German HDW Type 1500 submarines under license.

The success of India's military import-substitution programs, although more limited than Brazil's, is evident. India met just under half its arms acquisition needs through domestic production during the years 1960–90 (table 5-5). While the proportion of total acquisition attributable to domestic production is comparable with that in Brazil, the proportion of domestic production based on indigenous R&D capabilities rather than licensed production is somewhat less than in Brazil. From 1965 to 1990, over one-third of domestic production in India—compared with only 14 percent in Brazil—was the result of licensed production (table 5–4). A big difference between the two countries is that India's arms imports continue to dwarf its exports. Despite recurring announcements of export promotion drives, military import-substitution in India remains oriented toward the domestic market and shows little sign of enabling India to penetrate the international arms market.

Israel, South Africa and South Korea

Israel, South Africa and South Korea have also reduced their dependence on imported arms. Israel's technologically sophisticated and internationally competitive defense industry[64] is the source of an increasing share of its military requirements (table 5-5). Almost all Israeli defense production—99 percent during the years 1965–90 and 100 percent in 1990 (table 5–4)—is based on indigenous R&D. According to one assessment, "The story of

Israeli aerospace is almost a textbook example of the process of technology transfer through licensed production leading to indigenous development and production."[65] Israel's defense electronics R&D effort has been even more successful.[66] Another analyst has argued that the Israeli defense industry's capability to surge the manufacture of ammunition and spare parts and an ability to design and produce weapons tailored to Israel's military requirements have provided it a "significant degree of freedom."[67]

South Africa has become increasingly self-sufficient since the U.N.-sponsored voluntary arms embargo of 1963, even though its defense industry barely existed in 1963.[68] In 1966, South Africa acquired 70 percent of its arms from abroad.[69] One analyst, writing in the late 1980s, argued that "the local arms industry provides the South African Defense Force (SADF) with almost 100% of the equipment for its land forces, all the weapons for its airforce, and all the arms, ammunition and missile strike craft for its navy."[70] Arms imports averaged only $33 million annually from 1983 to 1993 (in constant 1993 dollars).[71] Licensed production accounted for only 18 percent of total domestic production from 1980 to 1984, 11 percent from 1985 to 1989 and 4 percent in 1990.[72]

South Korea, like Brazil, invested heavily in military import-substitution programs during the 1970s and 1980s.[73] As a result, the value of its arms imports has declined, while the level of weaponry produced domestically reached 50 percent by 1978[74] and, reportedly, 70 percent by the early 1980s.[75] In 1990, domestic production was the source of 80 percent of the country's total weapons acquisition (table 5-5). Production based on indigenous R&D accounted for 62 percent of the $6.7 billion worth of weapons manufactured in South Korea during the years 1965–90.[76]

Prospects for the Production and Export of Conventional Arms by Developing Countries

While countries such as Brazil, India, Israel, South Africa and South Korea have reduced their dependence on imported arms and, to a lesser extent, imported technology, not all developing countries that have turned to military import-substitution have been, or will be, able to do so. Argentina, with an across-the-board production capability, still found it necessary to employ imported weaponry in its attempt to wrest the Malvinas Islands from Britain in 1982. Egypt, which also has an across-the-board defense industrial capability, still relies heavily on arms from the United States and Western Europe. Other defense producers such as those classified as lesser and minor producers in table 5-2 have found it difficult to nationalize arms acquisition fully. Even the most successful of the developing world's defense manufacturers have not yet been able to eliminate arms imports— India is one of the developing world's leading arms importers, Israel still requires American arms, and South Korea still arms its troops, in part, with American weaponry.

Countries in a high-threat, technologically sophisticated military environment have found it extremely difficult to reduce dependence upon imported arms and technology substantially, even when, as in the cases of Israel, India and South Korea, they have devoted considerable resources to military import-substitution programs. Brazil has been able to develop significant self-reliance in part because it is situated in a relatively benign security environment. Militarily weak adversaries made the accomplishment of self-reliance less problematical for South Africa. India, Israel and South Korea have, however, found military import-substitution to be somewhat more arduous. Israel is located in what is arguably the most volatile region of the world. It is confronted with adversaries armed with some of the most advanced conventional weaponry available. India is confronted by China, which has one of the world's largest military establishments, and by Pakistan, which has been armed with advanced American weapons. South Korea faces a seemingly still implacable North Korea.

While India, Israel and South Korea have been able to reduce their dependence upon imported arms, it is inherently more difficult for them to become as self-sufficient as Brazil and South Africa. Brazil and South Africa were able to utilize intermediate-level technology in their military import-substitution programs. The threats confronting Israel, India and South Korea compel them to acquire relatively sophisticated weaponry. The products of their military import-substitution programs must be more able to counter the equipment supplied by advanced industrial countries.

In spite of the obstacles encountered by countries attempting to nationalize their arms acquisition through military import-substitution, the continuing diffusion of defense industrial capabilities has contributed to a convergence of arms acquisition patterns in the developing and advanced industrial worlds. Advanced industrial as well as developing countries import military equipment, and developing as well as advanced industrial countries manufacture arms. During the years 1983–93, every member of the North Atlantic Treaty Organization (NATO) and Organization for Economic Co-operation and Development (OECD), core Northern military and economic organizations, imported arms. Of the sixteen members of

NATO and twenty-four of OECD, only two (Iceland and Luxembourg) did not import arms every year during the period.[77] Brzoska and Ohlson estimated that the medium-sized advanced industrial countries produce 70–80 percent of their weapons inventories, the smaller advanced industrial countries such as Austria, Sweden and Switzerland 40–60 percent; they acquire the balance from abroad.[78] Advanced industrial countries, as in developing countries, also manufacture military equipment under license, incorporate foreign components in domestically designed weapons and collaborate on weapons development and manufacture.[79] The United Kingdom, for instance, depends upon American technology and hardware to maintain its "independent" nuclear posture. German, Italian, Japanese and Swedish military aircraft have incorporated imported avionics, engines and other components. As is evident from the deepening concern about the dependence of the American defense industrial base on foreign technology,[80] there are limits even to the military self-reliance of the United States. The difficulties encountered by arms manufacturers in the developing world have their counterparts in the advanced industrial countries. Arms procurement patterns in the developing world increasingly resemble those of the advanced industrial countries. Developing countries with military import-substitution programs are becoming less dependent on imports of arms and technology at the same time that advanced industrial countries rely more on imports of arms and foreign technology than is generally recognized.

While the diffusion of defense industrial capabilities from the advanced industrial countries of the North to the developing countries of the South has diminished differences in arms acquisition behavior and patterns, prospects for the developing world's arms manufacturers in the 1990s are somewhat less glowing than they appeared only a decade ago. As the experience of Brazil's struggling arms producers and exporters indicates, few of the developing world's defense industrial firms are as well positioned as those in the advanced industrial countries to weather continuing domestic and international declines in military spending and the consequent deterioration in the demand for their products. Evolving conditions in an international market that earlier enabled the developing world's defense industries to emerge, develop and even prosper now threaten the continued viability of these producers that have thrived without the security blanket provided by substantial domestic demand. Brazil, once held up as an example of the potential of military import-substitution, is now used to illustrate the fate of the developing world's producers and exporters whose success was highly export-dependent. Its once highly touted exporters are now struggling to survive.

Just as Brazil's international market performance in the past was exceptional, so too are its difficulties. The primary objective of military import-substitution was to reduce dependence on foreign suppliers by replacing imported weaponry with domestically manufactured arms. However, Brazil and other defense industrializers, especially, perhaps, Israel and South Korea, were quick to take advantage of export opportunities. In the process they become overly dependent on foreign demand for their products. Other producers, such as India and South Africa, whose primary markets were and are domestic, will also confront the dilemmas of declining defense spending. However, the problems they face are less severe than those of Brazilian producers, which must brave both a contracting international market and shrinking domestic demand.

It might seem that the retreat of the state evident in the rush to economic liberalization in many parts of the world could endanger military import-substitution as well. An activist state has, after all, been a feature of defense industrialization. In the realm of security, however, it is unlikely that states will abandon their mercantilist behavior. They will not leave national security entirely to the dictates, or whims, of "the market." The nationalization of arms acquisition, which is at the heart of military import-substitution, will continue to be a high priority. In the end, the fate of faltering defense industrialists in the developing world, such as those in Brazil, rests not with an invisible hand, but with a quite visible state.

The desire of developing countries to minimize their dependence on arms imports is no less strong now than during the Cold War. If anything, they will be even more wary of import dependence in a post-Cold War international arms market dominated by one country—the United States—than in a Cold War market dominated by two countries—the United States and the Soviet Union—that could be played off against each other. Consequently, they are unlikely to abandon the military import-substitution efforts initiated during the 1950s, 1960s and 1970s. At the same time, the economic and industrial prerequisites of defense industrialization will permit few other developing countries to begin that arduous process in the near future. Those already engaged in the process will attempt to consolidate their gains and build upon the foundation already in place. A domestic focus on nationalizing arms procurement—which is, it will be recalled, the point of military import-substitution—will, in all likelihood, be more pronounced in the future than it was during the height of the international arms market.

While forays into the export market by the likes of Brazil, Israel, South Korea and perhaps even India and South Africa will continue, the export successes of the past will be difficult to duplicate. Nevertheless, the developing world's arms producers and exporters are now a permanent feature of the international arms market.

Notes

1. See, for instance, Leonard Beaton and John Maddox, *The Spread of Nuclear Weapons* (New York, 1962); Lewis A. Dunn, *Controlling the Bomb: Nuclear Proliferation in the 1980s* (New Haven, 1982); Ted Greenwood, Harold A. Feiveson, and Theodore B. Taylor, *Nuclear Proliferation: Motivations, Capabilities, and Strategies for Control* (New York, 1977); George H. Quester, *The Politics of Nuclear Proliferation* (Baltimore, 1973); George Quester (ed.), *Nuclear Proliferation: Breaking the Chain* (Madison, 1981); Richard N. Rosecrance (ed.), *The Dispersion of Nuclear Weapons: Strategy and Politics* (New York, 1964).

2. Lewis A. Dunn, "Containing Nuclear Proliferation," *Adelphi Papers*, 263 (London, 1991); Mitchell Reiss and Robert S. Litwak (eds.), *Nuclear Proliferation After the Cold War* (Washington, D.C., 1994); Scott D. Sagan and Kenneth N. Waltz, *The Spread of Nuclear Weapons: A Debate* (New York, 1995); Leonard S. Spector and Mark G. McDonough, with Evan S. Medeiros, *Tracking Nuclear Proliferation: A Guide in Maps and Charts, 1995* (Washington, D.C., 1995).

3. See Michael Brzoska and Thomas Ohlson, *Arms Transfers to the Third World, 1971–85* (Oxford, 1987); Anne Hessing Cahn, Joseph J. Kruzel, Peter M. Dawkins, and Jacques Huntzinger, *Controlling Future Arms Trade* (New York, 1977); Philip J. Farley, Stephen S. Kaplan, and William H. Lewis, *Arms Across the Sea* (Washington, D.C., 1978); Amelia C. Leiss with Geoffrey Kemp, John H. Hoagland, Jacob S. Refson, and Harold E. Fischer, *Arms Transfers to Less Developed Countries* (Cambridge, Mass., 1970); David J. Louscher and Michael D. Salomone (eds.), *Marketing Security Assistance: New Perspectives on Arms Sales* (Lexington, Mass., 1987); Stephanie G. Neuman and Robert E. Harkavy (eds.), *Arms Transfers in the Modern World* (New York, 1979); Andrew J. Pierre (ed.), *Arms Transfers and American Foreign Policy* (New York, 1979); Andrew J. Pierre, *The Global Politics of Arms Sales* (Princeton, 1982); Uri Ra'anan, Robert L. Pfaltzgraff, Jr., and Geoffrey Kemp (eds.), *Arms Transfers to the Third World: The Military Buildup in Less Developed Countries* (Boulder, 1978); Stockholm International Peace Research Institute (SIPRI), *The Arms Trade with the Third World* (London, 1971); Robert E. Harkavy and Stephanie G. Neuman (eds.), *The Arms Trade: Problems and Prospects in the Post-Cold War World* in *The Annals of the American Academy of Political and Social Science*, 535 (1994); William D. Hartung, *And Weapons for All* (New York, 1994); William W. Keller, *Arm in Arm: The Political Economy of the Global Arms Trade* (New York, 1995); Edward J. Laurance, *The International Arms Trade* (New York, 1992); David Mussington, *Understanding Contemporary International Arms Transfers*, Adelphi Paper 291 (London, 1994).

4. See: Kwang-Il Baek, Ronald D. McLaurin, and Chung-in Moon (eds.), *The Dilemma of Third World Defense Industries: Supplier Control or Recipient Autonomy?* (Boulder, 1989); Michael Brzoska, "The Impact of Arms Production in the Third World," *Armed Forces & Society,* XV (1989), 507–530; Michael Brzoska and Thomas Ohlson (eds.), *Arms Production in the Third World* (New York, 1986); Renato Dagnino, "The Emergence of Military Industry in the Third World," in Joseph Rotblat and Ubiratan D'Ambrosio (eds.), *World Peace and the Developing Countries: Annals of Pugwash 1985* (London, 1986) 235–248; Saadet Deger, *Military Expenditures in Third World Countries: The Economic Effects* (London, 1986); Carol Evans, "Reappraising Third-World Arms Production," *Survival,* XXVIII (1986), 99–118; Luis Herrera-Lasso, "Economic Growth, Military Expenditure, Arms Industry and Arms Transfer in Latin America," in Christian Schmidt (ed.), *The Economics of Military Expenditures: Military Expenditures, Economic Growth and Fluctuations* (New York, 1987) 113–134; Chandran Jeshurun (ed.), *Arms and Defence in Southeast Asia* (Singapore, 1989); James Everett Katz (ed.), *Arms Production in Developing Countries* (Lexington, Mass., 1984); James Everett Katz (ed.), *The Implications of Third World Military Industrialization: Sowing the Serpents' Teeth* (Lexington, Mass., 1986); Robert E. Looney, *Third-World Military Expenditure and Arms Production* (New York, 1988); Stephanie G. Neuman, "International Stratification and Third World Military Industries," *International Organization,* XXXVIII (1984), 167–197; Robert M. Rosh, "Third World Arms Production and the Evolving Interstate System," *Journal of Conflict Resolution,* XXXIV (1990), 57–73; Ralph Sanders, *Arms Industries: New Suppliers and Regional Security* (Washington, D.C., 1990); Miguel S. Wionczek, "Growth of Military Industries in Developing Countries: Impact on the Process of Underdevelopment," *Bulletin of Peace Proposals,* XVII (1986), 47–58; Herbert Wulf, "Arms Production in the Third World, Effects on Industrialization," in Christian Schmidt (ed.), *The Economics of Military Expenditures: Military Expenditures, Economic Growth and Fluctuations* (New York, 1987) 357–383; Patrice M. Franko, "Small-Scale Competitiveness in the New International Arms Market," *Security Dialogue,* XXVI (1995), 449–462; José O. Maldifassi and Pier A. Abetti, *Defense Industries in Latin American Countries: Argentina, Brazil, and Chile* (Westport, Conn., 1994); David Mussington, *Arms Unbound: The Globalization of Defense Production,* CSIA Studies in International Security, 4 (Washington, D.C., 1994).

5. For examples and assessments of dependency theory, see Samir Amin, *Unequal Development* (New York, 1976); Douglas C. Bennett and Kenneth E. Sharpe, *Transnational Corporations Versus the State: The Political Economy of the Mexican Auto Industry* (Princeton, 1985); Fernando Henrique Cardoso and Enzo Faletto, *Dependency and Development in Latin America* (Berkeley, 1979); Eliana Cardoso and Ann Helwege, *Latin America's Economy: Diversity, Trends, and Conflicts* (Cambridge, Mass., 1992); Ronald H. Chilcote and Joel C. Edelstein (eds.), *Latin America: The Struggle with Dependency and Beyond* (New York, 1983); Theotonio Dos Santos, "The Structure of Dependence," *American Economic Review,* LX (1970), 231–236; Peter Evans, *Dependent Development: The Alliance of Multinational, State, and Local Capital in Brazil* (Princeton, 1979); Gary Gereffi, *The Pharmaceutical Industry and Dependency in the Third World* (Princeton, 1983); Robert A. Packenham, *The Dependency Movement: Scholarship and Politics in Development Studies* (Cambridge, Mass., 1992); John Sheahan,

Patterns of Development in Latin America: Poverty, Repression, and Economic Strategy (Princeton, 1987).

6. See Christian Catrina, *Arms Transfers and Dependence* (New York, 1988); Björn Hagelin, "Military Dependency: Thailand and the Philippines," *Journal of Peace Research,* XXV (1988), 431–447; Andrew L. Ross, "Arms Acquisition and National Security: The Irony of Military Strength," in Edward E. Azar and Chung-in Moon (eds.), *National Security in the Third World: The Management of Internal and External Threats* (Aldershot, 1988), 152–187; Panitan Wattanayagorn, "ASEAN's Arms Modernization and Arms Transfers Dependence," *The Pacific Review,* VIII (1995), 494–507.

7. See A. F. Mullins, *Born Arming: Development and Military Power in New States* (Stanford, 1987).

8. Speech by S. Rajaratnam, *Official Records of the United Nations General Assembly,* Tenth Plenary Meeting (New York, 1976), 149–150.

9. Military import-substitution is one of two counterdependence strategies adopted by developing countries. Many developing countries have attempted to mitigate the adverse consequences of military import dependence through acquisitions from multiple sources, which typically precedes efforts to engage in military import-substitution and continues as military import-substitution programs develop. See Edward A. Kolodziej, *Making and Marketing Arms: The French Experience and Its Implications for the International System* (Princeton, 1987), 306–315; Stephanie G. Neuman, *Military Assistance in Recent Wars: The Dominance of the Superpowers, The Washington Papers,* XIV (New York, 1986), 38–60; Frederic S. Pearson, "The Priorities of Arms Importing States Reviewed," *Arms Control,* IX (1988), 177–179; Ross, "Arms Acquisition and National Security," 164–166; Andrew L. Ross, "On Arms Acquisition and Transfers," in Edward A. Kolodziej and Patrick M. Morgan (eds.), *Security and Arms Control, A Guide to National Policymaking,* I (Westport, Conn., 1989), 107–108; Andrew L. Ross, "The International Arms Trade, Arms Imports, and Local Defence Production in ASEAN," in Chandran Jeshurun (ed.), *Arms and Defence in Southeast Asia* (Singapore, 1989), 22–25.

10. In actuality, there is a continuum between these two forms of the fifth stage. For alternative formulations, see Ron Ayres, "Arms Production as a Form of Import-Substituting Industrialization: The Turkish Case," *World Development,* XI (1983), 814; David J. Louscher and Michael D. Salomone, *Technology Transfer and U.S. Security Assistance: The Impact of Licensed Production* (Boulder, 1986), 4–7; David Saw, "The Emergence of the Third World Aircraft Industry," *Military Technology,* XII (1988), 44–45; Miguel S. Wionczek, "The Emergence of Military Industries in the South: Longer-Term Implications," *Industry and Development,* XII (1984), 119; Herbert Wulf, "Arms Production in the Third World," in Stockholm International Peace Research Institute (SIPRI), *World Armaments and Disarmament: SIPRI Yearbook 1985* (London, 1985), 330. Discussions of alternative formulations can be found in Keith R. Krause, "Third Tier Producers and the Structure of the Global Arms Transfer and Production System," a paper prepared for presentation at the 31st Annual Meeting of the International Studies Association (Washington, D.C., 1990), 4–9; Keith R. Krause, *Arms and the State: Patterns of Military Production and Trade* (Cambridge, U.K., 1992), 171–174.

11. See Neuman, *International Stratification,* 172.

12. According to data compiled by the Future Security Environment Working Group, Commission on Integrated Long-Term Strategy, from 1965 to 1984 the number of fighter aircraft producers increased from one to eight, the number of helicopter producers from one to six, the number of tactical missile producers from zero to seven, and the number of tank producers from one to six. Future Security Environment Working Group, Commission on Integrated Long-Term Strategy, *The Future Security Environment* (Washington, D.C., 1988), 49.

13. "Statement of Hon. William H. Webster, director of Central Intelligence," in *Chemical and Biological Weapons Threat: The Urgent Need for Remedies*, Hearings of the Committee on Foreign Relations, United States Senate, 101st Congress, 1st Session, 24 January, 1 March, and 9 May 1989 (Washington, D.C., 1989), 30.

14. U.S. Arms Control and Disarmament Agency (ACDA), *World Military Expenditures and Arms Transfers, 1988* (Washington, D.C., 1989), 17–20.

15. W. Seth Carus, *Ballistic Missiles in Modern Conflict,* The Washington Papers, 146 (New York, 1991), 1.

16. "The Proliferation of Ballistic Missiles," *Arms Control Today,* XXII (1992), 28–29.

17. Brzoska and Ohlson, *Arms Production in the Third World,* 8, 10.

18. U.S. Arms Control and Disarmament Agency (ACDA), "Arms Transfers" in *World Military Expenditures and Arms Transfers 1990* (Washington, D.C., 1991), 12.

19. U.S. Arms Control and Disarmament Agency (ACDA), *World Military Expenditures and Arms Transfers 1993–1994* (Washington, D.C., 1995), 14.

20. ACDA, *World Military Expenditures 1993–1994,* 17. ACDA's definition of "developing country" is broad. The six developing countries counted among the twenty leading exporters during the years 1985–89 included not only Brazil, China, Israel, and North Korea, but also Spain and Yugoslavia. Similarly, the eight developing countries counted among the twenty leading exporters in 1989 included Bulgaria, Chile, China, Egypt, Israel, North Korea, Spain, and Yugoslavia.

21. ACDA, *World Military Expenditures 1993–1994,* 14, 16.

22. Derived from data presented by Ian Anthony, Gerd Hagmeyer-Gaverus, Pieter D. Wezeman, and Siemon T. Wezeman, "Appendix 14A. Tables of the Volume of the Trade in Major Conventional Weapons, 1985–94," in Stockholm International Peace Research Institute (SIPRI), *SIPRI Yearbook 1995: Armaments, Disarmament and International Security* (Oxford, 1995) 510–512. These SIPRI data are in constant 1990 U.S. dollars.

23. Anthony, Hagmeyer-Gaverus, Wezeman, and Wezeman, "Appendix 14A," 493.

24. Richard F. Grimmett, *Conventional Arms Transfers to Developing Nations, 1987–1994,* Congressional Research Service (CRS), U.S. Library of Congress (Washington, D.C., 1995), 66.

25. It should be noted that while a reduction in dependence is the paramount objective, other political, military, and economic incentives reinforce decisions to initiate and develop military import-substitution programs. See Arthur J. Alexander, William P. Butz, and Michael Mihalka, *Modeling the Production and International Trade of Arms: An Economic Framework for Analyzing Policy Alternatives,* N-1444-FF/RC (Santa Monica, 1981); Brzoska, "The Impact of Arms Production"; Gregory Copley, "The Road to Self-Sufficiency," *Defense & Foreign*

Affairs, XV (1987), 24–30; Deger, *Military Expenditures*, 152–155; Evans, "Reappraising Third-World Arms Production," 99–101; Michael Moodie, *Sovereignty, Security, and Arms*, The Washington Papers (Beverly Hills, 1979), 23–30; Ilan Peleg, "Military Production in Third World Countries: A Political Study," 209–230, in Pat McGowan and Charles W. Kegley, Jr. (eds.), *Threats, Weapons, and Foreign Policy*, V, Sage International Yearbook of Foreign Policy Studies (Beverly Hills, 1980), 219–221; Andrew L. Ross, *Arms Production in Developing Countries: The Continuing Proliferation of Conventional Weapons*, N-1515-AF (Santa Monica, 1981), 20–22.

26. Declining market concentration reflected changes in the structure of the post-World War II international system as it evolved from the tight bipolar system of the 1950s and 1960s into the ever looser bipolar system of the 1970s and 1980s. It also reflected the increasing number of suppliers. On the relationship between market structure and system structure, see Robert E. Harkavy, *The Arms Trade and International Systems* (Cambridge, Mass., 1975).

27. For discussions of these changes in the international arms market, see Andrew L. Ross, "The International Arms Market: A Structural and Behavioral Analysis," in Andrew L. Ross (ed.), *The Political Economy of Defense: Issues and Perspectives* (New York, 1991), 113–133; Andrew L. Ross, "The International Arms Market: A Structural Analysis," *International Interactions*, XVIII (1992), 63–83.

28. Grant T. Hammond, "Offset, Arms, and Innovation," *The Washington Quarterly*, X (1987), 173–185; Michael T. Klare, "The Unnoticed Arms Trade: Exports of Conventional Arms-Making Technology," *International Security*, VIII (1983), 68–90; Signe Landgren-Bäckström, "The Transfer of Military Technology to Third World Countries," in Helena Tuomi and Raimo Väyrynen (eds.), *Militarization and Arms Production* (New York, 1983), 193–204; Stephanie G. Neuman, "Coproduction, Barter, and Countertrade: Offsets in the International Arms Market," *Orbis*, XXIX (1985), 183–213.

29. Bennett and Sharpe, *Transnational Corporations Versus the State;* Cardoso and Faletto, *Dependency and Development in Latin America;* P. Evans, *Dependent Development;* and Gereffi, *The Pharmaceutical Industry.*

30. Jeff Frieden, "Third World Indebted Industrialization: International Finance and State Capitalism in Mexico, Brazil, Algeria, and South Korea," *International Organization*, XXXV (1981), 407–431.

31. Alexander Gerschenkron, *Economic Backwardness in Historical Perspective* (Cambridge, Mass., 1962).

32. Gerschenkron, *Economic Backwardness*, 17.

33. On the role of the state in late industrializers, see Peter B. Evans, "Transnational Linkages and the Economic Role of the State: An Analysis of Developing and Industrialized Nations in the Post-World War II Period," in Peter B. Evans, Dietrich Rueschemeyer, and Theda Skocpol (eds.), *Bringing the State Back In* (Cambridge, England, 1985), 192–226; Leroy P. Jones (ed.), *Public Enterprise in Less-Developed Countries* (Cambridge, England, 1982); Atul Kholi (ed.), *The State and Development in the Third World* (Princeton, 1986); Atul Kholi, "The Political Economy of Development Strategies: Comparative Perspectives on the Role of the State," *Comparative Politics*, XIX (1987), 233–246; Gustav Ranis (ed.), *Government and Economic Development* (New Haven, Conn., 1971); Gautam Sen, *The*

Military Origins of Industrialization and International Trade Rivalry (London, 1984).

34. J. P. Nettl, "The State as a Conceptual Variable," *World Politics,* XX (1968), 564.

35. Sen, *The Military Origins.*

36. Keith Krause, "The Political Economy of the International Arms Transfer System: The Diffusion of Military Technique via Arms Transfers," a paper prepared for presentation at the 30th Annual Meeting of the International Studies Association (London, 1989).

37. Wulf, "Arms Production in the Third World," 381. See also Herbert Wulf, "Developing Countries," in Nicole Ball and Milton Leitenberg (eds.), *The Structure of the Defense Industry: An International Survey* (New York, 1983), 341.

38. Stephanie G. Neuman, "Arms Transfers, Indigenous Defence Production and Dependency: The Case of Iran," in Hossein Amirsadeghi (ed.), *The Security of the Persian Gulf* (London, 1980), 145; see also Anne Hessing Cahn and Joseph J. Kruzel, "Arms Trade in the 1980s," in Anne Hessing Cahn, Joseph J. Kruzel, Peter M. Dawkins, and Jacques Huntzinger, *Controlling Future Arms Trade* (New York, 1977), 78.

39. Moodie, *Sovereignty, Security, and Arms,* 31–32. See also Michael Moodie, "Defense Industries in the Third World," in Stephanie G. Neuman and Robert E. Harkavy (eds.), *Arms Transfers in the Modern World* (New York, 1979), 301; Michael Moodie, "Vulcan's New Forge: Defense Production in Less Developed Countries," *Arms Control Today,* X (1980), 2.

40. Peter Lock and Herbert Wulf, "Consequences of the Transfer of Military-Oriented Technology on the Development Process," *Bulletin of Peace Proposals,* VIII (1977), 135; Peter Lock and Herbert Wulf, "The Economic Consequences of the Transfer of Military-Oriented Technology," in Mary Kaldor and Asbjørn Eide (eds.), *The World Military Order: The Impact of Military Technology on the Third World* (London, 1979), 226; see also IPSH-Study Group on Armaments and Underdevelopment, *Transnational Transfer of Arms Production Technology* (Hamburg, 1980), 87–89.

41. Ayres, *Arms Production as a Form;* Nicole Ball, *Security and Economy in the Third World* (Princeton, 1988), 371–385; Landgren-Bäckström, *The Transfer of Military Technology;* Rosh, *Third World Arms Production;* Helena Tuomi and Raimo Väyrynen, *Transnational Corporations, Armaments and Development* (New York, 1982), 238–253.

42. An exception can be found in Kolodziej: "The capability of these states to free themselves increasingly of exterior dependency appears to be growing." Edward A. Kolodziej, "Re-evaluating Economic and Technological Variables to Explain Global Arms Production and Sales," in Christian Schmidt (ed.), *The Economics of Military Expenditures: Military Expenditures, Economic Growth and Fluctuations* (New York, 1987), 322. Kolodziej, however, did not directly engage those analysts who have subscribed to the analytical consensus.

43. "It is impossible to transfer all or part of the production process without transferring technology and technological and manufacturing know-how. Even the simple sale of military products has allowed some Third World countries to successfully imitate their production; this is now a well-tried path to indigenous

production . . . which is made all the easier by the transfer of production processes." Harbor, "Assessing the Scale," 379.

44. Ball, *Security and Economy,* 372.

45. As Martin Edmonds observed, "Once the knowledge of how to produce weapons has been transferred from one state to another, a major cause of dependence has been eliminated." Martin Edmonds, "International Military Equipment Procurement Partnerships: The Basic Issues," in Martin Edmonds (ed.), *International Arms Procurement: New Directions* (New York, 1981), 3.

46. Looney, *Third World Military Expenditure,* 112; see also Robert Looney, "Have Third World Arms Industries Reduced Arms Imports?" *Current Research on Peace and Violence,* X (1989), 15–26.

47. Similarly, it should be recognized that the ability to integrate imported components into locally designed and produced weaponry represents significant technological and political progress. Technologically, dependence on imported components is an improvement upon dependence on imported weapons systems. Politically, as Krause has noted, "dependency for system components rather than complete weapons systems has less of a political profile, and is less likely to incur political pressures." Krause, "The Political Economy," 22.

48. Ayres, *Arms Production;* Ball, *Security and Economy,* 354–385; Michael Brzoska and Thomas Ohlson, "Conclusions," in Brzoska and Ohlson (eds.), *Arms Production in the Third World,* 287; Harbor, "Assessing the Scale," 380; Michael T. Klare, "Deadly Convergence: The Perils of the Arms Trade," *World Policy Journal,* VI (1988–89), 146.

49. Roger Frost, "Pakistan's New Defense Minister on Missiles, Self-Reliance and Afghanistan," *International Defense Review,* XXII (1989), 427–428.

50. Surendra J. Patel, "The Cost of Technological Dependence," *Ceres,* VI (1973), 16; also quoted in Ball, *Security and Economy,* 351.

51. See Emanuel Adler, *The Power of Ideology: The Quest for Technological Autonomy in Argentina and Brazil* (Berkeley, 1987); Daniel Chudnovsky, and Masafumi Nagao, *Capital Goods Production in the Third World: An Economic Study* (New York, 1983); Charles F. Dolan, George Modelski, and Cal Clark (eds.), *North-South Relations: Studies of Dependency Reversal* (New York, 1983); Dieter Ernst (ed.), *The New International Division of Labour, Technology and Underdevelopment: Consequences for the Third World* (Frankfurt, 1980); Jorge M. Katz (ed.), *Technology Generation in Latin American Manufacturing Industries* (New York, 1987); Heraldo Muñoz (ed.), *From Dependency to Development: Strategies to Overcome Underdevelopment and Inequality* (Boulder, 1981); James H. Street and Dilmus D. James (eds.), *Technological Progress in Latin America: The Prospects for Overcoming Dependency,* (Boulder, 1979).

One analyst's comparison of the Australian and Singaporean defense industries is instructive:

> The Australian defense industry was asked by the Singapore Government to help establish a post-independence defense industry in that state. Some assistance was rendered, but now, with the exception of full aircraft production, the Singaporean defense industry completely outclasses its Australian godfather. That is true by whatever yardstick is applied: volume of production, type of production, and quality. And Singapore's aerospace industry is developing at such a rate that it will soon rival Australia's.

Copley, "The Road to Self-Sufficiency," 28.

52. See Alexandre de S. C. Barros, "Brazil," in James Everett Katz (ed.), *Arms Production in Developing Countries: An Analysis of Decision Making* (Lexington, Mass., 1984), 73–87; Clovis Brigagão, "The Brazilian Arms Industry," *Journal of International Affairs,* XL (1986), 101–114; Renata Dagnino and Domicio Proença, Jr., "Arms Production and Technological Spinoffs: The Brazilian Aeronautics Industry," a paper prepared for presentation at the Latin American Studies Association XIV International Congress, New Orleans (1988); Patrice Franko-Jones, "'Public Private Partnership:' Lessons from the Brazilian Armaments Industry," *Journal of Interamerican Studies and World Affairs,* XXIX (1987–88), 41–68; Patrice Franko-Jones, *The Brazilian Defense Industry* (Boulder, 1992); Peter Lock, "Brazil: Arms for Export," in Brzoska and Ohlson (eds.), *Arms Production in the Third World,* 79–104; Louscher and Salomone, *Technology Transfer;* William Perry and Juan Carlos Weiss, "Brazil," 103–117, in James Everett Katz (ed.), *The Implications of Third World Military Industrialization: Sowing the Serpents' Teeth* (Lexington, Mass., 1986); Dominco Proença, Jr., "Guns *and* Butter? Arms Industry, Technology and Democracy in Brazil," *Bulletin of Peace Proposals,* XXI (1990), 49–57.

53. U.S. Congress, Office of Technology Assessment (OTA), *Global Arms Trade: Commerce in Advanced Military Technology and Weapons* (Washington, D.C., 1991), 147.

54. Saw, "The Emergence of the Third World Aircraft Industry," 49.

55. Derived from data presented in table 5–5.

56. Andrew L. Ross, "Full Circle: Conventional Proliferation, the International Arms Trade, and Third World Arms Exports," in Kwang-Il Baek, Ronald D. McLaurin, and Chung-in Moon (eds.), *The Dilemma of Third World Defense Industries: Supplier Control or Recipient Autonomy?* (Boulder, 1989), 107–108.

57. ACDA, *World Military Expenditures 1993–1994,* 103.

58. OTA, *Global Arms Trade,* 144–145.

59. ACDA, "Arms Transfers," in *World Military Expenditures 1988,* 143.

60. SIPRI, *SIPRI Yearbook 1995,* 493. According to CRS data, among the leading suppliers of arms to developing countries, Brazil ranked eleventh from 1982 to 1989 (Richard F. Grimmett, *Trends in Conventional Arms Transfers to the Third World by Major Supplier, 1982–1989* [Washington, D.C., 1990], 56), tenth from 1984 to 1987 (Grimmett, *Conventional Arms Transfers to the Third World, 1984–1991,* 67), and eleventh from 1985 to 1988 (Richard F. Grimmett, *Conventional Arms Transfers to the Third World, 1985–1992,* 66). The financial difficulties currently confronting Avibras and Engesa call into question the continued success of the Brazilian defense industry's export promotion campaign (although not necessarily the continued success of military import-substitution). See Patrice Franko-Jones, "The Brazilian Defense Industry in Crisis," a paper prepared for presentation at the 31st Annual Meeting of the International Studies Association (Washington, D.C., 1990).

61. Derived from data presented in table 5–4.

62. See Thomas W. Graham, "India," in Katz (ed.), *Arms Production in Developing Countries,* 157–191; OTA, *Global Arms Trade,* 151–159; Raju G. C. Thomas, *Indian Security Policy* (Princeton, 1986); Raju G. C. Thomas, "India: The Politics of Weapons Procurement," in Katz (ed.), *The Implications of Third World Military Industrialization,* 151–163; Herbert Wulf, "India: The Unfulfilled Quest for Self-Sufficiency," in Brzoska and Ohlson (eds.), *Arms Production in the Third*

World, 125–146; Chris Smith, *India's Ad Hoc Arsenal: Direction or Drift in Defence Policy?,* SIPRI (New York, 1994).

63. Vivek Raghuvanshi, "India Adds Stealth to Top Fighters," *Defense News* (23–29 October 1995), 1, 44.

64. See Carus, "Israel"; Robert E. Harkavy and Stephanie G. Neuman, "Israel," in Katz (ed.), *Arms Production in Developing Countries,* 193–223; Aaron Klieman, *Israel's Global Reach: Arms Sales as Diplomacy* (Washington, D.C., 1985); Alex Mintz, "Military-Industrial Linkages in Israel," *Armed Forces & Society,* XII (1985), 9–27; Alex Mintz, "The Military-Industrial Complex: American Concepts and Israeli Realities," *Journal of Conflict Resolution,* XXIX (1985), 623–639; OTA, *Global Arms Trade,* 83–103; Stewart Reiser, *The Israeli Arms Industry: Foreign Policy, Arms Transfers, and Military Doctrine of a Small State* (New York, 1989); Gerald Steinberg, "Israel," in Nicole Ball and Milton Leitenberg (eds.), *The Structure of the Defense Industry* (New York, 1983), 278–309; Gerald Steinberg, "Technology, Weapons, and Industrial Development: The Case of Israel," *Technology in Society,* VII (1985), 387–398; Gerald Steinberg, "Indigenous Arms Industries and Dependence: The Case of Israel," *Defense Analysis,* II (1986), 291–305; Gerald Steinberg, "Israel: High-Technology Roulette," in Brzoska and Ohlson (eds.), *Arms Production in the Third World,* 163–192; Gerald Steinberg, "Large-Scale National Projects as Political Symbols: The Case of Israel," *Comparative Politics,* XIX (1987), 331–346.

65. Harbor, "Assessing the Scale," 375.

66. Harbor, "Assessing the Scale," 375.

67. Steinberg, "Indigenous Arms Industry," 302; see also Alex Mintz and Gerald Steinberg, "Coping with Supplier Control: The Israeli Experience," in Kwang-Il Baek, Ronald D. McLaurin, and Chung-in Moon (eds.), *The Dilemma of Third World Defense Industries: Supplier Control or Recipient Autonomy?* (Boulder, 1989), 137–151.

68. See Michael Brzoska, "South Africa: Evading the Embargo," in Michael Brzoska and Thomas Ohlson (eds.), *Arms Production in the Third World* (London and Philadelphia, 1986), 193–214; Michael Brzoska, *Shades of Grey: Ten Years of South African Arms Procurement in the Shadow of the Mandatory Arms Embargo,* Arbeitspapiere 13 (Hamburg, Ger., 1987); Landgren, *Embargo Disimplemented;* Ron Matthews, "The Development of the South African Military Industrial Complex," *Defense Analysis,* IV (1988), 7–24; James McWilliams, *ARMSCOR: South Africa's Arms Merchant* (London, 1989).

69. Matthews, "The Development of the South African," 7.

70. Ibid.; see also Landgren, *Embargo Disimplemented,* 15–19.

71. Derived from data in ACDA, *World Military Expenditures 1993–1994,* 130.

72. Derived from data presented in table 5–5.

73. See Louscher and Salomone, "Brazil and South Korea;" Chung-in Moon, "South Korean Defense Industry," *Journal of Defense and Diplomacy,* IV (1986), 2–27; Chung-in Moon, "South Korea: Between Security and Vulnerability," in Katz (ed.), *The Implications of Third World Military Industrialization,* 241–266; Chung-in Moon and Kwang-il Baek, "Loyalty, Voice, or Exit? The U.S. Third-Country Arms Sales Regulations and R.O.K. Countervailing Strategies," *Journal of Northeast Asian Studies* 4 (1985), 20–45; Janne E. Nolan, *Military Industry in Taiwan and South Korea* (New York, 1986); Janne E. Nolan, "South Korea: An

Ambitious Client of the United States," in Brzoska and Ohlson (eds.), *Arms Production in the Third World,* 215–232; OTA, *Global Arms Trade,* 129–140.

74. U.S. Congressional Budget Office, *Force Planning and Budgetary Implications of U.S. Withdrawal from Korea* (Washington, D.C., 1978), 15.

75. Shim Jae Hoon, "South Korea: Standing on Its Arms," *Far Eastern Economic Review* (23 October 1981), 26.

76. Derived from data presented in table 5–4.

77. Derived from data in ACDA, *World Military Expenditures 1993–1994,* 99–137.

78. Brzoska and Ohlson, *Arms Production in the Third World,* 27.

79. See Pauline Creasey and Simon May (eds.), *The European Armaments Market and Procurement Cooperation* (New York, 1988).

80. See Theodore H. Moran, "The Globalization of America's Defense Industries: Managing the Threat of Foreign Dependence," *International Security,* XV (1990), 57–99.

Part Three

ARMS SALES POLICIES
AND PRACTICES OF
MAJOR SUPPLIERS

SIX

United States

Janne E. Nolan

THE UNITED STATES has a long tradition of ambivalence in its policies toward trade in conventional armaments. This ambivalence is particularly evident in the enduring coexistence of policies that actively promote U.S. arms exports and the periodic initiatives to impose restraints on the transfer of advanced weapons or technologies. Even though the United States has dominated the international arms market for most of the post-war era, it has maintained one of the most complex and far-reaching policy apparatuses for reviewing and regulating arms sales. However, the agencies with authority over arms exports, such as the State Department's Office of Defense Trade, have a hybrid character—typically they are charged with both facilitating and constraining contracts for American arms. These institutional arrangements reflect the underlying tension in a U.S. policy that maintains the view that U.S. arms sales are a wholly legitimate instrument of diplomacy but also simultaneously the belief that arms exports can fuel regional instabilities and increase the likelihood of war.

To date, the Clinton Administration's arms transfer policies have, with modest exceptions, been consistent with the policies crafted during the preceding twelve years of Republican leadership. Presidents Ronald Reagan, George Bush and Bill Clinton share a relative permissiveness toward arms exports to all but a handful of pariah states. Even as a presidential candidate, Clinton publicly supported proposed sales of advanced aircraft to Saudi Arabia and Taiwan, and a major departure from past policies guiding arms transfers does not appear likely. The Clinton Administration issued the results of a two-year policy review in February 1995. The criteria it emphasized as guides for U.S. arms sales policy are the traditional rationales used for the past two decades. They include, inter alia, "consistency with U.S. regional stability interests," "the degree to which the transfer supports U.S. strategic and foreign

131

policy interests" and "the potential misuse of the export in question" by a recipient.[1]

The U.S.-led coalition war against Iraq was a sobering experience that indicated to many the need for a dramatic reorientation in thinking about this laissez-faire approach to the international trade in military technologies. Iraq's arsenal, which was made up of weapons and technologies provided mainly by the industrial countries, prompted the Bush Administration to reevaluate the wisdom of its past practices. For a time an international consensus for more effective controls over the diffusion of conventional military technologies appeared to be growing. Nevertheless, even under the succeeding president, U.S. non-proliferation policy remains predominantly focused on controlling the diffusion of nuclear, chemical, biological and missile technologies—not conventional arms.

This chapter discusses the major supply- and demand-related factors in the United States that help define the policies guiding U.S. sales of conventional arms and technology, and those that will have to be considered if a new approach of greater restraint were to be sought. The analysis assesses the relative feasibility and effectiveness of alternative approaches from the standpoint of U.S. policy and provides future recommendations.

The Historical Context

Control of conventional arms transfers to the Third World has long been the bastard child of international diplomacy, subordinate to all other forms of regulation of military activities. Unlike the international agreements to control the diffusion of nuclear or chemical weapons, the monopoly for all but the most advanced armaments is already shattered. Moreover, many dispute that conventional proliferation poses a danger, and the perceptions of the utility of arms transfers as an instrument for advancing some national interests overwhelm any moral opprobrium or often even long-term pragmatic military objectives. The one formal conventional arms restraint regime, the Missile Technology Control Regime (MTCR), restricts the sale of ballistic and cruise missiles, largely because of their association with the delivery of nuclear or chemical weapons. Even more problematical, Washington has always seen conventional weapons as a benign alternative to nuclear proliferation. Transfers of conventional weapons remain the most common instrument of dissuasion in the efforts to stop new states from going the nuclear route.

Additional formal efforts to control the arms trade have been plagued by a chronic inability to translate either ethical or pragmatic concerns into durable policy. Multinational efforts such as the proposals put forward at the 1977 United Nations (U.N.) Special Session on Disarmament or the Carter Administration's negotiations with the Soviet Union to limit sales of advanced weapons have failed. Few efforts by developing countries to establish regional restraint have endured.

Despite mounting evidence of the hazards the continued proliferation of military technology in unstable regions poses to U.S. and global security, at present there is no solid national or international consensus in favor of arms restraint. It is as if the sheer complexity of the problem not only defies political consensus, but also encourages a certain fatalism about efforts to alter current practices.

It is often overlooked that large-scale transfers of arms by the United States are a fairly recent phenomenon. The initial impetus for American military aid—to create a collective security apparatus in Western Europe in support of the policy of containment—evolved after the Korean War, to include bilateral security assistance arrangements with countries beyond the Western Hemisphere, such as the Republic of Korea (South Korea), Japan, Thailand and Pakistan.[2] Throughout the 1950s and early 1960s, the types and quantities of arms transferred to recipients consisted predominantly of surplus or obsolete equipment, most of which was exported as part of grant aid or highly concessionary programs.

Fundamental structural changes in the international environment beginning in the late 1960s resulted in a far more complex political-military calculus on the use of this instrument of arms transfers for advancing national interests. Foremost among these changes was the growing competitiveness of the arms market—a natural outgrowth of the economic recovery of the European nations, the growth of a robust Soviet arms industry and the ascendant economic capabilities of former colonial states. By the late 1970s and early 1980s the virtual U.S. and Soviet monopoly of the arms market, which had existed throughout the 1950s and 1960s, gave way to intense competition among the industrial countries, including some newly industrial countries such as South Korea and Israel, over arms sales to an ever larger number of countries (figure 6–1).

The advent of the Nixon Doctrine in 1969 marked a watershed in U.S. arms transfer policy. The declared objective was to promote self-reliance in defense among friendly countries in place of continued dependence on U.S. interventionary forces. This new policy made explicit a growing require-

Figure 6-1. *Supplier Shares of Arms Deliveries to the Third World, 1977–84*

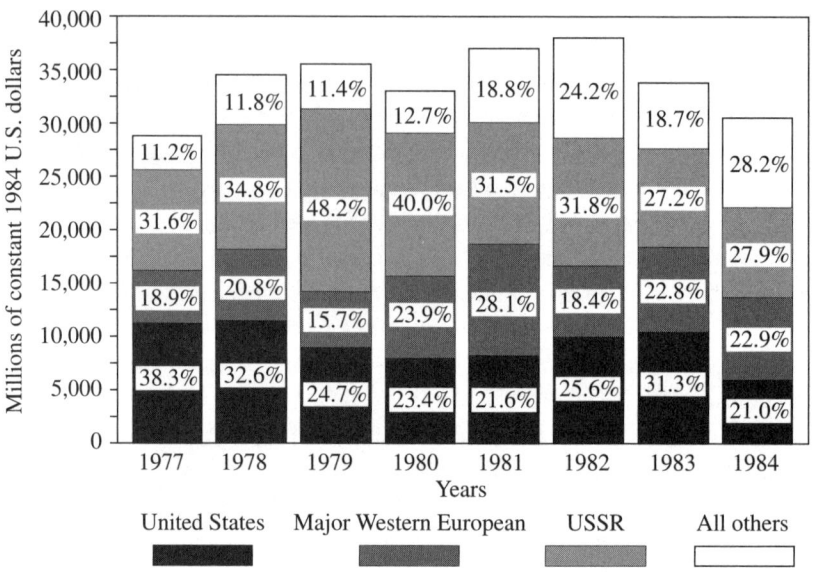

Source: Richard F. Grimmett, *Conventional Arms Transfers to the Third World, 1984–1991,* Congressional Research Service (CRS), U.S. Library of Congress (Washington, D.C., 1992).

ment that the demand of developing countries for greater independence in their own defense be accommodated. It also lent formal recognition to the decline of traditional instruments of diplomacy. At one time arms shipments were only one of several manifestations of a U.S. security commitment, usually accompanied by more tangible indicators such as formal security alliances, deployments of U.S. troops and a clear consonance of the political and military objectives of supplier and recipient. The Nixon Doctrine also coincided with a metamorphosis in the international defense technology market.

Arms transfers emerged in the 1970s as a leading instrument of U.S. and Soviet security relations with a wide range of developing countries. Increasingly, many of the countries receiving more liberal access to advanced military technology seemed remote from the traditional security perimeters of the superpowers. In some cases they exhibited little ideological or political compatibility. Although arms sales were a manifestation of U.S.-Soviet rivalry for international influence, by the end of the 1970s the two countries were equipping many of the same countries.

The new dynamic of interdependence among developed and developing countries over the decade changed the nature and volume of the arms supplied. A larger number of commercially competitive arms industries in advanced countries, accompanied by growing numbers of newly independent states with the financial resources to choose among suppliers, helped promote a rapid escalation in the volume and sophistication of the arms supplied globally. In the 1960s, the types of weapons transferred to the developing world consisted of limited quantities of relatively unsophisticated items such as MiG-17 or A-4 aircraft. By the mid-1970s, suppliers were competing for lucrative contracts with the Third World involving front-line, state-of-the-art technologies, including, in some cases, manufacturing technology and equipment diverted for export from the supplying countries' own forces.

In terms of volume, the number of aircraft, warships, missiles and armored fighting vehicles transferred to the Third World between the late 1960s and 1976 roughly equalled the total of the arms transferred between 1950 and 1970.[3] For the United States, exports to the Third World increased from an average of $250 million a year in the 1960s to over $10 billion a year in the mid- to late 1970s. A critical force driving this change was the redistribution of international wealth during this period, especially the emergence of the Middle East and Persian Gulf oil producers. Their near-monopoly over oil resources provided them with leverage over industrial suppliers. Although the overall volume of the arms sold declined in the 1980s, commensurate with global economic constraints and declining petroleum revenues, the sophistication of the arms exported from the industrial countries and the number of competing suppliers continued to grow.

As is evident from table 6-1, in the late 1970s the destination of U.S. arms exports began to focus on the Persian Gulf and the Middle East, a trend that continues today. By the time of the 1976 presidential campaign, arms sales were the focus of widespread public opposition in the United States. Although Soviet and Western European supplies of arms to Third World states had clearly grown significantly, the U.S. lead in this area— especially to countries of questionable stability, such as Iran—prompted public outcries against U.S. policy. Overall, the legacy of the Nixon Doctrine culminated in the late 1970s in a perception that arms sales were a policy out of control.

As a rhetorical rallying point, President Jimmy Carter's announcement of ambitious policies for unilateral and multilateral restraint, encapsulated in his statement in May 1977 that arms transfers were "an exceptional foreign policy implement, to be used only in instances where it can be

Table 6-1. *Regional Shares of the Value of Supplier Deliveries, 1977–84*
(Percent)

Country	East Asia/Pacific		Near East/ South Asia		Latin America		Africa (Sub-Saharan)	
	1977–80	1981–84	1977–80	1981–84	1977–80	1981–84	1977–80	1980–84
Non-communist								
of which:								
United States	16.57	21.11	80.90	75.06	1.44	2.56	1.09	1.27
France	3.39	2.34	72.28	84.78	15.37	7.49	8.96	5.39
United Kingdom	10.28	5.04	68.82	81.77	14.44	4.31	6.46	8.87
Germany, F.R.	4.59	9.99	69.51	37.08	10.66	41.91	15.25	11.02
Italy	5.88	5.50	58.40	56.91	16.39	26.73	19.33	10.87
All other	24.01	11.11	39.71	69.31	20.41	13.83	15.87	5.75
Total non-communist	13.61	12.77	71.56	73.31	8.42	9.32	6.41	4.61
(Major Western European)[a]	5.79	4.47	68.93	73.29	14.42	14.54	10.85	7.70
Communist								
of which:								
USSR	11.99	9.70	69.24	67.85	6.30	9.68	12.47	12.77
All other	5.18	3.43	76.99	88.90	.23	2.85	17.61	4.82
Total communist	11.22	7.92	70.12	73.84	5.61	7.74	13.05	10.51
Grand total	12.55	10.79	70.92	73.52	7.17	8.67	9.36	7.02

[a]Includes France, United Kingdom, Federal Republic of Germany and Italy.

Source: Richard F. Grimmett, *Conventional Arms Transfers to the Third World, 1984–1991*, Congressional Research Service (CRS), U.S. Library of Congress (Washington, D.C., 1992).

clearly demonstrated that the transfer contributes to our national security interests," at first met with general public approval. The Carter initiative imposed several specific controls over sales of advanced armaments to the Third World, at the same time that it excluded U.S. allies in the North Atlantic Treaty Organization (NATO) and the ANZUS (Australia, New Zealand and the United States) alliance from the restrictions. While providing for exceptions to the policy in the event of "extraordinary circumstances," the controls were intended to impose more careful scrutiny on pending arms requests.[4]

However, the increased domestic pressure to restrain arms sales coincided with a growing dependence on this instrument to achieve foreign policy objectives traditionally pursued by other means. The inherent conflict between efforts to appease domestic opposition to arms sales at the beginning of the administration and the subsequent pressures that arose from the dependence of the United States on arms exports created a chronic tension in policies that led to their constant revision. Damaged by the overblown rhetoric of early administration statements, the Carter policy was destined to incite opponents and disappoint supporters.

Among the first initiatives of President Reagan in 1981 was the abrupt dismissal of the Carter policies and resumption of arms sales as "routine diplomacy." They were to be considered "a vital and constructive instrument" that would assist "the state of preparedness of our friends and allies."[5] In contrast to the former administration's broad efforts to encourage negotiations with the Soviet Union in a number of key areas, the Reagan Administration, as it reaffirmed containment as a U.S. strategy, perceived arms sales, along with an overall U.S. military build-up, as critical to U.S. competition with the Soviet Union. A system of liberalized credits for arms recipients, reduced congressional oversight in reviewing sales and the establishment of a so-called Special Defense Acquisition Fund to stockpile weapons for rapid export in crises underscored the resurgence of the legitimacy of this instrument in U.S. diplomacy. The Reagan Administration also promoted sales of U.S. arms to covert and subnational groups in countries where it opposed the standing government, such as the *contras* in Nicaragua.

While international conditions and the growing competitiveness of the arms market actually reduced the overall U.S. share of the volume of weapons sold in the 1980s, important arms transfer decisions by the United States did affect the sophistication and destination of the arms sold. Advanced systems not previously approved for export to the Third World, such as F-16 fighters to South Korea and Venezuela and airborne warning and control system (AWACS) patrol planes to Saudi Arabia, were among the earliest decisions.

Two contrary elements of the Reagan legacy are worth noting. The first, which stemmed from the administration's overarching concern about technological competition with the Soviet Union, prompted the initiation of a sweeping "technology security" apparatus in the Pentagon aimed at preventing the diffusion of U.S. technology to the Warsaw Pact countries. In an effort spearheaded by then-Assistant Secretary of Defense for Policy Richard Perle, the Reagan Administration implemented a system of technology security reviews for military export licenses that transformed the Defense Department's Munitions Control Office, formerly an eight-person operation headed by an Air Force colonel, into a 100-person operation headed by a deputy under secretary of defense.[6]

The second was the crafting and announcement of the MTCR in 1987, a cartel-like agreement among the United States, Canada, the United Kingdom, France, the Federal Republic of Germany (West Germany), Italy and Japan to restrict exports of missile-related technologies.[7] Reached after four years of quiet negotiations conducted by mid-level officials, this agreement has become the cornerstone of efforts to manage the spread of space and other technologies relevant to missiles from North to South.

Ironically, it was President Reagan, not President Carter, who presided over the most significant decreases in U.S. arms sales in the last decade, a reflection of the extent to which policy is determined as much by exogenous factors as deliberate intent. While the growing structural dependence of the West and the Soviet Union on arms sales impeded efforts at restraint under the Carter Administration, the worldwide economic recession imposed constraints on the global demand for arms that even Reagan's policy of promoting U.S. military exports did not overcome.

From Bush to Clinton: Any Change?

After the war against Iraq, the United States became a leading voice in diplomatic initiatives on behalf of arms restraint. A series of negotiations prompted by the United States in 1991 led, in December of that year, to the U.N. General Assembly approving a proposal to establish an international registry of arms exports and imports under the auspices of the U.N. Secretary-General. The Permanent Five (P-5) members of the U.N. Security Council (the United States, Russia, China, the United Kingdom and France) also began discussions on procedures and guidelines for prior notification of arms contracts.

Whatever their limitations, these activities were the first formal efforts in many years to coordinate the North-South diffusion of technology through multinational agreement. They may presage a renewed commitment to managing the global trade in conventional weapons. At a minimum, the arms registry, if it succeeds, will be an important step toward greater transparency in international defense trade. Even modest steps such as the registry are an indicator of the higher profile countries are giving to non-proliferation objectives.

Still, skeptics, noting the concurrent effort to promote U.S. arms sales globally, have tended to dismiss U.S. involvement in such policy innovations as little more than belated, cosmetic efforts to respond to congressional or public pressure for even more serious action on this front. Decisions such as the announcement in 1991 of arms contracts valued at over $20 billion for the Middle East on the heels of the conclusion of Desert Storm, followed fairly shortly thereafter by decisions to pursue the sale of 72 F-15s to Saudi Arabia, 150 F-16s to Taiwan and dozens of the latest

M1-A2 tanks to Kuwait, seemed vastly to overshadow the diplomatic initiatives in the United Nations. In the final analysis, critics condemned the legacy of arms transfers under Bush as even worse than business as usual.

The Clinton Administration, as noted, has not produced any discernible change in U.S. arms transfer policy. In a declining global market, the United States continues to dominate sales to the Middle East and the Third World as a whole.[8] If anything, the pressure on the administration to treat arms exports as a domestic economic issue involving the preservation of jobs and production lines, rather than as a security or proliferation issue, became more pronounced in 1993–95.

Any effort to regulate arms contracts has to compete with the dependence of key U.S. defense companies on a high level of arms production at a time when domestic demand is contracting rapidly. With the dissolution of the Soviet Union and the end of the Cold War, the accelerated pace of the recession in the defense industry has helped generate powerful pressures for export promotion, arguably at the expense of longer term national security interests. While many thought the cessation of Cold War hostilities would make it far easier to control U.S. arms sales, for the immediate future the opposite seems to be true.

The perception of Clinton officials, consistent with that of their predecessors, that the United States has an obligation to reward the countries that supported it in the war against Iraq also militates against a change in U.S. arms export policy. The willingness of the Arab states, in particular, to support U.S. objectives in Iraq requires that they acquire better defense capabilities against future security threats. U.S. credibility, it is argued, requires that the United States demonstrate its commitment to the security of these states, a commitment that they would question if the United States denied them arms contracts.

Even as President Clinton declared before the United Nations in late 1993 that weapons proliferation is the greatest risk to U.S. security, his subordinates were taking steps to facilitate overseas sales, including assistance to U.S. defense firms in displaying their products at international arms exhibitions, the encouragement of embassies to become involved in the promotion of U.S. goods and attempts to speed up procedures such as the licensing of arms exports. As Robert Gallucci, assistant secretary of state for political-military affairs put it, "exports are essential to a strong economy . . . there's no question we hope to promote exports."[9] At the same time, the administration has spearheaded international efforts to create a successor regime for the Co-ordinating Committee on Multilateral Export

Controls (COCOM), the Cold War-era export cartel aimed at controlling militarily significant East-West trade. The new institution, known as the Wassenaar Arrangement, has twenty-eight members (including former U.S. adversaries such as Russia). It is focused on so-called rogue states such as the Democratic People's Republic of Korea (North Korea), India, Iran and Iraq. Despite U.S. efforts to achieve a more comprehensive and enforceable agreement, which was resisted by allies, the arrangement is wholly consensual and non-binding.[10]

The Conventional Arms Transfer (CAT) Negotiations: Lessons from the Past

If U.S. policy is to change in the near future, decisionmakers will benefit from an examination of the historical record. All too often, administrations have assumed office promising change without adequately appreciating the challenges involved. Inevitably, this leads to cynicism and fatalism when the policies fail, retarding the prospect for future initiatives.

The Carter Administration offers the one contemporary example of unilateral and multilateral efforts to restrain the sale of advanced arms globally. The Carter effort to develop arms restraint policies stemmed from a general political perception in the mid- to late 1970s that the international market in arms was outstripping the ability of policymakers to control it. The new administration wanted to find a basis for cooperation among suppliers and recipients of advanced military technology to limit trade in what were perceived to be destabilizing technologies. Some also saw cooperative efforts to reduce arms sales as a potential way to engage the Soviet Union in a dialogue about limits on the roles of U.S. and Soviet military power in the Third World and to reach mutual understandings about superpower "codes of conduct" in developing regions. As such, the motivating trends prompting a change in U.S. policy were similar to those currently prevailing.

The resulting Conventional Arms Transfer (CAT) talks—four rounds of bilateral negotiations on ways to limit arms sales—were held in 1977 and 1978.[11] Although some may argue the CAT experiment failed because of the politics of the Cold War, more subtle forces were at work that provide insight into today's challenges. First, the negotiations lacked clear objectives, backed by domestic political support. A precise definition of the purpose of an arms restraint initiative, with coherent institutional jurisdiction, is fundamental to sustaining a long-term commitment to such a difficult goal.

In turn, such support is only possible when strong and consistent presidential leadership ensures bureaucratic discipline and careful planning of strategy. The CAT talks were clearly a victim of competing domestic and international objectives and contradictory perspectives among senior officials that precluded the development of any reasonable foundation for their implementation. The political strains that plague current debates about controls over arms exports are, if anything, even more fractious today.

Third, the CAT talks demonstrated that sustainable, multilateral negotiations over an issue as controversial as arms sales are best served by beginning with modest objectives, which can be expanded over time. A diplomatic demarche should begin with incremental, technical measures that are minimally controversial within a region or industry to gain early support. For example, global bans on the export of weapons in which no government has a significant military stake—such as incendiary weapons, long-range surface-to-surface missiles or other weapons that pose particular risks to non-combatants—could be a reasonable starting point. A 1993 legislative initiative to extend a ban on the sale of anti-personnel landmines is one example of this kind of proposal.[12] The idea is to minimize the political resistance that arms restraint initiatives create at the outset in order to develop the fundamental infrastructure—both bureaucratic and international—that will later permit the institutionalization of arms restraint mechanisms.[13]

Prospects for Change

There are few reasons for optimism about the prospects for more active U.S. initiatives and diplomacy on behalf of restraint in conventional arms transfers. The tremendous structural impediments to a significant change in policy, regardless of the strength of the president's commitment, cannot be overlooked. Most notable is the growing political saliency of the issue because of the critical dependence of some of the larger U.S. defense companies on overseas sales. Until and unless other elements of a Clinton defense industrial strategy are implemented, particularly in the areas of industry diversification and conversion—as well as in the elicitation of support for a multinational restraint initiative—any major policy departures may prove politically impossible.

Still, in its struggle to find the right balance between domestic economic pressures and long-term security interests, the administration has tilted in the direction of the latter on several occasions. For example, the administration opposed legislation that would have created a $1 billion loan guarantee fund to subsidize U.S. arms exports, a proposal initiated by U.S. defense contractors and widely supported in the Senate.[14]

Conventional arms restraint has received some, albeit limited, attention in the Congress and among private commentators. One hundred and eleven members of Congress signed a letter in July 1993 urging President Clinton to consider a moratorium on arms trades to countries with "unelected or repressive governments or governments not complying with the new United Nations arms trade registry."[15] Still, the degree of congressional commitment to arms restraint remains to be proven.

Other constituencies may, however, favor restraint in arms sales. The U.S. military may be a new source of support for certain kinds of arms export controls, especially if the United States maintains an expanded military presence in the Persian Gulf region. Heavily armed countries that are politically unstable pose a direct risk to the security of U.S. personnel and military technology deployed overseas. The services have never been great fans of exports of advanced technology in any case, fearing that the weapons they use themselves could be captured by hostile forces. Technology is typically compromised when a well-defended client is transformed into an aggressively armed adversary, not an uncommon development in the volatile regions of the Third World.

Multinational efforts to share advanced technology among NATO countries or in Asia may also provide an avenue for greater coordination of guidelines on arms exports among participating countries. It is conceivable countries engaged in joint defense programs could be induced to accept codes of conduct restricting particular types of exports to volatile regions of the world, in return for access to high technology and assured outlets for exports in the developed world. In any event, given the common interest in preventing technology compromise of highly advanced systems, most new technologies developed for advanced countries' conventional deterrents should not be disseminated widely.

Another positive trend is the unprecedented stature currently being accorded international organizations and regimes, especially the United Nations, a trend that could not have been anticipated even five years ago. Most U.S. policymakers traditionally have viewed multilateral mechanisms as little more than symbols, often as nuisances, and certainly as powerless to effect significant change. Without these international arrangements to help guide countries toward even long-term restraint goals, the pressures on cash-strapped governments to sell arms could be even more difficult to manage.

Within the international lending institutions such as the International Bank for Reconstruction and Development, the International Monetary Fund and the more recently established European Bank for Reconstruction and Development, there is growing interest in linking international financial assistance to various norms of military behavior, including defense expenditures. This represents a marked shift in attitude among institutions that for decades formally promulgated the fiction that a country's defense sector should not be included in evaluations of its economic performance, political stability and other variables that go into decisions about credit or aid eligibility.

Additional examples of this linkage can be found in various Western arrangements to provide financial and material assistance to the countries of the former Soviet bloc. Aid packages can provide a form of leverage to induce countries of the former Soviet bloc to support and join in the international system of military and technology restraint. The threat that the central authority in Russia and other parts of the former Soviet Union will disintegrate, a situation that could result in uncontrolled diffusion of Soviet weapons in the open market, has compelled the Clinton Administration to devise a variety of initiatives to help forestall such a development. Current U.S. efforts to encourage the responsible disposition of military assets and promote successful economic modernization and defense conversion in the former Soviet Union are integral parts of a broader strategy to constrain the international diffusion of conventional arms.

In the last two years, the MTCR has elicited support from unexpected sources, growing from an initial list of seven in 1987 to twenty-seven members or adherents by the end of 1995.[16] Increasingly, the advanced country members of the MTCR seem to be recognizing that their ability to elicit new members and uphold the agreement's restrictions on missiles requires that they be more scrupulous in their own adherence to the guidelines. After much hesitation and dissembling over several years, France finally decided in early 1992 to cancel negotiations with Brazil for various rocket technologies pertinent to the development of ballistic missiles. In mid-1992, the United States imposed trade sanctions on the Russian space enterprise *Glavkosmos* and the Indian Space Research Organization (ISRO) for proceeding on a contract involving technologies specifically proscribed by the MTCR, including cryogenic materials and rocket components.[17] In addition, the contract would have involved the sale of MTCR target technologies to a non-MTCR country. For their part, the two countries counterargued that the equipment was to be used in India's civilian space program. Their protestations did not overcome the U.S. administration's interpretation of its legal obligations: the proposed contract invoked legislated sanctions that only the president could overturn. As a result of this U.S. pressure, the deal was cancelled in mid-1993.

Any renewed interest in international arms restraint will have to compete with a number of other factors that will make any initiative difficult. As has happened repeatedly, structural imperatives that favor permissive export policies may outstrip the political interest in restraint.

A critical factor militating against more stringent supplier controls is the changing and increasingly diffuse character of the international technology market. Protectionist instruments, such as the missile cartel, work only in proportion to the clout of the members and their relative monopoly over the products they are trying to control. Such arrangements also carry political costs, in that the countries being denied the technology typically perceive the arrangements to be discriminatory and illegitimate.

Over fifty countries produce weapons, and over a dozen of those can produce relatively advanced systems such as missiles. Some, including China, are unapologetic about their exports of proscribed technologies, including missile components, and have indicated they will not become full supporters of any restraint regime until they have a more equal share of the arms market. Moreover, in all areas of conventional weaponry other than missiles, including fighter aircraft or naval vessels, the number of potential exporters is larger and the consensus in favor of controls even weaker.

Trends in the technology market presage declining control by governments over the disposition of defense-related innovations. Critical technologies vital to defense, from supercomputers to biotechnologies and fiber optics, are increasingly commercial in origin. As developing countries establish their own weapons industries, they, too, are increasingly capable of tapping into new sources of commercial and dual-use goods without reference to superpower constraints. In the future, an ever-shrinking percentage of technology will be subject to any direct government controls. This situation will test the viability of cartels for all but a select number of the most advanced technologies.

The most difficult problem confronting limitations of arms sales is the absence of an agreed upon standard for judging what is a desirable and undesirable kind of export. Unlike nuclear or chemical weapons, which are difficult enough to control, there has never been an established procedure by which to calibrate what are defensive and offensive weapon acquisitions or force postures. Similarly, it is not easy to target particular technologies for "delegitimization," because the line between those weapons used for self-defense and those used for provocation must always be drawn within a particular politico-military context.

Conclusions

Several observations can be drawn from the current political environment about the prospects for and likely direction of U.S. arms export policy in the future. First, a selective system of restraints on the supply of certain key technologies may prove feasible at this point. Attempting a more comprehensive arms control regime at this stage would likely prove elusive and perhaps self-defeating. Second, to be realistic, any new initiative to regulate the international arms trade will increasingly require the cooperation of both developed and developing states. Third, those designing new policies will have to confront the primacy of the commercial pressures that are currently driving the quest for arms exports, domestically and internationally. Thus, they should involve industry in the conceptualization of the policies.

At the national level, a fundamental question is whether the U.S. government should remain in the business of promoting arms contracts globally. Providing advice to other governments and helping them to set defense priorities are legitimate official activities. Subsidies to contractors and other commercial activities are not, according to current Clinton policy. Further, it needs to be made clear that no amount of export promotion can save defense firms from the inexorable pressures of recession, declining levels of domestic procurement and an international market that is itself shrinking. The quest for short-term palliatives by industry, however understandable, cannot substitute for sound economic and security policy.

The pace of structural adjustment in the defense sector and the economy as a whole will influence the administration's ability to deemphasize arms sales in U.S. diplomacy. As candidate Clinton frequently urged, U.S. defense firms and their labor forces should not be unduly targeted as the scapegoats of change. An overall defense industrial strategy that helps American companies move away from excessive reliance on defense exports, as the administration has already put forward, would provide some degree of government assistance to the private sector. There will still be difficult problems of excess capacity and overemployment at the largest contractors, some of which do not have the ability or interest in diversifying out of defense production. Striking the appropriate balance between economic and security interests in arms export policy inevitably will be controversial.

No arms transfer policy can succeed without multilateral involvement and support. A major impediment to progress internationally continues to be the reluctance or inability of the advanced countries to renounce their own dependence on military exports and to move forward with economic restructuring. Inevitably, the continued embrace of exports as a way to mitigate the effects of domestic recession makes U.S. proposals suspect: they are viewed as supplier-centric, hypocritical and not politically acceptable in the eyes of key regional powers such as China or even allies such as France.

As discussed, at the outset the consultations with other arms suppliers should focus on modest objectives that impose the least cost on participants. The starting point could be technical discussions concerning weapon systems that are not central to any major power's foreign policy and that are widely considered as particularly dangerous—including missiles, antisatellite systems or weapons that can be diverted to terrorists, such as various kinds of biotechnologies. Weapons that have been the subject of international attention for their indiscriminate effects and that have marginal military utility, such as incendiary and fragmentation weapons, could also be considered. Such items could be the beginning of a list of types of weapons whose transfer would be banned globally or that would require prior consultation before a transfer took place.

Monitoring the flow of certain types of manufacturing technology useful in the production of advanced and destabilizing systems has, to date, been an area of particular weakness domestically and internationally. Potential types of restraints that could be considered multinationally range from outright prohibitions on particular classes of items to the elicitation of strict assurances for end-use and mechanisms for prior consultation among suppliers before transfer of particular systems. It should be anticipated, however, that many Third World countries will oppose technology denial arrangements such as the MTCR if they have no hand in devising them.

Whether an arms restraint initiative is to be a comprehensive multinational effort, engaging both suppliers and recipients at the outset, or should more properly begin with the major advanced countries, will depend on prevailing political conditions and the scope of the restraint proposals envisioned. A multinational regime could be pursued on several tracks, consisting of supplier negotiations and separate recipient negotiations, with the restraint regimes focusing on key regions over a period of time.

It is important to remember the embryonic nature of this area of diplomacy. Before the MTCR was initiated in 1987, there was no formal international apparatus to guide transfers of conventional technologies to developing countries. There is still very little interest in a regime to curtail exports of weapons or dual-use technologies. Governments, including those that adhere to the MTCR and the Nuclear Non-Proliferation Treaty, have so far resisted controls on transfers of advanced aircraft and non-ballistic missiles, despite their potential for the delivery of nuclear weapons and application in the development of ballistic missiles.

Discussions of technical issues—from global bans on destabilizing weapons to tighter restrictions on the disposition by clients of weapons received from the larger powers—need not be seen as either a substitute for, or a sacrifice of, the larger objective of developing more general codes of conduct for international exports. Given the novel nature of this kind of diplomacy, there is much to be learned even from technical discussions. Most importantly, they would demonstrate the practicality of negotiated restraint and create the procedures and institutions that could help implement more ambitious undertakings. As it stands today, conventional arms restraint is lacking even the most rudimentary diplomatic infrastructure or accepted politico-military vocabulary. Norms guiding even limited types of arms transfers could lend some measure of predictability to what is at present an area of great uncertainty.

The prospects for early or comprehensive success in such an endeavor should not be oversold. The experience of the Carter Administration shows how even relatively modest control policies resulted in intense opposition from those who saw them as a form of American withdrawal, from countries who feared a superpower condominium interfering with their security and from all who anticipated cataclysmic economic penalties, including U.S. industry and European suppliers.

Regardless of how much progress in formal arms restraint agreements is or is not achieved, the development of common understandings to guide crisis management and crisis resolution among nations will become increasingly important as Third World power alignments become more diffuse and clients more militarily capable. Bilateral and multilateral regional security talks, as have been pursued in the Middle East and South Asia in recent years, are in and of themselves a vital undertaking that in turn could serve as the foundation for agreements on arms shipments tailored to specific areas.

As a first step, the United States can take the lead in promoting unilateral and multilateral restraints. This would demonstrate U.S. commitment to such a goal. The United States has a leadership role to play in the United Nations and other international fora. It has played the role of catalyst to and chief promoter of a COCOM successor regime. It is well-recognized that the United States cannot possibly go it alone in this area, but it may help influence the receptivity of other countries to arms control goals.

Notes

1. "Fact Sheet: Criteria for Decision Making on US Arms Exports," The White House (Washington, D.C., 17 February 1995).

2. The following section on the evolution of the arms trade draws on previously published works by the author, including "U.S.-Soviet Conventional Arms Transfer Negotiations," in Alexander L. George, Phillip J. Farley, and Alexander Dallin (eds.), *U.S.-Soviet Security Cooperation* (London, 1988), 510–523; "Technology, Arms Races and the Distribution of Power," in Asa S. Clark, IV (ed.), *Defense Technology* (Baltimore, 1986), 49–63.

3. The methodology underlying the statistical sources on arms transfers varies, a condition that accounts for some disparity in the numbers given by different sources. Authoritative unclassified sources include: U.S. Arms Control and Disarmament Agency (ACDA), *World Military Expenditures and Arms Transfers* (Washington, D.C., annual); Richard F. Grimmett, *Conventional Arms Transfers to Developing Nations,* Congressional Research Service (CRS), U.S. Library of Congress (Washington, D.C., annual); Stockholm International Peace Research Institute (SIPRI), *SIPRI Yearbook: World Armaments and Disarmament* (Oxford, annual).

4. The Carter policy of unilateral restraints on U.S. arms sales included pledges not to be the first to introduce an advanced weapon system into a region, not to develop weapons solely for export, and not to retransfer weapons from one recipient to a third country. It also covered co-production agreements. See "Statement by the President on Conventional Arms Policy," 19 May 1977, reprinted in U.S. Library of Congress, Congressional Research Service, *Changing Perspectives on U.S. Arms Transfer Policy* (Washington, D.C., 1981), Appendix II.

5. "Text of the President's Directive on Arms Transfer Policy," reprinted in U.S. Library of Congress, Congressional Research Service, *Changing Perspectives on U.S. Arms Transfer Policy* (Washington, D.C., 1981).

6. Although no unclassified numbers have been compiled indicating licenses denied, American industry has complained vociferously that the system has entailed protracted delays and capricious denials that have resulted in de facto arms restraint. This development is important in that it illustrates how a clear and compelling objective—in this case, preventing communist nations from acquiring

U.S. technology—supported at the highest levels can lead to enforcement of restraint guidelines, commercial pressures notwithstanding.

7. Building on the London Nuclear Suppliers Group, the signatories agreed to voluntary guidelines for the control of all types of exports, military and dual-use, that can be identified as relevant to the production of civilian sounding rockets, space launch vehicles, and unmanned air vehicles with ranges in excess of 300 kilometers or capable of carrying more than 500 kilograms of payload.

8. For further discussion, see Richard F. Grimmett, *Conventional Arms Transfers to the Third World 1984-1991,* Congressional Research Service (CRS), U.S. Library of Congress (Washington, D.C., 1992).

9. As quoted in *Arms Trade News* (October 1993), 1.

10. See, for instance, "Successor to COCOM Agreed," *Financial Times* (20 December 1995), 4.

11. The decision to proceed with formal bilateral talks followed the conclusion of an initial round of discussions held in December 1977. The first meeting consisted largely of U.S. explanations to the Soviet side of the Carter arms restraint policy. The United States emphasized that sustaining the U.S. unilateral policy would require multilateral cooperation. Based on the surprising level of Soviet interest, a second and third round was scheduled for May and July 1978, both in Helsinki. A final round was held in Mexico City in December of that year.

12. Senator Patrick Leahy (D-Vermont) offered an amendment to the fiscal year 1994 Defense Authorization Bill that would impose a three-year moratorium on the sale, transfer, or export of landmines, a measure that passed the Senate by a vote of 100–0.

13. Progress in areas that incite a minimum of controversy could in turn lead to the development of agreed criteria for more far-reaching application—weapon systems of clear significance to countries' political/military policies, such as advanced fighter aircraft or ground combat vehicles. Although the emphasis in discussions of more sophisticated equipment would still be on the weapons themselves, these restraint criteria would in turn be applied regionally—with the definition of what constitutes technological sophistication and potential military consequences obviously varying region by region.

14. Known as the Kempthorne amendment after Senator Dirk Kempthorne (R-Idaho), the provision authorized $25 million in fiscal year 1994 for the establishment of a $1 billion loan guarantee program for U.S. arms exporters. Supporters stressed the need to strengthen the competitiveness of U.S. contractors in the international market in order to protect U.S. jobs. The measure gained fairly broad bipartisan support in the Congress, including from liberal Democrats who represented states dependent on defense business, including Massachusetts Sens. Edward Kennedy and John Kerry and Rhode Island Sen. Claiborne Pell.

15. Cited in *Arms Sales Monitor,* 22 (30 September 1993), 3.

16. New adherents include China, Israel, and numerous former Soviet states. For a full list of members, see *Arms Control Reporter,* XIV (1995), 183–193.

17. "Oversight of Rocket Technologies Urged," *Lituraturnaya Gazeta,* as translated in U.S. Department of State, Foreign Broadcast Information Service (FBIS), FBIS-URS-92–078 (26 June 1992), 2–3.

SEVEN

Western Europe

Lawrence Freedman and Martin Navias

DURING THE 1980s, European policies on defense production and exports began to adapt to a set of economic pressures that made weapons manufacture more costly and overseas sales more necessary. This shift involved efforts toward vertical and horizontal consolidation of defense manufacturing and a concomitant push to expand transfers of weapons to the Third World.

The pressures have intensified in the 1990s. It is unfortunate but unavoidable that this intensification is occurring at a time when Europe's broader political and security future is undergoing a major review. The decisions made by the Europeans in the 1980s concerning defense production and exports were difficult. Those to be taken in the rest of the 1990s and beyond will be incomparably harder. They will involve judgements about the future of national defense industrial bases and the size, shape and responsibility of a pan-European defense industry. Concomitantly, while pressures for continued exports of arms will strengthen, so will counter-demands that such sales be severely constrained. Dilemmas concerning the level—national or transnational—at which the transfer of dual-use technologies should be checked will also need to be resolved, lest more arms scandals embarrass already scandal-ridden political authorities.

In short, while Europeans may attempt to muddle through their quandaries regarding arms production and export, a steady buildup of economic and political forces could ensure that the various alternatives will have to be confronted decisively.

The Defense Industrial Base: The Erosion of National Independence

European governments have supported national defense industries to decrease their dependence on the United States for weapons manufacture

151

Table 7-1. *European Defense Production, 1989*

Country	Production (bil. ECUs)	Percentage of total of European Community	Percentage of gross domestic product
United Kingdom	19.0	34.0	3
France	16.8	30.0	2
Germany, F.R.	8–9.0	18.0	1
Italy	5.1	9.0	<1
Spain	1.7	3.0	<1
The Netherlands	1.5	2.0	<1
Belgium	1.0	1.8	<1

Source: Based on figures fromm David Foquet, Manuel Kohnstamm and Michael Noelke, "Dual Use Industries in Europe," a study prepared for the Commission of the European Community, Directorate-General for Industrial Affairs (Brussels, 1990).

and to ensure the greatest possible autarky in fulfilling their weapons requirements. The benefits were to be economic, political and military. Economically, the countries would save foreign exchange through decreased imports while increasing earnings through exports. Politically, dependence on the United States would be reduced and U.S. leverage over Europe in this crucial area lessened somewhat. Militarily, the particular defense requirements of European states would be addressed directly, while deployed forces could be supported more readily.

By the end of the 1980s defense-related industries had made substantial progress. Production in 1989 was calculated at 55 billion European Currency Units (ECUs) (table 7-1), while employment was estimated to be 1.5 million (table 7-2).

These figures demonstrate that production capabilities are not equally spread out across Western Europe. The United Kingdom, France and Germany dominated, accounting for three-quarters of Europe's defense production, with the other countries getting only a small cut of the pie. Only the dominant countries have succeeded in reducing their dependence on the United States and in meeting their defense requirements through domestic production. By the late 1980s Britain was importing only about 7 percent of its defense requirements from the United States, while France was in an even better position, importing only 1 percent. Germany imported a third of its defense equipment from the United States but could rapidly decrease that level if desired.

For smaller states with no significant industrial base, the picture is quite different. Italy and Norway depend on U.S. goods for over half their

Table 7-2. *Employment in the European Defense Industries, 1989*

Country	Defense employees (direct and indirect)
United Kingdom	620,000
France	400,000
Germany, F.R.	250,000
Spain	100,000
Italy	80,000
The Netherlands	35,000
Beglium	35,000

Source: Based on figures from David Foquet, Manuel Kohnstamm and Michael Noelke, "Dual Use Industries in Europe," a study prepared for the Commission of the European Community, Directorate-General for Industrial Affairs (Brussels, 1990).

military requirements, Denmark, Greece and Turkey for over three-quarters. While the major producers have succeeded in increasing their autonomy in defense production, overall the European-U.S. arms trade greatly favors the United States, and the level of U.S. trade in arms with Europe is greater than the intra-European level.

The result is that the benefits of a European military industrial base have always been concentrated in a few states and have not spread across Europe as a whole. Britain, France and Germany have benefitted the most, although not exclusively, from the decreased dependence upon the United States through the gains in foreign exchange resulting from exports of weapons.

At the same time, it is important not to underestimate the significance of domestic arms production and trade for the second tier of European producers. Italy, most notably, has a significant defense industry; Spain sees its defense industry as a technological driving force for the economy. Portugal and Greece are seeking to expand their defense production and to increase exports. In contrast, over the past few years Belgium has sold large segments of its defense industry, as has the Netherlands, although to a lesser degree.

No matter the size of each country's defense industry, they all face a common problem—national-based defense production is too small and fragmented to have a secure future in Europe. Unit costs for European goods have often been very high because of relatively small production runs, a reflection of the limited demand generated by Europe's small armies and defense budgets. As to the failure of Europe to establish significant external trade in arms, the reasons are the fragmentation of the market and duplication of efforts among the European nations, as well as cheaper foreign alternatives. Limited profits have constrained national investment in research and development (R&D) significantly. Although traditionally Europe has invested more than the United States in R&D as a percentage of revenue, even heavy national government involvement has not always kept European firms at the forefront of technological innovation. Significantly, this trend may strengthen over the next decade.

These problems were evident during the early 1980s, when the North Atlantic Treaty Organization (NATO) countries were attempting to implement the 1978 decision to increase defense expenditures annually by 3 percent in real terms. Since the collapse of the Warsaw Pact, defense expenditures have been declining steeply in virtually every European country. A steady decrease in defense purchases, on the order of 2–3 percent in real terms, can be expected for some years to come. The impact of these curtailments on the defense sector will be grave and will have significant multiplier effects on the wider economy. According to one estimate, approximately 300,000 European defense jobs were lost between 1988 and 1990.[1] Some suggest that 500,000 personnel, or a third of the total defense workforce in Europe, could become redundant by the middle of this decade.[2]

Concerns about employment naturally reinforce the desire to protect the defense industry base, so long as the original reasons for developing an independent capability still have validity, even if less pressing. Faced with small and declining domestic markets, difficulty obtaining economies of scale and the unpredictability of the overseas market, European planners are increasingly forced to seek new strategies to offset R&D costs. The trend has been to retain a defense production base at the expense of national independence. One method to reduce unit costs has been to pool research and production programs. European states have been collaborating on various defense projects, even though, during the 1980s, companies were not always glad to do so, and the record of successes and failures has been mixed, with some notable examples in the latter category.[3] Shifting economic realities have altered attitudes somewhat, and now, aside from the troubled European Fighter Aircraft (EFA) project, which has received so much attention, numerous, albeit smaller and more manageable, joint European projects are underway. Germany and France are building the Roland surface-to-air missile, the Hot 2 improved anti-tank missile, the Milan anti-tank system, the ANS anti-ship missile and the Tiger anti-tank helicopter. Together with the United Kingdom, France and Germany are constructing the Trigat anti-tank guided missile and a radar for the Cobra air defense system. The United Kingdom is involved in collaborative projects with other countries as well. It is developing, together with Canada, the Netherlands and Spain, an inertial navigation system; with France, the RTM 322 helicopter engine; with France, Germany, Italy and the United States, an improved multiple rocket launcher system; and with Italy, the Netherlands and Spain, the A-129 Tonal anti-tank helicopter. France, Germany, Italy and the Netherlands have also indicated their intention to proceed with the NH-90 tactical transport helicopter.

The problems besetting the EFA project raise the question of whether the pressure on public spending throughout Western Europe will allow even collaborative ventures on the most sophisticated systems. These problems also highlight the risks of interdependence in defense procurement when a key partner begins to have second thoughts. At one point Germany had stated its intention to withdraw from the project, but the German defense minister became persuaded that the political costs of pulling out were higher than the gains of conspicuously abandoning a "Cold War" aircraft. Although he was helped by a genuine reduction in projected costs, it was anger in Bavaria at the potential loss of defense jobs that convinced the minister. It is probable that future collaborative projects will focus on smaller, more manageable systems and subsystems.

A significant recent trend, prompted in part by the prospect of the European Community's (E.C.) single integrated market, has been a series of cross-border mergers and acquisitions, of which there are numerous examples. For some time the French Thomson-CSF conglomerate has sought to enter the British defense production sector by purchasing small U.K.-based companies such as Link Miles and Ferranti, Sonar Division. British Aerospace bought the German/Austrian company Heckler and Koch. Sieman's of Germany and GEC of the United Kingdom took over the U.K.'s Plessey. Thomson-CSF of France bought a subsidiary of Phillips (which had decided to opt out of the defense business). Alcatel of France acquired the Space, Defense & Telecom Division of ACEL of Belgium. Astra of the United Kingdom purchased MARC of Switzerland and Poudrière Réunie of Belgium.

At the same time, and perhaps more important in the short term than the intra-European mergers, is a trend toward consolidation within countries. In the United Kingdom and France the result has been the emergence of extremely large defense producers: in Britain, British Aerospace combined with Royal Ordnance; in France Aerospatiale and Thomson-CSF came together; and in Italy, GEC took over Ferranti Defense Systems. Also in Italy, Alenia was formed out of the merger of Aeritalia and Selenia; in

Germany, Daimler Benz took over MBB, TST and MBU and now clearly dominates the German aerospace and defense electronics industries; and in France, the aim of the formation of Sextant no doubt was to dominate the domestic aviation industry. These activities have resulted in greater economies in research costs and in a spreading out of production costs.

Collaborative projects had begun well before the onset of the new constraints in defense procurement. These constraints reinforce the importance of cooperation among defense manufacturers, and the trend toward collaboration will likely continue to develop (to some extent independently of the success of cooperation at higher political levels). In the final analysis, many European defense producers have little choice.

Recognizing the merits of joint ventures, during the late 1980s European governments sought to develop a framework that would abet joint ventures and enhance the capability of manufacturers to exploit the pan-European market. In 1976, they established the Independent European Programme Group (IEPG), composed of all European NATO countries except Ireland, which was not formally part of NATO. The IEPG was seen as a way to get France to cooperate on weapons production. More recently, following the 1985 Vredeling Report of the European Community, the IEPG is seen as a mechanism for developing an increasingly unified system of arms manufacture and procurement by liberalizing the European arms market, without, it should be emphasized, totally undermining national defense industries.[4] To this end a so-called action plan was drawn up in 1988. While some argued the plan would simply strengthen the stronger players in European defense production and not make manufacturing any more efficient, in theory the aim was to create a single European armaments market by ensuring managed access by European states into the markets of their neighbors. Unlike other European initiatives, it suffered from its dependence on national ministries for implementation. Nothing in the plan was binding, and no funds were available to ease adjustment.

A program based on *juste retour* might have succeeded had demand been increasing rather than decreasing. It would then have been easier for countries to balance self-restraint in the short term against benefits in the long term. In a period of severe pressures on national industries, that balancing will not always be easy. Future cooperation in production will likely be a function of narrow industrial perceptions of gain rather than a response to an overarching plan.

What of the present economic health of the firms themselves? Despite the recent cuts in employment, the vulnerability of the defense sector should not be exaggerated. European firms producing weaponry tend to be diversified enough not to suffer excessively from the decreases in domestic demand. According to figures of the U.S. Office of Technology Assessment (OTA), in 1988 the top twelve European defense companies earned only 17 percent of their revenues from weapons sales, compared with approximately 40 percent for equivalent U.S. firms.[5] A number of European defense companies—notable examples in the United Kingdom being British Aerospace and Rolls Royce—have over the past few years deliberately sought to decrease further the defense component of their sales and to expand their civilian products. Conversion of defense to civilian manufacture is notoriously risky and complex, however. The overall recession has already forced British Aerospace to accept that its best prospects lie with its role as a specialist defense contractor. Nonetheless, in general, greater diversification should translate into greater flexibility and an ability to survive reduced demands for defense products.

Despite these efforts to make European production more efficient, it is likely external producers will be allowed an increasingly large piece of the European defense procurement cake. For example, the British government's "value for money" weapons procurement policy may in a number of instances undercut British and other European producers. At the same time, while this outcome may apply in specific areas, European governments will continue trying to retain at least some autonomy in defense production and to continue R&D in important sectors. If they are to put greater emphasis on increased European cooperation in defense and foreign policy, it would be unwise—in a sense irrational—for them to abort all efforts in R&D and production of military equipment. Defense industries will likely continue to diversify into civilian production during the 1990s, at the same time the civilian technology sector will increasingly be employed to supply military requirements. This pattern has been increasingly evident since the 1970s, with civilian manufacturers moving into the defense field. There are, however, limits on this shift, as some items cannot be based on civilian technologies.

Governments are also becoming sensitive to the unemployment consequences of major cutbacks in the defense field. The continued downturn in British industrial production and growing unemployment may lead to greater intervention in support of local industry. The implication for defense producers is that even a much-downsized defense industry could expect some government input. One way to offset the costs of this intervention is continued exports. The 1992 devaluation of the pound may boost price competitiveness, although it is unclear how long this advantage will hold, especially if it leads to a resurgence in inflation.

The French government is committed to supporting its defense industry. During the 1990s France can be expected to continue to market its weaponry aggressively. The downturn in French sales in 1991–92 reflected the closure of the lucrative Middle East markets such as Iraq and the failure of its advanced fighter aircraft—the Mirage 2000—to compete with the U.S. F-16. The French were no doubt embittered by the U.S. decision to sell F-16s to Taiwan, as they believed they stood a chance of selling their aircraft. The United States justified the sale in part because China acquired Russian Sukhoi-27 aircraft but also because it wanted to maintain its influence over Taiwan by being the only supplier. Along with the United Kingdom, the French have expressed resentment at what they perceive to be a U.S. effort to dominate the Middle East market completely. Again, as with the United Kingdom, France is seeking to offset its losses in the Middle East with a focus on Asia.

In the early nineties, only one German company featured on the list of top ten European defense contractors. It has been presumed the strength of the German industrial base would enable it to challenge France and the United Kingdom in the area of high-technology defense production. However, this eventuality depends on a reversal of Germany's current plans to cut defense spending sharply and the use of its civil industrial base to push military developments forward. It is extremely difficult to forecast general economic developments in Germany for the mid- to late 1990s. It seems that the major structural change brought about by the high costs of German unification may lead to a decline in competitiveness over the long term. At any rate, Germany's growing indebtedness suggests that defense will not be a high priority in this decade. Most likely the government will simply seek to sustain existing military industries. It is also possible that the strength of the German machine tool sector may make that sector a target for countries seeking to acquire sensitive technologies; Iran could become a troubling example of this.[6]

Pressures to Export

To reduce their high unit costs, European defense producers have always depended heavily on external sales—far more than the United States has. European countries have long been interested in extending and enhancing their political influence through arms transfers, although this objective has been always subordinate to economic benefits—the reduction of unit costs and, more generally, the acquisition of foreign currency. Economics rather than politics has always given European arms sales their major impetus and rationale. With the costs of R&D and deployment of advanced weaponry so high, and with production runs relatively short, the major manufacturers have always been inclined toward export-led defense industrial policies. As a result, foreign sales have accounted for approximately a third of European defense production.

In the past European industries usually were able to rely on strong government support for exports. Governments often conducted intensive diplomacy in support of arms transfers. Good examples are the recent visits by senior French and British officials to the Gulf, where they argued the merits of their respective weapons systems. In October 1992 a number of British industrialists and politicians blamed the U.K. government for not doing enough to promote the Challenger tank to the Kuwaitis. In this instance, however, the greater political clout of the United States was the deciding factor, not the U.K. government's weak political or official will.

To reduce unit costs, European arms transfers have often included some of the most up-to-date weapons systems in Europe's arsenals. French Exocet missiles and U.K. Tornado aircraft are two examples. Sometimes the customers got their weapons before they were deployed in Europe. French forces have complained that some weapons are geared far more to the needs of overseas customers than to their own needs. It is in these high-technology items in specific market niches, such as air-to-surface missiles, frigates and corvettes, that European states have made their mark.

European producers have succeeded in complementing their limited trade across the Atlantic and within Europe by focusing on transfers to the Third World. From 1984 to 1991, 17 percent of sales to Third World states were by Europe. Nearly 20 percent of the total global trade in conventional arms is from Europe, with France, the United Kingdom and Germany supplying 90 percent of this total. According to one estimate, between 1987 and 1991 France transferred on average $2.4 billion annually, the United Kingdom $2 billion and Germany $1.3 billion.[7]

CRS figures suggest that while France, Germany and the United Kingdom dominated the European market during the 1987–91 period, other West European states supplied smaller but still significant amounts: Italy sold an average of $400 million a year, the Netherlands $400 million, Sweden $300 million and Spain $300 million. Although the Stockholm International Peace Research Institute (SIPRI) figures and

Table 7-3. *European Military Exports by Destination, 1987–91*

Country	Global Ranking	$ million 1990
A. World-wide		
France	3	11,220
United Kingdom	4	9,097
Germany, F.R.	6	6,115
Italy	8	1,878
The Netherlands	9	1,758
Sweden	11	1,524
Spain	13	1,128
B. Industrialized World		
Germany, F.R.	3	4,439
France	5	2,192
United Kingdom	6	1,498
Sweden	7	766
Italy	8	448
Switzerland	11	363
Spain	13	352
The Netherlands	14	345
C. Developing World		
France	3	9,028
United Kingdom	5	7,599
Germany, F.R.	6	1,676
The Netherlands	8	1,412
Italy	9	1,390
Spain	12	777
Sweden	13	758

Source: Stockholm International Peace Research Institute (SIPRI), *SIPRI Yearbook of World Armaments and Disarmament 1992* (London, 1992), 272–273.

country rankings differ from those of the CRS, the general overall picture does not (table 7-3).

Although there was a downturn in European arms exports at the turn of the decade, it did not occur across the board, and exports remained substantial. According to SIPRI, in 1991–92 Germany registered a large increase in arms transfers, emerging as the largest European exporter, with sales valued at close to $2 billion. France was the second largest European exporter of weaponry in 1992, with total deliveries of nearly $1.1 billion. The United Kingdom was in third position with $952 million—a sizable amount, although a 65 percent drop over 1989. Italy's exports doubled in 1992, from $163 million to $335 million. The smaller suppliers saw their sales fall in

1992. In comparison with 1991, the Netherlands exported arms valued at $305 million, Switzerland $83 million and Spain $37 million, decreases of approximately 12 percent, 70 percent and 26 percent respectively.[8]

It is important to note that the declines in sales were part of a global downturn in demand and not the result of a decision to export less. To the contrary, Europe is trying to revive sales. The efforts described above, which are aimed at making European defense production more cost-effective, are also intended to make European manufacturers more competitive in the global market. There is every indication that most of the weapons being produced collaboratively will be offered to overseas purchasers. In this way the overall costs can be reduced further and additional foreign exchange secured.

It can be argued that collaboration in defense production, consolidation of domestic producers, and acquisitions and mergers across borders will place European countries in a better position to export. The reason is that individually the countries might not have produced the goods they are developing jointly, or produced them as competitively. Coordination and cooperation are becoming the only options if European countries are to compete with the U.S. and other suppliers. However, collaborative production alone may not be enough to make the weaponry cost-effective. Exports will remain vital to offset the limited levels of European demand.

Pressures to Control Trade

Despite the necessity of exports to support the industrial base and yield economic benefits, European states have come under increasing pressure to curb transfers to the Third World. The Gulf War spurred new initiatives to control the arms trade, in the face of incontrovertible evidence that Saddam Hussein's determined efforts to construct an advanced military capability, including nuclear weapons, was assisted by European firms. The French government in particular had to confront the paradox that one of its greatest export triumphs in the Arab market was also a terrible embarrassment. During the 1980s, France had been Iraq's largest arms supplier after the Soviet Union. It sold Iraq about $5 billion worth of weapons, equivalent to about one-quarter of Iraq's total arsenal. Even before the invasion of Kuwait at the start of August 1990, this relationship was starting to trouble Paris. Reports suggested Iraq owed France 28 billion French francs (FF), of which FF15 billion was for military equipment. The French authorities had already imposed an informal arms embargo and were considering a more formal one. Coface, the French export credit agency, had stopped covering Iraqi credits, a step that frustrated efforts by Dassault to sell more than 50 Mirage 2000 jets to Iraq for FF22 billion.[9] When the war came, the French public was outraged that its armed forces were having to cope with an enemy that had been so substantially armed by French industry. French air strikes used AS-30 laser-guided missiles, of which there were 180, whereas Iraq had bought 240.

Revelations concerning the involvement of German companies in the upgrading of Scud missiles, the assembling of the "supergun," the chemical weapons plants in Samarra, the artillery and arms construction plant at Taji, and the Sa'd 16 research project on chemical and nuclear weapons influenced German attitudes toward the war.

The United Kingdom prided itself on having provided only a minuscule portion of Iraq's arms, and it had made a deliberate decision not to sell Hawk aircraft to Baghdad. Post-war investigations revealed, however, that decision-makers had been far less reticent in supplying dual-use technologies. In late 1992, a scandal broke out when a court case against the directors of a machine-tool firm, Matrix Churchill, was dropped. They had been accused of knowingly supplying dual-use machinery to Iraq for arms production. The case was dropped when it was revealed the government knew about much of what was going on and had relaxed its own guidelines, set during the Iran-Iraq war. This case revealed the complexities of decision-making—the managing director of the company was a major source of intelligence on the Iraqi procurement network, so that the government tolerated his firm's business. Trade with Iraq was bound up in assumptions about the imperatives of domestic reconstruction after the Iran-Iraq war. Only gradually did Saddam's true ambitions dawn on British policy-makers. Even after the revelations, the politicians directly responsible for promoting sales with Iraq, then based in the Department of Trade, were remarkably unrepentant.[10] They argued the need to pursue national interests and pointed to the problems of exercising restraint when others are prepared to break the rules.

These embarrassments and the need not to appear out of step with domestic and international opinion led Europe to declare a policy emphasizing the merits of restraining arms supplies to the Third World. Until now, at a declaratory level at least, European countries have not been outdone by American posturing on the desirability of international arms trade control regimes. French President François Mitterrand was influential in helping convene the first meeting in July 1991 of the five major arms suppliers (the P-5), which happen to be the five permanent members of the United Nations (U.N.) Security Council. British Prime Minister John Major was at the forefront in arguing for the creation of an international arms trade registry. In fact, the British and the French have been even more ambitious than their American counterparts. Where President George Bush wanted to focus the P-5 efforts on the Middle East in particular, Mitterrand wanted a global focus—not surprising, given the great importance of the Middle East market for French arms traders in the past. The July 1991 meeting supported both approaches, although in real terms the Middle East received the greatest attention. When it came to the arms register, the United States called for a more modest annual report system, while the United Kingdom pushed for a more expansive arms trade register system.

In practice, a large gap exists between these declaratory pronouncements and actual policy. By the time President Bush left office, the general view in Europe was that the control initiatives were casualties of Bush's reelection drive: his campaign suggested that Washington was only interested in curtailing the sales by others. The deliberate American undercutting of British sales of Challenger tanks to Kuwait further undermined London's views of the P-5 process. From a British perspective, this sale differed from some U.S. aircraft sales, for example, to Taiwan, in that the British suffered a direct loss as a result of the deal. Moreover, it was clear the Bush Administration had played an important role in undercutting Britain's bid. While the aircraft deals had weakened U.S. moral authority in calling for restraints on transfers of conventional arms, the tank deal severely harmed it.

The Character of Future European Arms Exports

Since the Gulf War, Europeans have suspected that American policy has been geared to gaining a lock on Middle East weapons purchases by forcing the United Kingdom and France out of the market. Within less than two years after the end of the Gulf War, the effort to control the arms trade seemed exhausted. The widespread view in Europe was that the reason was the aggressive American sales policy, which was incompatible with its professed interest in controls.

The truth is somewhat more complex. European governments and firms were not themselves following a policy of restraint. They were trying to compete with the Americans. The problem was that the United States had the comparative advantage. While the United Kingdom and other European countries will continue to contribute to discussions on limiting weaponry to the Middle East and elsewhere, this effort will go hand in hand with a steady commitment to ensuring European involvement in an increasingly competitive market.

The focus of European arms exports will remain the Middle East, complemented increasingly by East Asia. The Middle East has been an extremely lucrative market for the British, especially with regard to Saudi Arabia, where the 1988 Memorandum of Understanding ensures the continuation of the Yamamah agreement. The British believe that the failure to sell the Challenger tank to Kuwait was more the result of American strong-arm tactics than the advantages of the Abrams model. U.S. inroads into the Saudi air market also hurt. There is concern that pressure on the Saudi budget resulting from advance commitments to the United States will make it difficult for the British to secure pending contracts for the purchase of forty-eight Tornado combat and sixty Hawk trainer aircraft.[11] U.K. diplomats now feel strong pressure to support future sales to the Gulf states. Kuwait certainly has sought to balance its rejection of the Challenger by purchasing Warrior armored vehicles. If Gulf procurement decisions are based at least to some extent on political calculations, then, given Britain's residual influence in the region, some defense firms should not be too disappointed. Despite its traditional ties to the market in the Middle East, the priority in the United Kingdom is the expansion of sales to the Asian market. It has already concluded deals for Hawk trainer jets with Brunei, the Republic of Korea (South Korea), Malaysia and Indonesia. Further efforts in this arena can be expected in years to come.

As part of its commitment to support its defense industrial base, the French government should continue to market its weaponry aggressively during the 1990s. The French are even more unhappy than the British with what they view as an American desire to dominate the Middle East market. As with the United Kingdom, France is aggressively marketing its arms in Asia to make up for losses in the Middle East.[12]

The case of Germany is extremely interesting. It still works under self-imposed restrictions, but it has been steadily expanding its share of the overseas market. Its increased exports have involved primarily naval equipment—submarines, frigates and corvettes. Recipients have been Argentina, India, Norway, Poland, Portugal, Singapore and the United States. Germany also inherited thousands of pieces of equipment from the former German Democratic Republic (East Germany), amounts of which it has transferred to Belgium, Egypt, Finland, France, Hungary, Israel, the Netherlands, Norway, Poland, Spain, Sweden, Turkey, Uruguay and the United States. For historical reasons, Germany is very sensitive to weapons sales to the Middle East. However, it has circumvented this inhibition through collaborative projects such as the Tornado aircraft, whose sale is a joint one with its European partners rather than purely German. It is reasonable to assume that increasing European collaboration will mean increasing sales for German firms simply because of the role played up to now by Germany in collaborative ventures. This suggests that transfers of German weapons will remain high in the 1990s.

At some point the former members of the Warsaw Pact will seek to purchase arms from outside suppliers. Such sales are not imminent because of the large inventories left over from the communist years and the lack of capital with which to purchase new weapons. However, in the long term many of these countries will want to wean themselves away from old sources of supply. The European countries will be well-placed to meet this demand. How to handle transfers to the post-communist countries without sending unfortunate political signals will pose a challenge to Western European governments.

For Europe, as with all other suppliers, the arms trade has always involved an awkward mix of commercialism and strategic judgement. In the contemporary world—one no longer guided by the East-West confrontation—the basis for strategic judgement has become more complex. The natural customers of the European states have been different from those of Russia and China, but as commercial pressures grow in the West, competition for those natural markets will grow. If actual U.S. sales policy continues to differ radically from declared policy, if the Chinese sell arms and if the Russians prove unable to control the flow of arms, Europe will find fewer reasons to control its flows.

In any event, the question of controls only touches this export drive at its extremities. It is notable that even the theorizing about the new measures proposed in 1991 presumed the purpose was not so much to dampen the trade in arms per se as it was to guard against the recurrence of an Iraq-type situation, in which a single country acquired far more advanced weaponry than it could conceivably have needed for its defense.

Export Controls: The Surrender of National Autonomy

While Europe's major arms exporting governments may be ready to participate in collaborative projects and willing to tolerate cross-border acquisitions, they have, as noted, traditionally been extremely wary about giving up full control over their weapons production capabilities. Not surprisingly, they have also been slow to agree to surrender their aegis over overseas sales and control of those sales. They have found support in Article 223 of the Treaty of Rome, which places national arms industries and the prerogative for arms sales and arms control decisions firmly under national governments. Article 223 explicitly says that no member country is obliged to disclose information it deems vital to national security, nor is it prevented from adopting steps it considers essential to protect its trade in armaments.[13] The transfer of weapons remains subject to registration with national customs authorities even if they are going to other EC member states.[14] Military manufacture and export controls are thus generally differentiated from the EC's other trade and commercial policies.

The future of this national autonomy will be increasingly debated during the 1990s. The reasons are the continuing integration of European defense industries (which place such industries beyond the control of single nations); the desire by some in Europe to use the Maastricht Treaty to move toward the development of a common European foreign and security policy (with coordination and control of arms sales a significant component); and, more immediate and pressing, the response to the movement toward a single market and the removal of internal customs borders. The removal of such borders has implications not only for weaponry but also for dual-use technology, which is not exempt from EC control under Article 223. Existing differences among European states in the areas of product and country control lists, of legislative practices and of approaches to end-user certificate requirements create major loopholes in any pan-European effort at controls.

Two issues are increasingly coming to the fore as a result of the movement toward integration: (1) whether items exempted by Article 223 from EC control could be surrendered to broader community supervision; and (2) how Europe can ensure dual-use items—whose control will only be monitored once they leave European borders—do not exit the continent by way of the member state with the most lax export laws.[15]

While theoretically the control of arms and dual-use technologies should be treated similarly, in many respects the implications of Article 223, with its imposed limits on EC authority, mean the two present very different sets of problems. Because of Article 223, control of arm transfers originally had to be discussed outside EC institutions. These transfers came under European Political Co-operation (EPC), an institution set up to discuss foreign policy issues within the EC. The EPC provided the framework for discussions on unified EC arms control actions, for example, the arms embargoes during the 1980s against Argentina, Libya, South Africa and Syria and during 1990 against Iraq. Only in 1986 did the Single European Act integrate the EPC into the EC. Soon thereafter, in 1991, the EPC's Political Committee established a working group to draw up criteria for use in establishing control over arms exports. In June of that year seven criteria were agreed upon and reported to the European Council, which accepted the report and added an eighth criterion. The criteria are:

(1) Respect for the international commitments of member states, in particular, for resolutions of the U.N. Security Council, EC sanctions and multilateral non-proliferation regimes.

(2) Respect of human rights in the destination country.

(3) The internal situation in the destination country.

(4) The preservation of regional security and stability.

(5) The national security of member states and allied countries.

(6) The behavior of the buyer in relation to the international community.

(7) The possibility of re-export from the destination country.

(8) The compatibility of arms exports with the technical and economic capacity of the recipient state.[16]

It was the Council's hope that acceptance of these criteria would form the basis for coordinating national control efforts.

Acceptance, however, is only one thing. Agreement on interpretation by member states is another, and efforts within the EPC to secure a common interpretation of these criteria have been unsuccessful. A common arms export policy depends on common interpretations by all members, and this situation does not seem likely in the short term. Efforts continue in this area, and some criteria will likely have to be refined and others dropped before unanimity is achieved.

It should be noted that a common policy on controls over exports of conventional armaments was first put forward in the Inter-Governmental Conference (IGC) on European Political Union. This was supported by the European Commission and European Parliament, but some member countries quickly rejected it. The Maastricht Treaty, which emerged from the IGC, did not deal directly with arms control, but it allowed for the coordination of arms exports in the context of a common foreign and security policy. Movement here depends on broader political processes, including ratification of the Maastricht Treaty by all EC member states.

In the future, Article 223 could be discussed and deleted, but at the moment this action seems unlikely. The political climate in Europe will have to change significantly, control mechanisms will need to improve substantially, and acceptance of most of the Council's criteria for arms exports would have to have taken place.

Although major difficulties will need to be overcome, there is room for optimism in the area of dual-use technologies. In August 1992 the Commission published a "Proposal for a Council Regulation (EEC) on the exports of certain dual-use goods and of certain nuclear products and technologies."[17] This proposal includes: authorization of exports of specific dual-use items identified on a complementary list (based on the Non-Proliferation Treaty [NPT], Missile Technology Control Regime [MTCR], and Co-ordinating Committee on Multilateral Export Controls [COCOM] lists); the setting up of end-user clauses to ensure non-diversion; and the establishment of safety clauses ensuring national control for goods not identified on the complementary list. For the proposal to be successful the Commission recognized there would have to be: agreement on lists of products to be curtailed and on a "negative" list of countries that could not receive certain technologies; the establishment of common criteria for issuing export licenses; the establishment of coordination mechanisms for ensuring pan-European cooperation; and improved control procedures among the various European states.

While there do not appear to be many problems with regard to agreement on a product list (based on dual-use items on the COCOM and NPT regime lists), significant differences remain with regard to criteria for granting export licenses and to the country lists. France and the United Kingdom have expressed strong opposition to the latter. It is also evident that, within Europe, capabilities to monitor dual-use exports vary greatly. Nevertheless, it is possible that some of these issues may be resolved in the not too distant future and that a pan-European system for controlling the export of dual-use technologies may soon emerge, albeit one with a limited scope.[18]

Conclusions

In individual European countries the importance of exports to the defense industries, to employment and to foreign exchange earnings militates against any quick surrender of authority over weapons production and marketing. It is one thing to support collaboration. It is another to surrender full autonomy over defense production. It can be expected that throughout the 1990s states such as the United Kingdom and France will resist Euro-cratic efforts to control exports and domestic military production. These countries will continue to protect their defense industry bases, or at least significant elements thereof, even when doing so runs counter to economic and technological logic.

Conditions for manufacturers in Europe are harsh. They face both de-creased domestic demand and increased overseas competition. Military equipment is one of the few items many post-Communist states can export successfully. Large amounts of second-hand equipment are flowing out-ward in the confusion following the collapse of the Soviet bloc. Whatever interest China had in supporting the concept of control was undermined by the American and French sales of aircraft to Taiwan. Sales such as these highlight the basic philosophical questions that have always dogged efforts to control conventional arms transfers. Why should friends be denied the means of self-defense? If they are to be denied, will they be compensated by security guarantees? To what extent must nuclear non-proliferation take priority even if it means allowing conventional arms as an alternative to a nuclear strategy?

The balance between economic and strategic considerations has always shaped conventional arms control. When the strategic considerations are confused, it is not surprising the economic predominate. More than in the United States, in Europe economic imperatives have driven the sales of weapons. The major restraint on West European trade is less likely to be self-imposed than to be a function of decreased overseas demand, especial-ly if the recent hard-sell by American manufacturers continues. In the meantime, former Soviet and Third World producers will gather strength. The domestic pressures within Europe associated with the decline in domestic military procurement and generally troubled economies will force arms dealers to redouble their export efforts.

Controls are not a high priority on the West European agenda. Certainly, progress toward the establishment of more effective pan-European control mechanisms will depend on progress on the broader political issues related to the establishment of more unified EC policies. Attempts to establish joint mechanisms to control unwanted transfers of arms and technology out of the EC have already fallen afoul of the general problems associated with the Maastricht Treaty and the move toward European integration.

At some point an issue will arise over sales to former members of the Warsaw Pact. How transfers to post-communist countries are handled with-out sending unfortunate political signals will challenge Western European governments. For the moment the main problem in post-communist Europe has been the arms embargo imposed against the former Yugoslavia. An embargo is the strictest form of arms control, and it raises all the traditional problems. It penalizes the side without large stocks or indigenous forms of production, and so favors Serbia. It is extremely difficult to police, especially given the surplus equipment readily available from within the Commonwealth of Independent States. Wars do not need the most sophisticated weapons to keep them going. Tremendous damage can be inflicted with small arms and mortars.

As noted, the arms trade has always involved an awkward mix of commercialism and strategic judgement. It is likely that arms exporters will seek to move beyond their own natural trade partners to the natural partners of competitors. If the Chinese continue to be unwilling, and the Russians unable, to control the flow of arms, Western states will find fewer reasons to control their flows. If the Western states are unwilling to provide security guarantees to potentially vulnerable states or to confront those states, such as Iran, that are clearly building up their military power, it will be hard to argue that friends should not have the means to defend themselves.

These are familiar conundrums. If a "New World Order" had emerged out of the wreckage of the old, with the permanent members of the United Nations able to act in harmony, ready to take on rogue states at the first signs that clearly defined international norms were being violated, then a policy for arms transfer controls might have followed naturally. Instead the strategic environment has become more confusing than ever, a reality that is inevitably reflected in the efforts to control arms transfers.

Notes

1. "Memento on Defence Disarmament 1992," European Institute for Research and Information on Peace and Security, as quoted in *Defense News* (7–13 September 1992).

2. Ian Anthony, Agnes Courades Allebeck, and Herbert Wulf, *West European Arms Production: Structural Changes in the New Environment* (Stockholm, 1990), 2.

3. Francis Tusa, "Euro Industries Take the Lead in Multinational Collaborative Efforts," *Armed Forces Journal International,* CXXXI (1992), 15.

4. Ian Gambles, "Prospects for West European Security Cooperation," *Adelphi Papers* (International Institute for Strategic Studies, 1989), 34.

5. U.S. Congress, Office of Technology Assessment (OTA), *Global Arms Trade: Commerce in Advanced Military Technology and Weapons* (Washington, D.C., 1991), 76–77.

6. "Bonn Technology Exports to Iran Soar," *International Herald Tribune* (7 December 1992).

7. Richard F. Grimmett, *Conventional Arms Transfers to the Third World, 1984–1991,* Congressional Research Service (CRS), U.S. Library of Congress (Washington, D.C., 1992).

8. Stockholm International Peace Research Institute (SIPRI), *SIPRI Yearbook 1993: World Armaments and Disarmament* (Oxford, 1993), 444.

9. For background on West European sales to Iraq, see Kenneth Timmerman, *The Death Lobby: How the West Armed Iraq* (London, 1992).

10. See, for example, the article by Alan Clark, the minister at the center of the Matrix Churchill scandal, in *Daily Telegraph* (11 November 1992).

11. *The Independent* (13 December 1992).

12. See Anthony, Allebeck, and Wulf, *West European Arms Production.*

13. Clause 1 of Article 223 of the Treaty of Rome states:

The provision of this Treaty shall not preclude the application of the following rules: (a) No Member State shall be obliged to supply information the disclosure of which it considers contrary to the essential interests of security; (b) Any Member State may take such measures as it considers necessary for the protection of or trade in arms, munitions and war material; such measures shall not adversely affect the conditions of competition in the common market regarding products which are not intended for specifically military purposes.

14. Interpretations of the article vary among the members, with France and the United Kingdom interpreting it broadly to include many goods and Germany and the Netherlands expressing a preference for a limited national scope.

15. See Wolfgang Reinicke, "Arms Sales Abroad: European Community Export Controls," *The Brookings Review,* X (1992), 22–26.

16. "Conclusions of the Presidency of the European Council Held in Luxembourg on 29 June 1991," *EPC Press Release* (29 June 1991).

17. European Commission, "Proposal for a Council Regulation (EEC) on the exports of certain dual-use goods and of certain nuclear products and technologies," Com (92), 317 final (Brussels, 31 August 1992).

18. *Arms and Dual Use Exports from the EC: A Common Policy for Regulation and Control* (Bristol, 1992).

EIGHT

Russia

Julian Cooper

UNTIL THE END of 1991, the Soviet Union was one of the world's major exporters of conventional weapons. It had pursued this trade for political ends rather than commercial gain, within the framework of a global division of labor shaped by the East-West rivalry, which dominated the post-war epoch. The simultaneous collapse of communism and the Soviet empire abruptly fractured this familiar set of relations and patterns of behavior. The USSR, once a monolithic, large-scale supplier of armaments, was fragmented into fifteen independent states, each inheriting a portion of the vast Soviet military-industrial complex. This chapter explores the implications of this revolutionary transition for arms proliferation.

From Stalin to Gorbachev

While some arms transfers took place during the 1930s, only in the late 1940s did the Soviet Union begin to trade in weapons on a regular, organized basis. Josef Stalin established a specialized Engineering Directorate within the foreign trade ministry to handle arms sales, initially to other countries in the socialist camp. From then on the Soviet government undertook arms transfers in accordance with a set of guiding principles that remained virtually unchanged until the *perestroika* of President Mikhail Gorbachev.

The export of conventional arms was a monopoly of the Ministry of Foreign Economic Relations (MFER) of the USSR Council of Ministers. Before Gorbachev, two directorates of the ministry were responsible for all questions of arms transfers and military technical cooperation with foreign powers. The Chief Engineering Directorate (*Glavnoe Inzhenernoe Upravlenie,* or GIU), the successor to Stalin's Engineering Directorate, handled the sale and delivery of weapons to foreign clients. The Chief Technical Directorate (*Glavnoe Tekhnicheskoe Upravlenie,* or GTU) was concerned with the transfer of licenses and the building of factories for the production of Soviet weapons abroad and the repair of transferred equipment. These directorates were staffed predominantly by military personnel, with an especially strong representation of naval officers. Work in arms sales, which provided opportunities for foreign travel and favorable remuneration, was regarded as a highly desirable posting. Not surprisingly, many of the staff of the directorates were children of the country's military and political elite.[1]

The MFER worked in consultation with the Ministry of Defense, the USSR Committee on State Security (KGB) and the Foreign Ministry, although the latter appears to have had a subordinate role. Decisions on which countries were to receive Soviet armaments and which types of weapons could be sold were a prerogative of the country's political leadership, specifically, the Communist Party Central Committee Secretariat and, ultimately, the Politburo. All export deals had to have the formal approval of the chairman of the Council of Ministers.[2]

Arms transfers were regarded as an important instrument of policy in the Soviet Union's strategic drive for global influence; commercial considerations were secondary. Some clients identified as especially deserving received arms as free gifts or were expected to make no more than symbolic payments on a barter basis. More common were transfers at highly favorable terms of credit, usually extending to fifteen years at zero or minimal rates of interest. In many cases repayment was in goods or the soft currency of the recipient country, without expectation of much hard currency reimbursement, if any. During the time of President Leonid Brezhnev, sales for hard currency grew, but maximization of such earnings was not the primary goal.

The economic arrangements for Soviet arms transfers were highly centralized with the enterprises, with the development organizations of the defense industry playing a largely passive role, as they had no material interest in export sales. As with any other product, arms for exports were planned. MFER requirements for exports for the coming year were incorporated into the annual plans of production plants, alongside equipment for the domestic forces. Arms marked for export were transferred to the MFER for payment in rubles at domestic ruble prices. The MFER then charged whatever price it felt appropriate to the foreign client. Any hard currency receipts went to the state budget. Neither the design organizations responsible for the development of an exported weapon nor the manufacturing enterprises received any share of the hard currency earnings. As far as they were concerned, exports were almost identical to deliveries to the Soviet armed forces. The only real difference was that they had to meet special export specifications. It is not clear the additional costs of these special requirements were always covered by the prices the MFER paid.

A striking feature of Soviet arms sales prior to the Gorbachev period was the extraordinary secrecy surrounding all aspects of the business. No statistics were published on the scale of the trade, and Soviet arms export policy could not be discussed in the press. Probably only a handful of top party, military and government leaders knew the details. As has been acknowledged since the demise of the USSR, this impenetrable secrecy was highly convenient to the workers of the MFER system, who could readily conceal their errors and privileges.[3]

Change Under Gorbachev

Under Gorbachev's perestroika the traditional arrangements for arms transfers began to change. The new foreign policy orientation was associated with a progressive depoliticization of the arms trade, with a corresponding growth in emphasis on commercial considerations. With *glasnost* and democratization, for the first time Soviet arms transfers became a matter of public debate, with open criticism of the lack of information on the volume and terms of sales. One of the first to raise the issue of secrecy was the then-head of the foreign ministry's international organizations department, Andrei Kozyrev, the future foreign minister of Russia, who noted the absurdity that even he had no choice but to use data from the Stockholm International Peace Research Institute (SIPRI).[4]

Beginning in 1987, the Soviet Union began to decentralize its foreign trade, giving individual enterprises more autonomy, including the right to retain a share of hard currency earnings from exports. This policy put organizations in the defense sector at a disadvantage in relation to their civilian counterparts. Taking advantage of the new openness, from early 1990 on some prominent figures in the defense industry began to campaign openly for an equivalent material reward for their military exports. Prominent in this lobbying were the leading designers in the aircraft industry, Mikhail Simonov, general designer of the Sukhoi organization, and Rostislav Belyakov of MiG.[5]

An additional impetus to greater freedom in external economic activity for defense sector organizations was the policy of unilateral force reductions the Soviet Union announced at the end of 1989. According to the new policy, the defense budget was to be reduced by 12 percent by 1991 and would include a reduction in expenditures on the procurement of weapons and other military hardware of almost 20 percent. Capacity released from military production was to be converted to civilian use, with priority to goods of direct benefit to the welfare of Soviet citizens. As the cuts in military expenditure progressively deepened, many enterprises in the defense sector found themselves in increasing economic difficulty, unable to organize new civilian activities on short notice. Efforts to draw up a comprehensive national plan for conversion were protracted and increasingly divorced from reality, as the traditional structures and mechanisms of the planned economy disintegrated. Growing tensions and fears within the defense industry contributed to the febrile conditions that sparked the abortive coup attempt of August 1991.[6] It is not surprising that in the months preceding the demise of the USSR, many within the defense sector were beginning to regard increased arms exports as a relatively benign alternative to reduced military production.

The collapse of the communist regimes in Eastern Europe had an immediate impact on Soviet arms sales. Until then, the volume and pattern of transfers had been broadly similar to those in the first half of the 1980s, although deliveries to Afghanistan partially offset a decline in sales to the Middle East and Africa. The overall trends are shown in table 8-1. Former Soviet statistics are still woefully inadequate, but according to Russian sources, 1989 was the peak year for exports of arms and other military equipment, with sales of 14.6 billion foreign trade rubles. The value declined by 25 percent in 1990 to 10.8 billion.[7] It has been claimed that during the 1986–90 period, 32 percent of arms sales were for cash and 54 percent for credit, with the remaining 14 percent being transferred free of charge.[8]

Both Western and Russian sources confirm the dramatic decline in arms exports in 1991, although the precise magnitude is uncertain. According to Russian statistics, exports of arms and other military equipment amounted to 4.3 billion foreign trade rubles, which means a decline of 60 percent, a level identical to that indicated by SIPRI data.[9] However, in September 1992 the MFER released data claiming that USSR sales in 1991 of basic categories of armaments were only U.S. $1.55 billion (including free deliveries of $20 million). This understatement may be the result of using

Table 8-1. *Soviet Transfers of Major Weapons, 1981–90*

Years	Total (mil. 1985 U.S. prices)	Warsaw Pact	Non-Warsaw Pact socialist countries[a]	Mideast	India	Africa	Afghanistan
1990	6,397	31.1	13.1	7.0	19.3	8.7	17.0
1989	12,220	23.4	11.9	14.6	22.9	5.0	17.6
1988	12,559	24.8	10.2	20.9	21.2	8.7	7.4
1987	14,916	24.9	4.1	31.1	22.5	9.0	4.3
1986	14,731	27.5	10.8	29.9	16.6	7.9	4.4
1986–90	60,823	25.8	9.5	22.5	20.5	7.8	9.0
1981–85	60,389	28.9	11.4	34.8	10.0	9.1	1.9

[a]Cuba, Cambodia, Laos, North Korea, and Vietnam.

Source: Calculated from SIPRI data, presented in Ian Anthony, ed., *Arms Export Regulations* (Oxford, 1991), 181–182.

an inappropriate exchange rate.[10] The MFER declared the information was being released in accordance with Russia's obligations to the United Nations (U.N.). More revealing than the dollar total was the breakdown by type of equipment, presented according to the categories to be used in the U.N. Register of Conventional Arms: tanks, 553 units; armored combat vehicles, 658; large caliber artillery systems, 381; combat aircraft, 40; attack helicopters, 1; warships, 3; missiles, 1,783; and air defense missile systems, 1. Of total sales in 1991, 61 percent went to countries in the Middle East, 12 percent to Europe and a mere 1 percent to Africa.[11] Thus, even before the collapse of the Soviet Union, the volume and destination of arms exports were undergoing profound change.

The Russian Federation as an Independent State

In the newly independent Russian Federation, the post-communist regime under President Boris Yeltsin took over most of the former Soviet defense industry. Facilities located in Russia accounted for at least 80 percent of end-product weapons production and 90 percent of military research and development (R&D).[12] It also took over all former Soviet institutions for the management and control of arms exports. From the outset there was a fundamental break with the past: the Communist Party, now banned, could no longer exert any influence. While the party's role had already diminished under Gorbachev with the development of the presidential system, it seems likely that until August 1991 it retained some ability to shape arms export policy.

The new Russian government soon made clear its determination to make a rapid transition to a market economy, beginning with harsh measures to stabilize an economic system in severe disequilibrium. Efforts to balance the budget included a sharp reduction, by at least two-thirds, in expenditures on military procurement. The defense industry, already disorganized as a result of the earlier cuts, the ill-considered conversion, the breakup of the former industrial ministries and the supply relations disrupted by the fragmentation of the union, now faced a collapse in domestic military orders, with little prospect of being able to take up the slack through more civilian work. To make matters worse, the newly created Russian armed forces lacked a military doctrine and had no clear idea of their future requirements for military equipment. These immediate circumstances of the first few months of the post-communist regime helped shape the emergent state policy on arms exports.

Uncertainty and Local Initiatives

In the initial period of confusion from late August 1991 to early 1992, the Russian government had no clear policy for the conduct and control of arms exports. Yeltsin soon made clear he had no objection to efforts to sell arms abroad. Speaking to members of the industrial group of the Russian Parliament at the end of October, he argued that arms exports could be profitable for the country. Yeltsin also revealed he had suggested to the U.S. president that they should divide the arms market between the two countries, but President Bush had deemed such a measure inadvisable.[13] At this time, Yeltsin appeared to have been under the influence of Mikhail Bazhanov, president of a shadowy commercial organization, Protek, and formerly chief of a directorate of the General Staff of the Soviet Armed Forces. While retaining his company presidency, Bazhanov was appointed chairman of a State Committee of the Russian Federation for Conversion attached to Yeltsin's presidential apparatus. In this capacity he was given the right to license and control the export of weapons. It was envisaged the State Committee would work closely with the Ministry of Defense and a specially created directorate of the Foreign Ministry, with the latter advising on the political dimension of potential sales.[14] Concern was expressed in Russian parliamentary and government circles when it emerged that Yeltsin had issued a secret presidential order to form a state foreign trading company for weapons sales, with the company to be overseen by Bazhanov's State Committee. The concern was understandable: at this time Protek was rumored to be involved in potential large-scale deals with intermediaries in Spain, Liberia and the United States, the speculation being that military technology was to be sold off at giveaway prices.[15]

At the same time, anxiety was rising in the West as evidence accumulated in the autumn of 1991 suggested that Russia was adopting an aggressive arms export policy. A range of Russian combat aircraft was displayed at the Dubai air show, with prices far below normal market rates. Iran and Syria were both rumored to be looking at deals with Russia to acquire the latest aircraft and tanks. These developments prompted U.S. warnings to Moscow that Western aid could be threatened.[16]

By the end of 1991 Yeltsin appeared to have realized the necessity of establishing an effective state licensing system to control sales of weapons, but it took time for appropriate structures to be created. Meanwhile, arms producers and regions dominated by military production were pushing for decentralization of trade. As noted, demands that arms developers and producers have a more active role predated the collapse of the USSR. From early 1992 on, the sharp cut in domestic defense orders and fears of unemployment and social unrest prompted the political leadership to relax its opposition to local initiatives.

In January 1992 the Speaker of the Supreme Soviet, Ruslan Khasbulatov, headed a government delegation to Udmurtiya in the Urals, one of the regions in Russia most heavily dominated by military production. Khasbulatov suggested that arms exports could be used to raise money to fund defense industry conversion, with a substantial share of the export earnings generated by local enterprises to be retained by the Udmurt Republic.[17] This idea was developed in further discussions with the Udmurt authorities by Mikhail Malei, appointed state advisor on conversion issues in November 1991. Later in the spring Yeltsin approved this local initiative. This act was the origin of what Malei later termed "economic conversion."[18] At about the same time, in late January, two arms-producing enterprises in Tula were granted the right to sell to customers directly. To underline that this move had Yeltsin's backing, the President himself gave permission for a plant of the Nizhnii Novgorod automobile association (GAZ) to sell 300 armored personnel carriers independently.[19] This move established what became a characteristic pattern of presidential patronage: a succession of decisions by Yeltsin granting individual enterprises the right to find customers for military goods and to retain a large portion of the hard currency earnings.[20]

A new lobby of the defense industry, the League of Defense Enterprises, under its president, Alexei Shulunov, took up the idea of granting an independent role to arms producers. Writing in January 1992, Shulunov argued the new possibilities of expanded sales were not being exploited because of excessive centralization of export activity. However, he acknowledged the need to retain a system of overall state control.[21] In March the government approved an extension of the Tula initiative: on an experimental basis, until the end of 1992, a group of defense industry enterprises was given permission to deal directly with foreign customers.[22] Also in March, the leader of the government, Egor Gaidar, approved in principle the establishment of a state arms trading company to sell weapons produced in the Nizhnii Novgorod region.[23] By the summer more enterprises had been granted the right to trade on their own account, including the Moscow aviation production association, MAPO, permitted by presidential order to find customers for its MiG-29s, and the Almaz association, which was building the S-300 (SA-10) anti-missile defense system.[24] MAPO was the first arms plant to advertise its product in the Russian press; others followed, including the Tula machine-building works, which has actively promoted its aviation guns and other systems.[25]

By the autumn of 1992, local initiative in the trading of weapons had taken new forms. The Nizhnii Novgorod Trade Fair for the first time featured a large display of arms available to foreign clients.[26] Arms began to be offered for sale through the commodity exchange system. In mid-October, for example, a large consignment of equipment, ranging from MiG-29s to Kalashnikovs, was offered for hard currency on the Universal Ukrainian-Siberian Exchange by a seller who chose not to be identified. Fears were expressed that the weapons could find their way to regions in conflict in the former Soviet Union.[27] In January 1993 the same exchange offered weapons for sale valued at $2 billion, but an investigation by Russian journalists cast doubt on the reality of this offer, hinting that the Russian security service may have been using the exchange for intelligence purposes to expose illegal arms traders.[28]

New Commercial Agencies for the Arms Trade

In all cases, enterprises and regions were limited as to the independent action they could take because: the MFER had to approve any deal and issue a license. The ministry also had the right to fix the prices at which arms were exported, a regulation that generated considerable discontent at the local level.

As discussed, the MFER's traditional approach to selling weapons abroad was not commercial in character. The GTU and GIU were concerned above all with the implementation of intergovernmental agreements. Initiative to find customers and to close contracts in the face of competition was not a quality in strong demand: in the words of Vladimir Shibaev, deputy minister for arms exports in 1992, the agencies' main activity was the "formulation of countless documents."[29]

During the first year of the new Russian regime, the agencies of the MFER underwent progressive commercialization. First to be transformed was the GIU, which in January 1992 was converted into *Oboroneksport,* the Russian state foreign economic association for the export of military goods and services, under the leadership of Major General Sergei Karaoglanov. One of the tasks of the new organization was to assist enterprises in their efforts to export.[30] In April the GTU was transformed into *Spetsvneshtekhnika,* a state trading company charged with finding customers for weapons and concluding deals. Headed by Major General Valerii Brailovskii, it could work as an intermediary on behalf of industrial clients, charging a commission that could not exceed 5 percent of the value of the contract. An activity of this new company was the organization of commercial exhibitions of Russian weaponry. One of the first, held in July at the Moscow International Trade Center, featured missile and artillery systems.[31]

During the Gorbachev period a third organization, the Chief Directorate for Collaboration and Cooperation (*Glavnoe Upravlenie po Sotrudnichestvo i Kooperatsiya,* or GUKS), was created within the MFER to work with socialist countries in the area of license transfers and cooperative projects for weapons production. In the autumn of 1992 GUKS underwent a radical change, becoming Konvimeks, a joint stock corporation for sales of weapons. The shareholders were arms producers belonging to the League of Defense Enterprises. The league, jointly with *Spetsvneshtekhnika,* was involved in the founding of yet another joint stock company to supply spare parts and services and to modernize Russian weapons in the hands of foreign clients.[32] These new organizations will provide a degree of commercial interest in arms sales hitherto unknown for former Soviet defense plants.

In addition to these new organizations in Moscow, some design organizations and firms have been creating their own commercial structures for sales promotion and service. One of the first was MiG-Servis, created by the MiG design organization and the Moscow production plant to provide foreign clients with spare parts and maintenance services.[33] In addition, a number of regionally based arms export firms have been formed, including Kalashnikov in Udmurtiya, Siberia Conversion in Novosibirsk,

Bol'shaya Volga in Nizhnii-Novgorod, and Zolotye vorota Rusi in Vladimir. They are joint stock companies with local defense plants as the
principal shareholders.[34] The Vladimir-based firm is organizing auctions
at Suzdal of unsold stocks of weapons produced originally for state
export orders. Foreign buyers are permitted, provided they can get MFER
approval.[35]

Toward a New Control Regime

In the face of vigorous pressure to decentralize arms export activities,
during 1992 the MFER under Minister Petr Aven and Deputy Minister
Shibaev consistently struggled to maintain a regime of strict central control.
While new organizations emerged and struggled for influence, the MFER
structures for military-technical cooperation retained most of their former
staff and, partly through inertia and partly for reasons of professional pride,
continued to act in a manner that must be judged, from the viewpoint of the
outside world, as highly responsible. In maintaining their essentially conservative approach they appear to have had the firm backing of the Ministry
of Foreign Affairs under Andrei Kozyrev.

Early in 1992 it was revealed that a law on military-technical cooperation between Russia and foreign states would be prepared before the end of
the year.[36] Action was more rapid than envisaged: during a legislative
vacuum, Yeltsin on May 12 approved a decree on military-technical collaboration and regulations on procedures for state control of arms exports
and imports.[37] The president was to exercise ultimate control, with advice
from relevant government bodies. The president would make decisions on
the general policy for arms exports, the establishment of relations with
states with which Russia had not previously had cooperation, the sale of
weapon systems not previously traded, the license of sales for foreign
production of Russian weapons, and the termination of sales to individual
foreign states. All activity in the field of arms exports was to be coordinated
and monitored by the Interdepartmental Commission on Military-Technical
Collaboration between Russia and Foreign Countries (known in Russia by
the acronym KVES).

The KVES was created in May 1992. At first it was chaired by the acting
Head of Government, Egor Gaidar, but he was soon replaced by Georgii
Khizha, Deputy Head of Government for industry. The ministers of all the
government departments concerned with arms transfers, including the
foreign intelligence agency, the domestic security service and the finance

ministry, sit on the commission. It rarely meets as a whole because its members are too busy. Instead, it has a permanent secretariat and is supported by an Interdepartmental Expert Group.

The May decree delineated the role of each major actor. The Ministry of Foreign Affairs monitors Russia's observance of international obligations and the political implications of arms sales. The Ministry of the Economy, together with the Ministry of Industry—later replaced by the new State Committee for the Defense Branches of Industry—and the armed forces, are charged with drawing up proposals on the volume and type of weapons to be exported and imported. The Commission for Export Control of the Ministry of the Economy is responsible for regulating militarily sensitive technology transfers. The Ministry of Defense assists with the delivery of arms and with training in their use, repair and general backup. It can also supply surplus equipment. Many of the practical issues of arms trade are the concern of the MFER, including implementation of intergovernmental agreements, monitoring of prices and other financial aspects of arms sales, and, of particular importance, the monopoly right to prepare and issue licenses to participants in the arms trade authorized by the Russian government. The Ministry's Chief Directorate for Military-Technical Collaboration grants the licenses.[38]

This control structure was not to the liking of all parties involved. The Ministry of Industry and, possibly, the Ministry of Defense, together with the leaders of many arms-producing enterprises, wanted to weaken the control functions of the MFER so that arms producers could conclude contracts with the maximum degree of independence.

Before the May decree, Parliament had demanded involvement in the regulation of arms transfers. Yeltsin's unpublished decree of autumn 1991 appears to have included reference to Supreme Soviet approval being necessary for the export of weapons of types listed in a so-called "red" book.[39] The formal structures for control do not in fact embody a role for the Parliament. However, a parliamentary intercommittee group for military-technical collaboration has been established under Vitalii Vitebskii, the Supreme Soviet's deputy chairman for industry and energy and a former defense industry design engineer. This group of nine senior parliamentary deputies is charged with monitoring Russia's arms export policy and practices. They are supposed to have access to relevant documentation to permit a review of individual contracts at all stages through completion.[40]

The basic regime for controlling arms sales has been supplemented by additional measures to regulate transfers within the Commonwealth of

Independent States (CIS) and the export of strategic technologies. A government decree issued in July established a licensing procedure to control deliveries to other CIS member states of components for military hardware. Deliveries of weapons to CIS states involved in conflicts was to be governed by the procedures for foreign powers set up according to the May decree.[41] Earlier in the year, in April, Yeltsin issued a decree on measures to create an export control system to regulate transfers of equipment, material and know-how with potential use in the manufacture of weapons, including missile, nuclear, biological and chemical systems. An Export Control Commission attached to the Ministry of the Economy, chaired by Georgii Khizha, is responsible.[42] In July it issued the first list of items to be controlled.[43] To assist this process, a Council for Export Control, chaired by Academician G. A. Mesyats, was created by the Presidium of the Russian Academy of Sciences.[44] Later in 1992 new control regimes were adopted for nuclear-related exports and exports of dual-use technologies.[45] These legislative initiatives, designed to create a Western-style control regime, culminated in the adoption at the end of January 1993 of a statute on the licensing procedure for the export and import of listed armaments and other military goods and services. This statute confirms that the MFER must issue licenses. Companies violating the license procedure risk being blacklisted, with individuals subject to possible imprisonment of three to twelve years.[46]

By the spring of 1993 the Russian government had made real progress in establishing a system of control of arms exports and militarily sensitive technology. It remains to be seen whether this system can operate effectively in the face of powerful pressures for a more lax control regime and in an environment characterized at all levels by a chronic weakness of executive power and authority. Working in the direction of more effective control has been international pressure for improved regulation of conventional arms transfers.

International Obligations

Under Gorbachev, the position the Soviet Union was beginning to adopt on the arms trade was more responsive to concerned international opinion. After the Gulf War, during which Soviet-supplied weapons were deployed in breach of the U.N. Charter, Russian Foreign Ministers Edvard Shevardnadze and Andrei Kozyrev were both active in promoting improved regulation of conventional arms transfers.[47] The Soviet Union backed the creation of a U.N. Register of Conventional Arms and began to argue for a strengthening of the control regime for missiles. While the USSR was not a formal signatory of the Missile Technology Control Regime (MTCR), Gorbachev and Shevardnadze on a number of occasions made clear its provisions would be observed.[48]

Since the collapse of the USSR, Yeltsin, Kozyrev and other Russian government representatives have reaffirmed support for international efforts to regulate trade in conventional arms. Russia participated in the P-5 process, reaffirmed its support for the U.N. Register, made its annual submission and declared allegiance to U.N. embargoes on sales to individual countries.[49] According to Shibaev, Russia is strictly observing the principles adopted by the London meeting of the permanent members of the Security Council in October 1991. As a result, he stated, Russia is observing embargoes on sales to Afghanistan, Angola, Cambodia, Iraq, Kuwait, South Africa and the former Yugoslavia. He also indicated Russia will no longer sell arms to Ethiopia and Libya.[50] This commitment to the observance of international rules indicates the Russian administration's awareness of the importance of not allowing arms export activity to complicate relations with the West at a time when assistance and political support are needed. The need for this assistance remains the strongest card in the hand of Kozyrev and the foreign ministry in their effort to maintain an effective regulatory regime for arms transfers.

Tensions exist, however, and they surfaced in early May 1992 when Russia pursued the sale of cryogenic rocket engine technology to India. This deal, worth $200–250 million, had been concluded by the former Soviet Union. The United States challenged it as a breach of the MTCR. Although Russia has not signed the MTCR, as legal successor to the USSR it had agreed to abide by its provisions. The U.S. move, which was ultimately successful in preventing the sale, caused considerable bitterness in Russia. The Russian authorities have been adamant that the technology transfer is purely civilian in nature. It has become an article of faith in Russian governmental circles that the U.S. opposition was motivated by a desire to stifle Russian competition in the space technology field.[51] The Indian deal fed a growing suspicion that Russia was facing an informal embargo a U.S.-inspired attempt to prevent Russia from gaining access to new advanced technology markets. Notwithstanding these reservations, in January 1993 the Russian government adopted a new regime to control the export of equipment, materials and technology associated with missile production; it included publication of a detailed list of controlled items.[52]

The Search for New Markets

With the collapse of the Warsaw Pact, a large captive market for former Soviet arms effectively disappeared. When these countries eventually re-equip their forces, some may decide to purchase from Russia, but those orders will have to be won in the face of Western competition. The Commander of the Hungarian Air Force, Lieutenant Colonel Balogh, probably spoke for many when he declared that Russian aircraft could not be ruled out, but were unlikely, when finding a replacement for the country's aging fleet of MiGs. It is important, he declared, to find a supplier in a country with a strong industrial base.[53] There have also been reports that Poland has decided not to buy more MiGs because of doubts about the availability of spare parts.[54]

Since the breakup of the Soviet Union, Russia has been attempting to enter new markets for arms sales. Ideally, those responsible for the development and manufacture of high technology weaponry would like to find clients in the advanced industrial countries. Since the MiG-29 was first shown at Farnborough in 1988, it has become normal for the principal air displays of the West to feature prominently a range of Russian combat aircraft and, more recently, missile systems.

This sales drive has met with only modest success. Various countries have expressed interest, including Switzerland and Finland, and Yeltsin is reported to have offered Italy MiG-29s and the latest tanks during his December 1991 visit to Rome, but orders have failed to materialize. Following the announcement of Germany's intention to withdraw from the European Fighter Aircraft (EFA) project, there were rumors in Moscow of German interest in the MiG-29 as a possible cheaper alternative. After all, it was argued, German Air Force representatives had praised the qualities of the twenty-four MiGs taken over from the former forces of the German Democratic Republic (East Germany). Again, nothing practical has arisen from these heightened expectations. The one exception is the deal concluded between the Kamov design organization and the Geneva and Washington-based Group Vector company for joint work on the Ka-50 (Hokum) combat helicopter. Group Vector is fronting an attempt to sell almost 120 Ka-50 helicopters to the British Army, at a price almost one-third below that of competitors. It is thought the ability of the company to provide adequate spare parts and service backup could prove a crucial factor in securing this £2 billion deal.[55]

While seeking to penetrate advanced markets, Russia has been trying to export to a number of somewhat less developed countries with little or no previous experience with Soviet armaments. There were reports in August 1992 that Russia had proposed to Taiwan the sale of fifty MiG-29s original-ly intended for Iraq.[56] It later emerged that the Russian government had decided not to pursue arms sales to Taiwan because of a desire not to upset China, considered to have a better market potential.[57] It has also made approaches to the Philippines, offering to sell MiG-29s and naval vessels, including minesweepers and corvettes, in exchange for bananas and other tropical fruits.[58] For over two years starting in October 1991, Russia sought to conclude a deal with Malaysia for the sale of MiG-29s. This episode revealed starkly Russia's relative inexperience in trading on a competitive, commercial basis. Having obtained Malaysian agreement in principle to buy the Russian plane, including consent to a price of $24 million per plane, the MFER prepared a contract for signing. At one stage eighteen Russian middlemen, including a popular singer, suddenly appeared demanding a commission from the MFER. To complicate matters further, the main MiG-29 production plant attempted its own independent approach, offering the aircraft it had previously attempted to sell to Iran. By this time Malaysia had also received proposals from the United States for the F-16 or F/A-18, on very favorable terms. In March 1993 the Russian Vice-President, Alexander Rutskoi, visited Kuala Lumpur in an attempt to conclude the deal, offering as an additional incentive for Malaysia part of the payment in goods instead of full payment in the hard currency previously requested. Eventually, in 1994 a deal was closed in which Malaysia spent $1 billion, half of which bought eighteen MiG-29s and half of which purchased eight F/A-18s. Altogether, Russia's performance in this sale raised questions about its commercial skill.[59]

Israel is another country cited as a target of the Russian sales drive. As early as June 1991 it was reported that Russia had offered to sell a variant of the MiG-31 interceptor, and later in the year it was claimed it had also offered the S-300 (SA-10) air defense missile system Russia's equivalent to the Patriot with a proposal by its manufacturer, the Almaz Association, for joint work to enhance its capabilities.[60]

Russia's relations with South Africa have improved dramatically since the collapse of communism. At the time of President Frederik W. de Klerk's visit to Moscow in June, there was speculation, officially denied, that arms transfers were on the agenda. It was reported, however, that South Africa was interested in acquiring technology for aircraft, lasers and other research-intensive products.[61] There has also been interest in a Russian proposal to use a modified SS-25 missile to launch South African communications satellites from the Plesetsk launch site.[62] A senior representative of the Russian Defense Ministry has acknowledged that South Africa has shown interest at an official level in acquiring Russian military equipment, but the Foreign Ministry has been adamant that Russia will respect U.N. sanctions.[63]

Representatives of the MFER have also made reference to efforts to secure sales in Singapore and Thailand; the latter has confirmed officially that it will not purchase MiG-29s.[64] In this category of countries Turkey provides one of the few examples of success: according to the Turkish Ministry of Internal Affairs, helicopters, armored transporters and other equipment worth $300 million are being purchased.[65]

In the past the USSR had had little success in selling arms to Latin American countries, although it supplied some tanks and aircraft to Peru in 1974. Now there is active Russian interest in finding new markets in that region, with Brazil as the most favored potential partner. In February 1992 it was reported that a Brazilian company had acquired monopoly rights on MiG-29 sales in Latin America, and organizations in the Russian space program have been attempting to develop cooperation agreements with the Brazilian aerospace industry.[66] Military contacts with Chile were developed but did not lead to arms sales. The visit of Argentine Defense Minister Gonzales to Moscow in October 1992 prompted speculation that arms supplies were under consideration. Gonzales denied this report, but he did acknowledge Argentine interest in acquiring new technologies made available by Russian defense industry conversion, in particular satellite technology. He also hinted that at a later date cooperation with Russia on an Argentine nuclear submarine might be explored.[67]

The collapse of the Soviet Union has inevitably affected arms supply relations with India, traditionally one of the USSR's closest partners outside the Warsaw Pact. It has been estimated that approximately 70 percent of the weapons of the Indian armed forces are of Soviet origin. Russia's willingness to continue the relationship has not been at issue. The problem is the commercial terms on which future cooperation is to be based. In the past, arms supplies to India and the licensing of Soviet weapons for local production were undertaken on little more than a barter basis. The Soviet Union granted credit for up to seventeen years at an annual rate of interest of 2.5 percent, with repayment in rupees, which were then used to import goods from India. Before the breakup of the USSR, there were reports that India was interested in the creation of joint ventures with the Soviet Union for the manufacture of aircraft and other weapons for export. Relations have now been soured by a disagreement over the terms of payment and the scale of indebtedness for past arms transfers. In 1990 the USSR had extended the traditional payments mechanism to 1995, but the new Russian government indicated it wanted terms that were more acceptable from a commercial point of view, including at least partial settlement in hard currency. India has also had serious problems obtaining spare parts for former Soviet military equipment, as Russian producers have been insisting on settlement in cash. To maintain the traditional close relations, Russia granted India a long-term credit of $830 million for arms purchases, and by the end of 1992 an agreement had been reached on the supply of thirty MiG-29s and the organization in India of a joint venture for the production of spare parts for aircraft.[68] Meanwhile, as indicated below, India has been seeking closer relations with Ukraine as an alternative source of supply of armaments and spare parts.

As economic problems have complicated the traditionally close relations with India, Russia has made tentative efforts to interest Pakistan in its military equipment, with rumored Pakistani interest in the Su-27 and MiG-29. However, at an official level in Russia it is recognized that this issue is very sensitive, with potential to damage relations with India seriously. Yeltsin has declared that Russia will not give Pakistan any military technical assistance.[69]

Russian observance of U.N. embargos and the switch to trade on a strictly commercial basis have resulted in a sharp decline in arms sales to the Middle East, one reason for the strenuous efforts to cultivate new partners. Sales to Iraq and Libya have been halted; trade with Syria has been disrupted by Russian insistence on hard currency payments for equipment supplied previously and the settlement of outstanding debts. Russia inherited from the Soviet Union contracts for arms sales to Iran worth some $3.5 billion, including MiG-29s, Su-24s, T-72 tanks and diesel-powered submarines. Delays in settling mutual debts between the two countries have complicated fulfillment of these contracts. Russia is also aware of U.S. displeasure, clearly manifested in relation to the delivery of Kilo-class diesel submarines. Russia has denied press reports of a major new deal with Iran, although Foreign Ministry officials asserted Russia's right to supply Iran provided the regional strategic balance is not upset.[70] Perhaps the greatest success in opening up a new market in 1992 was the sale of BMP-3 infantry combat vehicles to the United Arab Emirates, the target of a concerted export promotion drive that included a proposal by the Russian Defense Minister Pavel Grachev for the joint manufacture of weapons.[71]

China was a notable exception to the very limited success the new Russian regime had in finding alternative outlets for its arms exports during its first year. The Russian armed forces were anxious to restore good relations with their Chinese counterparts, and the arms deals concluded appear to have been a direct product of high-level military contacts during 1991 and 1992. The Komsomol'sk-na-Amure aviation works is supplying twenty-four Su-27s under a $1.4 billion contract apparently concluded directly by the Sukhoi general designer, Mikhail Simonov, after efforts by Minister of Foreign Economic Relations Pyotr Aven failed.[72] Sales of the MiG-31 interceptor and T-72 tanks are also expected, and there have been reports that China has decided to purchase more than 100 S-300 air defense missile systems.[73] However, the Russian press has voiced concern over the extent to which unofficial contacts have developed between Russian defense sector enterprises and their Chinese counterparts.[74]

In the case of the Su-27 deal, 65 percent of the value of the contract is being settled on a barter basis, with the remaining 35 percent by payment in hard currency.[75] Notwithstanding the Russian government's desire to switch to normal financial terms for all arms deals, barter is likely to remain a major feature of sales to China.

Selling off the Surplus

With a reduced military budget, deployable equipment constrained by Conventional Forces in Europe (CFE) limitations and the clearly stated intention of Russia's political leadership to scale down the armed forces located on its territory, Russia finds itself in possession of a vast stock of surplus weaponry. At a time when budgetary pressures make it impossible to provide even the most basic housing and social needs of officers and conscripts, especially those returning to Russia from duty in Eastern Europe, the Baltic states and other parts of the former Soviet empire, not surprisingly the country's military leaders have been anxious to raise additional financing by exporting surplus equipment.

As early as October 1991 the air force was complaining that sales of surplus aircraft had to be conducted through the MFER, which kept more than 90 percent of the proceeds.[76] From then on the armed forces campaigned for the right to export surplus arms independently, with a minimum of interference by the MFER. They got the support of some regions in Russia that were holding large stocks of equipment. The local authorities in Omsk, for example, sought permission to sell old T-55 tanks, more than

1,000 of which were at a nearby army base.[77] In February there were rumors that Yeltsin's advisor on military affairs, General Konstantin Kobets, had drafted a decree for the president's approval allowing the armed forces to find customers, negotiate deals and conclude contracts for the sale of surplus equipment, with only a minimal system of control through licenses.[78] Soon thereafter Yeltsin authorized that the air force find customers for 1,600 surplus combat aircraft, with the proceeds to be used to finance housing and pay increases for personnel. This action gave credence to the rumors. The initiative was shortlived, however. Yeltsin's order was rescinded in July, apparently in response to charges of high-level intrigue and corruption.[79]

There has been much speculation in the Western press that, notwithstanding the control regime of the MFER, Russia has exported arms on an unofficial basis. Perhaps the most dramatic claim was a report in early 1993 in the London *Observer* that $360 million worth of arms were sold to Serbia in a secret deal involving the Russian army. The Russian Foreign Ministry vigorously denied this report.[80] There is little doubt some surplus equipment of the armed forces has found its way to the Caucasian republics. Occasional arrests and interceptions in the West indicate that intermediaries have been attempting to sell arms to foreign clients. In early October 1992, for example, the Bavarian police arrested someone attempting to sell former Soviet weapons, including eight MiG-25s, to Iran and Nigeria.[81] The Estonian authorities have intercepted consignments of armored vehicles and infantry weapons destined for clients in Western Europe.[82] The absence of effective controls and customs services at the borders of Russia and its new neighbors has facilitated such unofficial arms trading. Such controls are now being created, and in time it should become more difficult to export illegally. There is also evidence Russia's security and intelligence services are taking the problem more seriously. The Russian Foreign Intelligence Service has created an export controls directorate to guard against the proliferation of sensitive military technologies, in particular nuclear, biological, chemical and missile systems. In early 1993 the security service intercepted sixty specialists in missile technology about to leave Russia to work for the Democratic People's Republic of Korea (North Korea).[83]

Russian military leadership is increasingly assertive, and it is possible the armed forces will succeed in getting a relaxation of the requirement that all sales of surplus equipment be licensed by the MFER. That development could open the door to aggressive efforts to find clients for relatively advanced weaponry at low prices. However, this issue is bound up in the broader issue of the fate of Yeltsin and the Russian government.

Export Hopes Frustrated

During the initial period of the new Russian state, many in leadership positions in the presidential apparatus, bureaucracy, armed forces and industry have backed an active arms export policy in the expectation of earning enough hard currency to help alleviate the acute transition problems of the hypertrophied military-industrial complex. On a number of occasions Yeltsin gave his personal backing to efforts to sell weapons, describing the trade as a potential "shock absorber" for Russia. At the same time he has emphasized that international obligations must be met.[84] Vice-President Rutskoi has been more outspoken in his support and has openly criticized the MFER for the strict control it exercises.[85] The most committed public advocate of sales, however, has been Yeltsin's advisor on conversion issues, Mikhail Malei. Malei has championed what he terms "economic conversion": the financing of defense industry conversion from the proceeds of arms exports during a transitional period of three to four years. The total cost of conversion, in his view, is some $150 billion over fifteen years. Exports could earn at least $5 billion a year toward that sum. Malei has expressed confidence that up to 40 percent of the enterprises in the defense industry possess the potential to compete in the world market.[86]

This strong official backing for arms exports, coupled with the partial decentralization of authority to negotiate deals, raised expectations in industry and the armed forces that more sales would soon be achieved. These hopes have been frustrated. After more than a year it is apparent that non-communist Russia has had little success in finding customers for its new or old weaponry. Arms exports in 1992 amounted to a mere $2 billion.[87] In September 1992 Economics Minister Andrei Nechaev forecast 1993 earnings at approximately $3 billion, but the state's actual order for arms production for export was set at only $2 billion.[88] This low level of orders can be explained at least in part by the large stocks of unsold weaponry: in early 1993 it was claimed that enterprise stocks included more than 200 fighters and bombers ready for sale and 1,000 tanks.[89]

In a manner all too familiar from Soviet times, the limited success in expanding arms sales has provoked a search for scapegoats. Many in Russia's industrial and governmental circles believe the United States is at least partly responsible, insofar as it is perceived as attempting to block

Russia access to markets, both new and old. The principal culprit, however, is widely held to be the MFER. Its critics argue it is too restrictive and not sufficiently enterprising in its approach to arms sales. It attempts to maintain relatively high prices and appears reluctant to allow producers to retain export earnings. The critics may have a point that the staff of the ministry is not accustomed to working in a competitive environment, but they are wrong to attack its efforts to keep arms transfers under effective state control. Some critics also see the hand of the Foreign Ministry under Andrei Kozyrev, regarded by critics as too subservient to Western foreign policy positions and too ready to back U.N. resolutions, even those resulting in the loss of traditional, valued customers for Russian arms. The removal of Aven from his post as minister of foreign economic relations at the end of 1992 was a victory for those campaigning against him. Aven's successor, his former first deputy Sergei Glaziev, appears to be more prepared to pursue an active arms export policy. Soon after taking up his post he raised the issue of the alleged loss of sales resulting from Russian observance of the U.N. arms embargoes.[90] In response to this concern, in February 1993 Yeltsin signed an order setting out a new procedure for the imposition of an embargo on arms sales, a procedure that will make it difficult to take for granted Russian support for new U.N. sanctions.[91]

Although the MFER may have been somewhat responsible for Russia's relatively poor performance, there are clearly other reasons of a less subjective nature. Recognition has been growing within Russia that while certain types of armaments may be competitive from the point of view of technical specifications and combat performance, they are not always attractive to potential customers looking for systems viable over a normal period of service life. A particular case in point is the technologically complex weapon systems that Russia is so anxious to sell. Russia is simply not in a position to offer the kind of long-term service, spare parts and modernization now taken for granted in the advanced markets Russia is seeking to enter. Russia's ability to maintain a strong domestic manufacturing and development capability is also in doubt: customers risk becoming owners of systems produced by a country no longer able to undertake their further technical development. Furthermore, not all weapons now being offered are fully modern, especially with regard to their electronic systems and between-service life of major assemblies. There is no dispute that Russia manufactures some highly competitive infantry weapons and other equipment for ground forces. The problem is that in the new post-Cold War, post-Gulf War environment, such equipment is widely available at low prices, including items built outside Russia under Soviet licenses. In these circumstances, to growing domestic dismay, Russia is finding it difficult to generate export earnings from the very systems best suited to foreign customers.

Recognition within Russia that it faces major obstacles to expanding conventional arms sales was reinforced by its participation in February 1993 at the Abu Dhabi Idex-93 arms exhibition. At the beginning of February, following a visit to India, Yeltsin had informed the government that he had doubts about the wisdom of cutting back arms production so drastically. Instead, arms exports should be expanded. He gave the go-ahead for an unprecedented Russian export campaign at Idex-93, fronted by the Defense Minister Grachev and with the participation of many of the country's leading weapons designers. Some of the latest systems not previously offered for export were displayed. Before the event Russia boldly claimed it expected sales worth $11 billion. However, no deals were concluded. Malei, ever the optimist, made vague reference to possible future deals worth $2–3 billion.[92]

The experience in Abu Dhabi was sobering. Russia had to offer competitive terms of credit and a dependable system of service backup. However, what was required above all was political stability at home. Serious customers would not be found, it was acknowledged, while the political future of Russia remained so uncertain.[93] Against this background Foreign Minister Kozyrev began to explore the possibility of an agreement with the Clinton Administration on the opening up of arms markets to Russia to aid conversion.[94] This initiative, echoing Yeltsin's earlier approach to President George Bush, can be interpreted as a sign that the Russian government realizes there is little chance economic restructuring can be financed through large-scale arms transfers without Western assistance. In making this proposal, however, Kozyrev must also have been aware that in return the West would expect nothing less than Russia's maintenance of a truly effective control regime over all sales of conventional arms.

Ukraine and the Other Successor States

Ukraine has taken possession of a potentially powerful defense industry and substantial stocks of surplus military equipment. End-product weapon systems produced in Ukraine include tanks, missiles and naval surface ships. In addition, enterprises manufacture radar systems, military optical equipment, communications systems and conventional munitions. However, production is heavily dependent on supplies of materials, systems and components from Russia. Creation of a viable national defense industry will require considerable investment and restructuring. It is not clear the Ukrainian government will adopt this course of action. While Defense Minister Morozov appears to favor the development of a strong domestic arms industry, Minister for Machine-building, the Military-Industrial Complex and Conversion, Viktor Antonov, takes a more sober view, envisaging a modest defense sector accounting for no more than 4 percent of industrial output.[95] Meanwhile, as in Russia, defense plants are experiencing a collapse in orders while endeavoring to keep their personnel employed.[96]

Antonov is opposed to an active arms export policy and does not believe conversion of the defense industry should be financed by sales of weapons. His ministry operates the control regime for foreign sales of both new weapons and surplus equipment of the armed forces. Antonov chairs a fifteen-person expert commission that is responsible for issuing licenses; its membership includes representatives of the defense and justice ministries and the security service.[97]

Given the difficulties Ukrainian arms producers face and Antonov's lack of enthusiasm for exports, it is not surprising that sales in 1992 appear to have been extremely modest. The sole reported deal of any size was a contract with India to supply arms and spare parts for the country's stock of Soviet military equipment. In return, Ukraine will receive medicine and cloth for military uniforms for the country's new armed forces, plus some hard currency.[98] Rumors of sales have abounded, however, including reports, officially denied, that arms were supplied in partial payment for oil delivered from Iran. In October the Russian press reported a story in a South African newspaper that that country was contemplating the purchase of MiG-29s from Ukraine. The Ukrainian Defense Ministry promptly refuted the story, declaring it had no intention of supplying arms to South Africa.[99] The most persistent rumor, however, one denied officially on several occasions in both Ukraine and Russia, concerned an alleged sale to China of the Varyag aircraft carrier under construction at the Nikolaev shipyard.[100]

The only other successor states with sizable military production are Belarus and Kazakhstan. The defense industry in the former is more concerned with high-technology systems—radar, radio, communications and optical equipment—than with end-product weapons. The government of Belarus, however, has made clear its intention to export surplus military equipment, in particular aircraft and tanks. It sold tanks to Italy, probably for scrap, and is reported to have offered Finland surplus military aircraft.[101]

The military production of Kazakhstan is closely integrated with that of Russia. Weapons manufactured include heavy machine guns, missile transport erector-launch vehicles and a range of naval armament, including torpedoes and ship-to-shore missiles. The Nazarbaev regime has been anxious to maintain good relations with Russia, including continuation of the traditional cooperation between plants of the two countries involved in military work. Early in 1992 news that the Metallist plant in Ural'sk had sold, with President Nursultan Nazarbaev's reported approval, 500 heavy machine guns to a company in Germany caused a stir. Transfer of shore-to-sea missiles to Libya was also reported to be under consideration. These developments, together with a report that Kazakhstan was proposing to sell arms through the commodity exchange system, gave rise to fears that uncontrolled arms transfers were in the works.[102] There is no evidence the sales have happened; Nazarbaev appears to have maintained strict control.

Before the country became torn by internal strife, it appeared for a while that Georgia might be contemplating an active export policy to find foreign buyers for the Su-25, built at a plant in Tbilisi. This facility is so heavily dependent on Russian supplies for engines and other basic systems, however, it is difficult to see how Georgia could pursue an independent export policy beyond the sale of any stock of completed aircraft.[103]

Details of the formal structures and procedures for the control of exports of conventional arms from Kazakhstan, Belarus and other independent states are not available. It seems likely, however, that, following Russian practice, systems of high-level government control are being operated, at least for sales of newly produced military equipment. As with Russia, the most worrying issues are the lack of clarity in the control regimes for exports of surplus equipment of the former Soviet armed forces and the possibility of illegal transfers where effective border controls are lacking.

Conclusions

Western fears of an uncontrolled flood of weapons from Russia and the other successor states have so far not materialized. A combination of the inertia of former Soviet institutions, an awareness of the potential economic and political costs of provoking serious Western concern and the inability of Russia and other states to respond readily to the new competitive market conditions have limited their ability to sell in new markets to compensate for the dramatic loss of traditional clients. However, Western concern has prompted the Russian government to adopt appropriate legislation to control sales of conventional arms and exports of nuclear, missile and dual-use technologies.

Russia's lack of success in exporting weapons has generated serious domestic tensions. In circumstances of continuing political uncertainty and instability, these tensions could yet lead to the adoption of policies genuinely unsettling to the rest of the world. Russia's future stance on arms exports is likely for some time to be uncertain as the country evolves politically and economically. It is difficult to envisage, however, that anything like the traditional Soviet system of arms transfers can be restored.

Notes

1. See *Nezavisimaya Gazeta* (19 February, 11 July, 26 August, and 30 September 1992); *Krasnaya Zvezda* (24 July 1992).

2. Nikolai Ryzhkov, *Perestroika: istoriya predatel'stv* (Moscow, 1992), 258, 263. Ryzhkov observed this when he was prime minister, with his signature needed in all cases.

3. According to Vladimir Shibaev, then deputy minister of MFER for arms sales, *Nezavisimaya Gazeta* (30 September 1992).

4. See *Izvestiya* (20 February 1992).

5. These developments under Gorbachev are discussed in more detail in Julian Cooper, "Soviet Arms Exports and the Conversion of the Defense Industry," in Luciano Bozzo (ed.), *Exporting Conflict: International Transfers of Conventional Arms* (Firenze, 1991), 135–142.

6. Conversion and the defense sector's involvement in the August coup are discussed at greater length in Julian Cooper's Chatham House paper, *The Soviet Defence Industry: Conversion and Reform* (London, 1991).

7. *Rossiiskii Ekonomicheskii Zhurnal* (1) (1993), 62. These ruble data appear to be in current prices, so that the decline is understated.

8. Calculated from *Rossiiskii Ekonomicheskii Zhurnal* (1) (1993), 62. See also *Nezavisimaya Gazeta* (29 September 1992).

9. *Rossiiskii Ekonomicheskii Zhurnal* (1) (1993), 62; Stockholm International Peace Research Institute (SIPRI), *SIPRI Yearbook 1992: World Armaments and Disarmament* (Oxford, 1992), 272. Free transfers amounted to only 100 million foreign trade rubles.

10. *Nezavisimaya Gazeta* (29 September 1992). The equivalent SIPRI estimate is $3.93 billion.

11. *Nezavisimaya Gazeta* (29 September 1992).

12. *Ekonomiki i Konversii* (4) (1991), 7.

13. British Broadcasting Corporation (BBC), *Summary of World Broadcasts,* SU/1219 B/3 (2 November 1991).

14. *Kommersant'* (46) (25 November–2 December 1991).

15. *Glasnost'* (49) (1991).

16. *Izvestiya* (19 November 1992); British Broadcasting Corporation (BBC), *Summary of World Broadcasts,* SU/1236 C2/4 (22 November 1991).

17. British Broadcasting Corporation (BBC), *Summary of World Broadcasts,* SU/1283 B/3–4 (21 January 1992).

18. *Rossiiskaya Gazeta* (29 January 1992).

19. British Broadcasting Corporation (BBC), *Summary of World Broadcasts,* SU/1291 B/6 (30 January 1992), and SU/W0215 A/1 (31 January 1992).

20. The most striking example is probably the Nizhnii Tagil Uralvagonzavod plant, the world's largest tank producer. During a visit to the factory in June, Yeltsin told the workforce to sell their T-72 tanks abroad and keep 80 percent of the foreign currency earnings (*Izvestiya* [8 June 1992]).

21. *Rossiiskie Vesti* (3) (1992).

22. *Komsomol'skaya Pravda* (31 March 1992); *Kommersant'* (23) (1–8 June 1992); *Izvestiya* (27 April 1992).

23. *Nizhegorodskie Novosti* (13 March 1992); *Komsomol'skaya Pravda* (13 May 1992). The Bol'shaya Volga arms export joint stock company was created later in the year (British Broadcasting Corporation [BBC], *Summary of World Broadcasts,* SU/1607 C1/1 [8 February 1993]).

24. *Kommersant'* (23) (1–8 June 1992); *Nezavisimaya Gazeta* (26 August 1992).

25. The first advertisement for the MiG-29 appeared in *Krasnaya Zvezda* (16 June 1992). A few days earlier (11 June), in the same paper, an article appeared under the heading, "Want a MiG-29? Give Us a Ring!"

26. *Izvestiya* (9 September 1992).

27. *Nezavisimaya Gazeta* (27 October 1992).

28. *Komsomolskaya Pravda* (5 February 1993); *The Financial Times* (6–7 February 1993); *Kommersant'* (6) (8–14 February 1993), 5.

29. *Nezavisimaya Gazeta* (30 September 1992).

30. *Vneshnyaya Torgovlya* (3) (1992).

31. *Kommersant'* (29) (13–20 July 1992).

32. *Nezavisimaya Gazeta* (1 October 1992).

33. *Krasnaya Zvezda* (11 August 1992).

34. British Broadcasting Corporation (BBC), *Summary of World Broadcasts,* SU/1607 C1/1 (8 February 1993).

35. *Moscow News* (45) (1992), 10, and (49) (1992); *Finansovye Izvestiya* (12) (21–27 January 1993).

36. V. Shibaev, *Rossiiskie Vesti* (3) (1992).

37. Published in *Rossiiskaya Gazeta* (15 May 1992).

38. *Rossiiskaya Gazeta* (15 May 1992); *Nezavisimaya Gazeta* (11 July 1992).

39. *Argumenty i Fakty* (49) (1991).

40. *Izvestiya* (12 June 1992).

41. *Rossiiskaya Gazeta* (6 August 1992).

42. *Rossiiskaya Gazeta* (6 August 1992).

43. *Rossiiskie Vesti* (37) (1992). A later decree of December 21, 1992, established a new export and import regime for nuclear-related materials and technologies (published in *Rossiiskaya Gazeta* [28 January 1993]).

44. *Poisk* (28) (1992).

45. Published in *Rossiiskie Vesti* (28 January and 3 March 1993). .

46. *Kommersant'* (1–7 February 1993); *Rossiiskaya Gazeta* (17 February 1993); *Moscow News* (25 February 1993).

47. *Krasnaya Zvezda* (3 October 1990).

48. See Ian Anthony (ed.), *Arms Export Regulation* (Oxford, 1991), 224–225; Stockholm International Peace Research Institute (SIPRI), *SIPRI Yearbook 1991: World Armaments and Disarmament* (Oxford, 1991), 220–222, 334–335.

49. Yeltsin underlined Russian support for the P-5 process in February 1992 (*Krasnaya Izvestiya* [22 February 1992]).

50. *Rossiiskie Vesti* (2) (1992); *Nezavisimaya Gazeta* (19 February 1992); *Izvestiya* (12 June 1992).

51. See *Izvestiya* (6 May 1992); *Rossiiskie Vesti* (8) (1992) (article by Sergei Glaz'ev, then first deputy minister of foreign economic relations).

52. Published in *Rossiiskie Vesti* (17 March 1993).

53. *East European Markets* (21 August 1992).

54. *Kuranty* (23 June 1992).

55. *Financial Times* (30 October 1992); *The Independent* (20 October 1992).

56. *Sovetskaya Belorussiya* (29 August 1992).

57. *Nezavisimaya Gazeta* (1 October 1992).

58. *Izvestiya* (14 September 1992); *RFE/RL Research Report* 1 (25 September 1992), 56; *Izvestiya* (21 January 1993).

59. *Nezavisimaya Gazeta* (30 September 1992); *The Financial Times* (3 March 1993); *Jane's Defence Weekly* (13 March 1993), 12; *Izvestiya* (5 March 1993); *Defense News* (4 and 11 October 1993, 14 February 1994).

60. *Izvestiya* (21 June 1991); *Missile Monitor* (2) (Monterey Institute of International Studies 1992), 66. Early in 1993 there were reports of a possible Russian-South Korean joint venture to manufacture the S-300 (British Broadcasting Corporation [BBC], *Summary of World Broadcasts,* SU/1585 C2/4 [13 January 1993]).

61. *Krasnaya Zvezda* (17 June 1992); *Izvestiya* (15 July 1992).

62. *Finansovye Izvestiya* (28 January–3 February and 11–17 February 1993).

63. *Nezavisimaya Gazeta* (11 July 1992); British Broadcasting Corporation (BBC), *Summary of World Broadcasts,* SU/1576 A1/4 (1 January 1993).

64. *Krasnaya Zvezda* (2 March 1993).

65. *Izhenernaya Gazeta* (102, 103) (1992).

66. British Broadcasting Corporation (BBC), *Summary of World Broadcasts,* SU/1312 A1/2 (24 February 1992); *Delovoi Mir* (1 October 1992); *International Affairs* (Moscow) (February 1992), 37.

67. British Broadcasting Corporation (BBC), *Summary of World Broadcasts,* SU/1516 C3/1 (20 October 1992). See also *Moscow News* (7) (1993), 4.

68. *Izvestiya* (29 December 1992).

69. *Izvestiya* (8 April 1992, 13 October 1992); British Broadcasting Corporation (BBC), *Summary of World Broadcasts,* FE/1601 C1/3 (1 February 1993).

70. See *Izvestiya* (28 July 1992); *Nezavisimaya Gazeta* (30 July 1992); British Broadcasting Corporation (BBC), *Summary of World Broadcasts,* SU/1516 A1/3 (20 October 1992).

71. *Literaturnaya Gazeta* (13 May 1992); British Broadcasting Corporation (BBC), *Summary of World Broadcasts,* SU/1407 A4/2 (15 June 1992), SU/1421 A4/1 (1 July 1992); *Izvestiya* (5 January 1993).

72. *Nezavisimaya Gazeta* (1 October 1992); British Broadcasting Corporation (BBC), *Summary of World Broadcasts,* SU/1451 C3/5 (5 August 1992).

73. *Izvestiya* (5 March 1993).

74. British Broadcasting Corporation (BBC), *Summary of World Broadcasts,* SU/1470 A1/3 (27 August 1992); *Izvestiya* (3 December 1992); *Komsomolskaya Pravda* (11 December 1992).

75. *Komsomolskaya Pravda* (11 December 1992).

76. *Rabochaya Tribuna* (15 October 1992).

77. British Broadcasting Corporation (BBC), *Summary of World Broadcasts,* SU/1285 A1/2 (23 January 1992), SU/W0214 A/3 (24 January 1992); *Izvestiya* (5 February 1992).

78. *The Independent* (13 February 1992).

79. *Financial Times* (2 March 1992); *Komsomolskaya Pravda* (17 March 1993). Later in 1992 there were reports that a group of top officials of *Spetsvneshtekhnika* had been arrested for obtaining hard currency bribes. It is possible this corruption related to the sale of surplus aircraft (British Broadcasting Corporation [BBC], *Summary of World Broadcasts,* SU/1563 C1/2 [14 December 1993]).

80. *The Observer* (28 February 1993); *Izvestiya* (2 March 1993).

81. *Izvestiya* (6 October 1992).

82. *RFE/RL Research Report 1* (32) (14 August 1992), 48; *Izvestiya* (16 June 1992).

83. *Rabochaya Tribuna* (11 February 1993).

84. See, for example, *Izvestiya* (22 February 1992).

85. *Moscow News* (35) (1992).

86. British Broadcasting Corporation (BBC), *Summary of World Broadcasts,* SU/1304 C1/5 (14 February 1992), SU/W0230 A/16–17 (15 May 1992). Malei's estimates of potential export earnings vary over time, ranging from $4 billion to $12 billion.

87. According to the MFER, arms sales in 1992 were $1.8 billion (*Finansovye Izvestiya* [4–10 February 1993]). Malei has claimed $1.9 billion against a "plan" for the year of $6 billion (*Izvestiya* [23 February 1993]; *Kommersant'* [7] [1993]); other sources claim sales of $3 billion (for example, *Rosiiskie Vesti* [4 March 1993]).

88. *Izvestiya* (29 September 1992); *Rossiiskie Vesti* (4 March 1993).

89. *Rossiiskaya Gazeta* (17 February 1993).

90. British Broadcasting Corporation (BBC), *Summary of World Broadcasts,* SU/1592 i (21 January 1993). According to Glaziev, the losses amounted to $16 billion.

91. *Rossiiskie Vesti* (24 March 1993).

92. British Broadcasting Corporation (BBC), *Summary of World Broadcasts,* SU/1615 C2/6 (17 February 1993); *Izvestiya* (23 February 1993).

93. *Komsomolskaya Pravda* (23 February 1993); *Nezavisimaya Gazeta* (26 February 1993).

94. *The Financial Times* (16 February 1993), *Finansovye Izvestiya* (18–24 February 1993); *Moskovskie Novosti* (10) (1993).

95. See *Nezavisimaya Gazeta* (23 October 1992); *Krasnaya Zvezda* (26 February 1993).

96. The defense industries of Ukraine and the other successor states are discussed in more detail in Julian Cooper's "The Defense Industries of the USSR Successor States," in Herbert Wulf (ed.), *Arms Industry Limited* (Oxford, 1993) 87–108.

97. *Nezavisimaya Gazeta* (23 December 1992).

98. British Broadcasting Corporation (BBC), *Summary of World Broadcasts,* SU/1516 A1/3 (20 October 1992).

99. British Broadcasting Corporation (BBC), *Summary of World Broadcasts,* SU/1516 C3/2 and i (20 October 1992). The original South African press report referred to MiG-29 engines and radar equipment, not complete aircraft (*The Weekly Mail* [16–22 October 1992], 6).

100. British Broadcasting Corporation (BBC), *Summary of World Broadcasts,* SU/1451 C3/5 (5 August 1992); *Pravda Ukrainiy* (23 October 1992); *Krasnaya Zvezda* (26 February 1993). The latter reports that the Varyag is 70 percent built, but cannot be completed without substantial Russian involvement.

101. British Broadcasting Corporation (BBC), *Summary of World Broadcasts,* SU/1445 C3/3 (29 July 1992), SU/1601 C2/3 (1 February 1993); *RFE/RL Research Report* 1 (14 August 1992), 44.

102. *Izvestiya* (5 February 1992).

103. In early 1993 the Russian press reported that Iran was interested in buying Georgian-built Su-25s or spare parts for the seven Su-25s Iran obtained from Iraq (*Izvestiya* [21 January 1993]).

NINE

China

Gerald Segal

THE MAJOR ARMS exporters have distinctive mixes of policies and practices. China is distinct in part because it appears to be the only major arms exporter other than Germany that did not see a decrease in the percentage and volume of its transfers in the early 1990s, despite a rapidly shrinking market. In 1994 it ranked third among exporters of major conventional arms to the developing world and fourth overall.[1]

China is also distinctive because, by virtue of its place in the heart of Asia, its indigenous arms production affects the arms transfer policies of other states in East and South Asia. Virtually all other supplier states discussed in this chapter are outside the main recipient regions being assessed. China needs to be understood as an arms producer in the widest sense: both as exporter and as stimulus to the imports of its neighbors.

China is distinct in a third way. As a developing country, its arms industry is far less sophisticated than that of other major suppliers. In terms of exports, China has focused on cheaper and less sophisticated technology and has thereby limited itself to certain markets. China is unique among major exporters in not being able to sell to most developed countries.

Nevertheless, these distinctive features are subject to major changes that must be assessed. China's ability to increase its market share depends to an important extent on the volume of arms Russia sells from its bargain basement and the extent to which it also produces inexpensive weapons for export. For much the same reason that China buys weapons from Russia, Russia may be able to cut deeply into China's market niche. China may also discover that its concentration on the export of lower technology systems will not succeed in the new marketplace, especially because the Gulf War demonstrated the utility of smarter weapons. China may find itself losing market share or entering into closer collaboration with other states such as Russia in order to co-produce more sophisticated weapons.

Finally, China may come under increasing pressure to constrain its arms transfers. China is treated as a great power in part because of its arms transfers (another is its possession of nuclear weapons). If the other great powers become more serious about limiting arms transfers, they will put greater pressure on China to join them. Internally, China's policy on arms transfers is part of a broad debate over whether to pursue a more independent or interdependent foreign policy. That choice is tied to other choices about how far and fast to pursue domestic reform and about interconnections with the international market economy. Thus, China's choices about arms transfers are very much a part of broader decisions about openness in foreign policy and its commitment to reform.[2]

The Character of Chinese Exports

The main features of China's arms export trade are summarized below.[3] First, as noted, in 1994 China ranked third among arms exporters to the developing world and fourth in arms exports overall. In 1988 China held these same rankings, but in 1989 it ranked fifth and fifth, and in 1990 it ranked fourth and sixth in these categories. In the period 1988–92 as a whole, China ranked third among exporters to the developing world and fifth overall (table 9-1 and figure 9-1). These swings in China's export levels were also reflected in its shifting market share. In 1988 its market share of exports to developing countries was 8.9 percent, but in 1989 that share fell to a low of 4.4 percent, while its share of total arms transfers was just 2.6 percent. By 1992, its share of the Third World market had reached a level almost double that of 1988. In light of these export levels, China joins the four major European arms exporters— France, Germany, Russia and the United Kingdom—as a middle-ranking power.

The second major feature of Chinese transfers is that they go almost entirely to markets in the developing world. In this regard, China is less like Germany and more like France and the United Kingdom, whose arms exports also depend on the markets of the developing world. In the 1987–91 period, over 98 percent of Chinese arms transfers went to the developing world, by far the highest percentage of any major arms exporter (table 9-2). The reason is the lower quality of Chinese arms exports and China's

Table 9-1. *Exporters of Major Conventional Weapons, 1988–92*
(Millions of constant 1990 U.S. dollars)

Exporter	1988	1989	1990	1991	1992	1988–92
			A. To the developing world			
United States	4,494 (18.97)	3,662 (16.97)	4,622 (26.14)	4,147 (31.32)	3,075 (32.99)	20,000 (23.39)
USSR/Russia	10,280 (43.40)	10,348 (47.86)	6,615 (37.41)	3,987 (30.11)	1,904 (20.43)	33,135 (38.73)
China	2,097 (8.85)	945 (4.37)	1,249 (7.06)	1,705 (12.88)	1,535 (16.47)	7,531 (8.80)
France	1,668 (7.04)	2,051 (9.49)	1,794 (10.15)	724 (5.47)	351 (3.76)	6,588 (7.70)
United Kingdom	1,505 (6.35)	1,993 (9.22)	1,163 (6.58)	697 (5.26)	658 (7.06)	6,016 (7.03)
Germany	284 (1.20)	208 (0.96)	857 (4.85)	425 (3.21)	296 (3.18)	2,069 (2.42)
			B. To all states			
United States	12,204 (30.48)	11,848 (31.07)	10,822 (36.11)	11,666 (47.67)	8,429 (45.80)	54,968 (36.40)
USSR/Russia	14,658 (36.61)	14,310 (37.52)	9,724 (32.44)	4,448 (18.18)	2,043 (11.10)	45,182 (29.91)
China	2,161 (5.40)	1,009 (2.64)	1,249 (4.17)	1,705 (6.97)	1,535 (8.34)	7,658 (5.07)
France	2,403 (6.00)	2,846 (7.46)	2,129 (7.10)	820 (3.25)	1,151 (6.25)	9,349 (6.19)
United Kingdom	1,704 (4.25)	2,710 (7.10)	1,456 (4.86)	801 (3.27)	952 (5.17)	7,623 (5.05)
Germany	1,241 (3.10)	814 (2.13)	1,677 (5.59)	2,530 (10.34)	1,928 (10.47)	8,190 (5.42)

Note: The figures in parentheses are percent of market total.

Source: Stockholm International Peace Research Institute (SIPRI), *SIPRI Yearbook 1992: World Armaments and Disarmament* (Oxford, 1993).

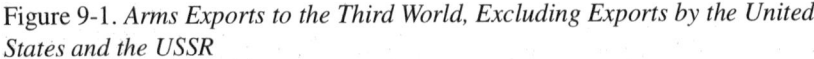

Figure 9-1. *Arms Exports to the Third World, Excluding Exports by the United States and the USSR*

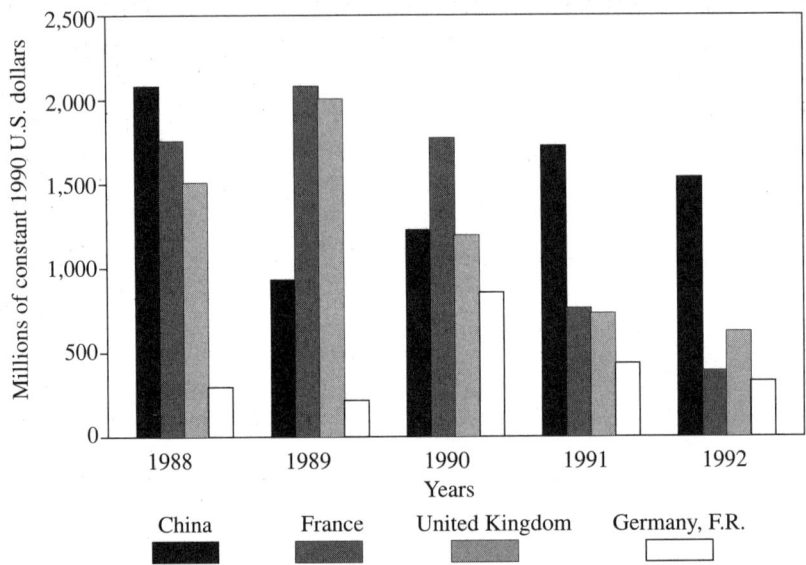

Source: Stockholm International Peace Research Institute (SIPRI), *SIPRI Yearbook 1993: World Armaments and Disarmament* (Oxford, 1993).

concomitant inability to cater to the richer states seeking more sophisticated technology.

A third point is that China, like most arms exporters, depends heavily on markets in the Middle East. In the 1987–91 period, nearly half of Chinese arms exports went to that region, with East Asia and South Asia splitting most of the rest. Of the five countries that dominated the Chinese export ledger from 1988 to 1992, two were from the Middle East (Iran and Saudi Arabia) and three from Asia (Pakistan, Thailand and Bangladesh).

The fourth major feature of Chinese transfers is that Beijing is rarely the largest arms supplier to its major markets. In 1988–92, China followed the USSR as the market leader in Iran, with 26.9 percent of the Iranian market. It accounted for only 9.8 percent of the Saudi market and 4.7 percent of the Iraqi market but 45.3 percent of the Thai market. The Democratic People's Republic of Korea (North Korea) acquired just under 10 percent of its arms from China (it ranked sixth among China's export markets). On the other hand, China had 55.5 percent of the Pakistani arms market, while Bangladesh (ranked seventh), Myanmar (ranked eighth), Zimbabwe (ranked ninth) and Sri Lanka (ranked tenth) each received approximately 80 percent of

Table 9-2. *Chinese Exports of Major Conventional Weapons by Importing Region, 1987–91*

(Millions of constant 1990 U.S. dollars)

Region	Total	Percent of Chinese exports
East Asia	1,726	22.0
South Asia	1,812	23.1
Middle East	3,850	49.2
Africa	333	4.1
Americas	136	1.6

Source: Stockholm International Peace Research Institute (SIPRI), *SIPRI Yearbook 1992: World Armaments and Disarmament* (Oxford, 1992).

their arms from China. While China rarely opens up markets, in many cases it is a significant supplier, and its policies are important to the management of arms races in the region.

Fifth, China has avoided undue dependence on exports to any one state. It was thought that following the end of the Iran-Iraq war China would fade as an arms exporter. However, it found markets in Thailand and Myanmar and retained its market in Iran. Iraq disappeared as a customer because of the United Nations (U.N.) embargo in 1990, but China is well-placed to return when Iraq is allowed to re-arm. Unusual deals such as its sale of the CSS2 medium-range ballistic missile (MRBM) to Saudi Arabia suggest China is prepared to be flexible in its dealings and is willing to sell to various sides in a complex zone of conflicts.[4] As shown below, Chinese motives are multifaceted and changing.

Finally, China has transferred a wide range of weapons. The image of China as primarily a supplier of small arms has not been accurate for some time, especially when considering its exports to Asian states. Chinese exports to the Gulf region involved large numbers of aircraft and tanks. The MRBM deal with Saudi Arabia demonstrated that China was willing to sell weapons that no other power would provide. In brief, China has shown its willingness to sell virtually any conventional weapon in its arsenal.

Chinese weapons are generally less modern and effective than those sold by the developed states. However, although Chinese equipment is clearly not first rate, it is more than third rate. Against modern arms, as Iraq discovered in 1990–91, Chinese equipment is not much use. However, as the Iran-Iraq war demonstrated, Chinese arms sufficed against a less modernized enemy. China has apparently won its market share by offering relatively low prices to compensate for the poor quality. Little hard evidence is available on the actual prices China charges, but Thai officials complain about the poor quality despite bargain prices and probable heavy pay-offs to corrupt colleagues. The deals with Thailand suggest that because China's arms are low quality, it may be hard for it to win contracts with armed forces seeking prestige weapons. On the other hand, Chinese equipment will do when bribery is a motivator.

It should be noted that China has an important impact on the pattern of arms exports simply because it is the dominant power in Asia. Even if China did not export weapons, its domestic arms production would be a major factor in Asian arms races. As Andrew Mack points out, it is impossible to understand fully the pattern of the arms trade in Asia without understanding Chinese policy.[5] Its size alone makes China a factor in the decisions of nearly every Asian power that acquires weapons. Pakistan and India must consider how China might use its power in the event of renewed fighting on the subcontinent. The states in the Association of South-East Asian States (ASEAN) are especially concerned about the extension of China's naval and air power. How a crisis is managed on the Korean peninsula depends to a large extent on Chinese capabilities and intentions. One of the most critical factors in Japan's post-Cold War defense policy is its relationship with a China that continues to increase the size of its defense budget.

The fact that China continues to modernize its military, to use force in the South China Sea, to increase its defense spending and to test its nuclear weapons is a vital issue in Asian arms races. While it is true that China's arms are not top of the line, it could overwhelm most of its neighbors with sheer numbers reinforced by a few relatively modern weapons systems. China's acquisition of modern equipment from Russia has led to increased tension in the South China Sea and was a factor in the sale of U.S. F-16s to Taiwan. While China is now putting greater stress on quality than quantity in its weapons procurement, recent trends also indicate greater interest in weapons with longer range and greater lethality. Various reports suggest China is seeking closer cooperation with the Russian defense establishment on possible co-production of newer weapons.[6] China has apparently acquired in-flight refueling technology from Iran but is having major problems deploying the system because of the complex skills required.[7] As the range and lethality of China's domestic arsenal increase, concern over its exports will grow.

Why China Exports

China's motives for exporting arms are perhaps harder to explain than those of the other great powers.[8] The Communist Party still sets much of Chinese foreign policy, and aspects of Chinese national security are more tightly wrapped in secrecy than is true in most other countries. Study of China of necessity must take into account the highly productive rumor mills, many of which originate in Hong Kong, a fact that makes analysis difficult. Caution is necessary even when using basic data about the scale of China's arms transfers. Even more caution is needed when suggesting clear motives for Chinese behavior. This uncertainty over China's policies and motives makes it difficult to formulate any initiative to control Chinese behavior.

These problems are evident when considering what is said to be a major motive for Chinese exports—the desire of Chinese officials to make money for themselves and their families.[9] It is often argued the so-called Red Princes and Princesses—the children of the aging leaders of China—are key players in China's complex network of arms-exporting corporations. As China's economy has become more open to market forces, Chinese leaders and their children have created vast empires and amassed equally large fortunes selling arms abroad. The anecdotal evidence of the involvement of the Red Princes is indisputable, although it is impossible to know how much money they have made and how specific decisions on arms transfers are arrived at.

It is far more difficult to assert that this nepotism makes it impossible to control China's arms transfers. The network of corporations exporting arms is more complex than in the former Soviet Union, although, as China becomes more of a market economy, the comparison is best made to such middle-level powers as France and the United Kingdom. The number of Chinese arms exporters is comparable to that in France and the United Kingdom, although the familial nature of the links to the ruling elite in China makes the power of Chinese corporations harder to assess. There is undoubtedly pressure from the corporate level to increase exports and for other parts of government to lift restrictions, but this pattern pertains in market economies as well. Competition between corporations, whether in China or the capitalist world, adds to the pressure to increase sales. However, whether analysts speak of a military-industrial complex in the West, or Red Princes in the East, there are limits to personal avarice as a motive for arms sales.

The key to answering the question of whether the arms trade can be controlled is whether the government will choose to. Ultimately the motives of the government are the key factor, with the motives of corporations and individuals contributing ones. China is just as able to control the practices of its corporations and individuals as are the governments of France, the United Kingdom and the United States. In fact, its control over the internal and external means of production and communication is probably tighter than that in the West, despite the growing emergence of a market economy. Tanks and aircraft are at least as hard to make and get to the market in China as in the West. The necessary raw materials are found within the state sector. The government provides support to fledgling Chinese firms without which they could not successfully negotiate with foreign customers.

In short, the Red Princes are exporting arms because the state desires it. No doubt pressure from well-connected family members makes the government push harder for some deals or makes it more difficult for the state to stop them. However, this situation is little different from, for example, American electoral politics fueling the F-16 exports to Taiwan and corporate contributions to election war chests. Nevertheless, the bottom line is that China's government can halt transfers of arms whenever it desires. A case in point is the arms trades with Iraq. The Red Princes, like many others, were making vast amounts of money selling arms to Iraq. However, after the members of the U.N. Security Council agreed on an arms embargo (with China abstaining), China, along with the other powers, stopped supplying arms to Iraq. Suggestions that Chinese arms exports should be treated more leniently because of the role of the Red Princes are tantamount to indulging in ethnic-chic.

Shifting Chinese Motives

The presence of the Red Princes encouraged the greater emphasis China placed on economic motives for arms transfers in the 1980s. In the days before the opening of China to economic reform in the late 1970s, China was said to export arms primarily for political reasons. Weapons were "sold at friendship prices" to support revolutionary causes or friends in coalitions against whatever superpower was deemed the major threat. In short, China made little money from its arms transfers. In the 1980s, as China's economy became increasingly market-oriented, there was an oft-noted shift from simple transfers of arms to sales of arms. Analysts have found it difficult to assess how much money was being made on individual deals and who was earning it. There has been much speculation the Chinese People's Liberation Army (PLA) was obtaining important extrabudgetary allocations because the arms exporters it controlled were able to retain a large share of the profits. Some of these funds were said to have been plowed back into research and development (R&D) of new equipment for the PLA.[10] The rise of the Red Princes seems to have been both part of this process and part of the reason revenues to the PLA might have been falling in recent years. There is no way to get an accurate picture of this problem.

Even if the PLA continues to earn money from arms sales, the figures are unlikely to be large enough to offset the cuts in the defense budget in the early and mid-1980s. If the state as a whole pocketed the profits, the figures would be minuscule as a percentage of China's rapidly rising foreign trade. Marketing Chinese arms certainly brought more revenue than giving the equipment away at friendship prices. Nevertheless, caution is needed when suggesting the change from friendship prices to "market" prices was crucial for Chinese foreign or defense policy.

Of more significance was that the search for profit took China into new markets, most notably the Middle East. The vast majority of Chinese sales, as opposed to "friendship" transfers, were to Iran, Iraq and Saudi Arabia. As some of these markets have dried up in recent years, China has been able to make money from Thailand and Myanmar, two countries to which China usually sold at friendship prices. Even if the Middle Eastern markets re-open later in the 1990s, it is unlikely the scale of the transactions will yield sufficient profits to be the major motive for arms transfers. At a minimum, China will remain interested in deals that enable it to obtain samples of foreign technology, such as its agreement with Iran to receive in-flight refueling technology. China will also welcome the opportunity to see its weapons tested on the battlefield and learn how best to modernize its equipment.

While profit will be a factor in China's arms transfers in the 1990s, its relative importance has lessened in recent years, above all because the end of the Cold War has revolutionized the strategic environment in which China operates. Arms transfers have taken on added importance as a political tool to increase China's international standing. China recognizes that in today's fluid international environment it will be far harder to play great powers off each other. In an atmosphere of increased paranoia and uncertainty, China can better justify arms transfers as being in the national interest, although the specific type of national interest may vary.

China's most important national security goal is to support friendly neighbors. In the 1988–92 period, the most unambiguous examples are the 4.0 percent of Chinese transfers that went to North Korea and the 25.2 percent to Pakistan, although for differing reasons. North Korea is of decreasing importance as China no longer sees it as a necessary component in a competition for influence with the Soviet Union. In fact, North Korea has complicated China's effort to normalize relations with the Republic of Korea (South Korea). North Korea has become an embarrassment to China, and its nuclear weapons program risks drawing China into unwanted conflict in the region. Chinese arms transfers to North Korea will likely decline even further. However, should Sino-American relations deteriorate, China may feel that in the short term it makes sense to defend North Korea as a communist state, at least until the shift to a more moderate and reformist regime can be arranged. North Korea is also useful as a channel for Chinese arms exports that otherwise might be disruptive of China's relations with the West. Persistent reports of such use of North Korea for missile sales to the Middle East muddies the waters of Chinese policy and Western policy responses.[11]

Chinese transfers to Pakistan are likely to hold up far better. China sees Pakistan as a key pressure point on India. Despite the rapid changes in the power balances in the region, this rationale still holds. India is no longer closely aligned with the Soviet Union, and Sino-Indian relations have been improving.[12] China has no reason for wanting a conflict in Kashmir, but it also does not particularly wish India to gain its objectives. India and China share a concern over the uncertainty posed by a potentially Islamic Central Asia, but China knows that Pakistan will be an important actor in this region, and therefore close ties are important. China would strongly oppose any dismemberment of Pakistan that might result from a free-for-all in Central Asia, as it would threaten the stability of China's own Central Asian territory. As Pakistan finds it harder to obtain arms from a United States that is wooing India and worried about Pakistan's nuclear weapons program, China is likely to want to enter into supplier arrangements for continuing arms transfers to Pakistan. The reported sale of M-11 missiles may be only part of a much more complex set of weapons being provided, including fighters, main battle tanks (MBTs) and anti-tank missiles. Co-production deals, for example, for a type 2000 MBT or a K-7 jet trainer, are likely to continue as both China and Pakistan see the benefit of cooperation in the face of efforts by advanced Western countries to isolate and pressure them.[13]

Chinese transfers to neighboring countries are much smaller and more complicated. During the Afghan war of the 1980s, China sold weapons to the U.S. Central Intelligence Agency (CIA) and donated others directly to Pakistan for the *mujahideen* in the largest covert operation since World War II.[14] The 0.6 percent of Chinese transfers that went to Afghanistan consisted mostly of small arms. The Afghan situation does not threaten China's security. Rather, it is part of a more general concern over the fate of Central Asia.

The 5.3 percent of Chinese transfers that go to Myanmar, a pariah regime, are also not the result of concern over national security. They have far more to do with making money and helping control the flow of arms and drugs in China's southwest. China has little interest in Myanmar's battles with Thailand or Bangladesh, made possible by the provision of arms. China might have some interest in the victory of a more reformist regime in Rangoon, and consequently support for the present junta is counterproductive. However, China's greater fear is that Myanmar will disintegrate ethnically under a less ruthless military regime, which would have less control of the drug barons. As it stands, Chinese officials are said to profit to some extent from the illicit drug trade in the frontier region and are confident that the scale of the problem there is under control. Nevertheless, although most of China's deals with Myanmar were transacted more for money than for national security, concerns about the future persist.[15] Reports that China is helping Myanmar build a naval base on the coast of the Andaman Sea that might eventually serve Chinese ships in a blue-water role lead to worries that China has major, long-term political objectives in South Asia.[16]

If China's national security in the Asian region is defined more broadly, then transfers to such countries as Thailand (19.3 percent in the 1988–92 period) must be included. However, as suggested, Chinese deals with Thailand in the early 1990s have had a great deal to do with the corruption of the Thai armed forces. It is unlikely those weapons will be used except against internal dissent. If Thailand were planning to use them against external enemies, it would have to worry about the poor quality and difficult maintenance of the Chinese weapons. With the easing of tension in Indochina, the Thai armed forces should settle down to the more prosaic interest of earning their share of illicit trade with Cambodia and Laos. Although the diminution of concern with Vietnam means China needs Thailand less than it did in the 1980s, China still sees Thailand as a friend in ASEAN. In sum, Thailand is a case where the motives underlying China's arms transfers are a mixture of political and economic, with the latter of increased importance in recent years.

A Global Power

It seems that, as the geographical distance from China increases, political and national security motives would decrease in importance. While that conclusion was correct when China saw its national security primarily in Asian terms, it no longer takes such a parochial view. The disintegration of the Soviet Union has freed China of its obsessive worry about Soviet strategy and made it even more aware of the need to be a player on the international stage. Chinese fears of a single American superpower and the deterioration in Sino-American relations reinforce that attitude. China sees a need to stress its global role as a way of deterring American pressure and compelling Washington and the West to take it seriously, particularly since they no longer need China as a foil to the USSR.

China's attempt to prove it is more, rather than less, important in the post-Cold War world depends on its ability to demonstrate it is a global player. China has quickly learned that threats to proliferate nuclear weapons in the Islamic world (Algeria or Iran) make its policy a major concern in the West. Threats to block the use of the U.N. Security Council to sanction collective action against the likes of Iraq, Libya or Serbia make the wooing of China more important in the post-Cold War world. Similarly, a desire to see Chinese adherence to the Missile Technology Control Regime (MTCR) or willingness to cooperate with the P-5 to limit arms transfers to the Middle East gives China leverage. China's conventional arms transfers are, as with its nuclear weapons, key reasons China must be treated as a great power.

For these reasons, China can rightly see its arms transfers to the Middle East as an important political means of ensuring it is taken seriously and its concerns are not trampled in the pursuit of an American and Western-defined new international political order. Chinese transfers to countries such as Iran may not have begun for primarily political reasons, but in the post-Cold War world the rationale has become primarily political. This is not to say China necessarily feels that, as the superpowers did in the Cold War, they could manipulate arms sales to support allies against the ideological enemy. China, a one-time arms seller to both Iran and Iraq, is far too pragmatic for such illusions of direct influence. The reported sales of M-9 missiles to Syria were not made in the expectation of more influence in the Middle East peace process. At the same time, China is fully aware that such sales cause concern in the United States and the West.[17] Beijing may hope that arms transfers will lead to contracts for civilian construction and win some extra votes in international institutions. China also knows the threat that it might resume arms sales to Iraq and sell more indiscriminately in the Middle East as a whole is useful in helping deter the United States from revoking most-favored nation (MFN) status and allowing a sharp deterioration in Sino-American relations. Should Sino-American relations take a serious turn for the worse, China will feel freer to sell what it wants in the Middle East.

Thus, it seems that China is motivated to sell arms primarily to worry the United States and the West. Its threat to break up the P-5 process on limiting arms transfers to the Middle East did not stop the United States from selling F-16s to Taiwan. The broader concern with China as a geopolitical power, however, helped keep President George Bush anxious to avoid a rapid deterioration in Sino-American relations. The American decision to allow the sale of communications satellites to China in September 1992, just after the announcement of the F-16 deal, suggested that China could not win all the battles but that its general strategy was working. The fact that China quit the P-5 conventional arms talks over the Middle East but was generally restrained in its reaction to the F-16 sale for fear of harming President Bush in the November election suggested China is able to play a subtle great-power game when it wants.

There is little reason to expect China to change course. Therefore, Russian arms sales to China and American and Russian sales to the Middle East could be interpreted to mean that efforts to control the transfer of conventional arms are in serious trouble. Just as China was not the first to open most arms markets, so it will not be the first to pull out. China will sell arms where it can make money at a low political cost. Where political gain is evident, even in the form of making others worry, China will be reluctant to curb its arms transfers. It understands that it is part of what Europeans might term a concert of great powers, involving a complex balance of power among several key actors. The trick for China, and the rest of the international community, is to maintain credibility. This mainly political basis for transferring arms requires that China act credibly about joining in international efforts to control arms proliferation. If China is never seen as likely to join the international arms control arrangements, there will be less reason to bargain or cooperate with China. For the same reasons that China transfers arms, it must cooperate with efforts to control the spread of conventional arms.

Dealing With the China Problem

China is a middle-sized power and middle-sized problem for those wishing to control the proliferation of conventional arms. If arms control is to succeed, it must involve China. It is unlikely, however, that China will be primarily to blame if arms control fails. China has dragged its feet as attempts have been made to strengthen the MTCR, control arms transfers by the P-5 or implement the U.N. Arms Register. None of these efforts failed simply because China held back. It is fair to say that China held back because the other powers were also undermining these international accords.

China can and does take part in international arms control efforts when it is pushed sufficiently hard, when the risks in agreeing are low and when the benefits of agreeing are high. China eventually agreed not to block the sanctions and arms embargo against Iraq in 1991, signed the Nuclear Non-Proliferation Treaty (NPT) in 1992, agreed to abide by the terms of the MTCR in 1992 and agreed to sanctions and an arms embargo against Serbia later that year. In 1992, however, China remained skeptical of the chemical weapons pact and, as noted, withdrew from the Middle East P-5 process.[18] These agreements and expressions of doubt, as with policies related to the management of the international economy, are part of a gradual transition in China toward acceptance of international interdependence. Still, the transition is obviously painful and fitful, as China's internal debate about reform staggers forward. As the succession struggle in China approaches, these debates about reform and the degree of interdependence become more acute and difficult.[19] The Chinese seem to have a particular problem with the concept of a "regime" in international affairs (as in the MTCR). One reason may in part be explained by the different approach of the Chinese to international law in general.[20] China's willingness to test the limits of definition in the MTCR may therefore be something more complicated than sheer duplicity. China's ambivalence and caution about conventional arms control essentially are yet another manifestation of the domestic debates about reform and interdependence, at a time of uncertainty about the post-Cold War pattern of international relations.

A wide range of efforts is underway to contain the conventional arms race. However, to understand how to deal with China, the starting point must be the problems and not the solutions. There are two main regions that need to be addressed, the Middle East and Asia.[21] In the latter, China is both a major supplier to states in the region and, by virtue of its presence and power, a major source of insecurity.

With respect to Asian security, what will be the likely reaction of China to efforts to achieve conventional arms control? In the 1980s China undertook major unilateral efforts to reduce the size of its armed forces. This process was crucial to the development of Sino-Soviet detente and was enhanced when the Soviet Union itself disintegrated. The result was that China is more secure in its frontiers than it has been for several hundred years. It is now moving troops away from the northern frontier and is paying far more attention to the need for more modern and mobile forces. Possible threats include unrest in Central Asia, naval action in the South China Sea, a conflict with Taiwan and a possible arms race with Japan. The lessons of the 1990–91 Gulf War taught China the virtues of modernity and mobility in war, and it is now streamlining and modernizing its forces. It is filling major gaps in equipment through selective imports from Russia, assistance from Israel and a long-term R&D program. Should Russia and China agree to co-produce weapons, China will have filled a major gap in its defense industry and seriously strengthen its position in the arms export business.[22]

These developments are not necessarily antithetical to conventional arms control if they are undertaken more slowly, the defense budget is kept stable and less emphasis is placed on air and naval force projection. However, as long as China continues to develop this more dangerous defense policy, and as long as it does so in a region mostly devoid of multilateral arms control and replete with latent and active tension, China will be a major obstacle to regional arms control. The fact that China remains largely a revanchist power with intensely held desires to acquire territory, particularly Hong Kong, Taiwan and South China Sea, as well as other minor disputed territories, means it has less reason to support arms control. Chinese arms purchases from Russia and discussions about more co-operation on defense in the long term are one reason the United States decided to sell F-16s to Taiwan. That step in turn fueled the arms race, as it gave fodder to those in the Chinese system determined to resist Western pressure on arms sales and supported the hardline factions opposing reform and interdependence.[23]

To be fair to China, arms control has not featured prominently on the Asian agenda—the region is better known for its nationalism than for its multilateralism. Often Chinese irredentism and suspicion only appear to be a large problem because China is such a major player in Asia and involved in so many subregional discussions. Arms control in Northeast Asia is only on the forefront in academic fora. For a long time it was the United States more than China that blocked progress. Chinese detente with South Korea was crucial in shaping the possibilities for a relaxation of tensions in the region. In Central and South Asia, not even this minimal level of discussion of conventional arms control has taken place.

Where China looms as a larger problem for conventional arms control is in Southeast Asia. China's determination not to discuss sovereignty has hampered efforts to contain disputes in the South China Sea. China has also used oil exploration by an American firm and a policy of aggressive naval patrols to extend its claims when political and military conditions are propitious. China does not seem inclined to surrender its objectives of acquiring territory unless it meets direct and powerful counterpressure. The acquisition of in-flight refueling and new aircraft from Russia suggests that China is prepared to use force in this dispute and eschew effective arms control.

The Taiwan problem also involves contested sovereignty. China's military modernization has provoked Taiwan, as has been true elsewhere in Southeast Asia, to acquire its own modern weaponry. China is highly unlikely to accept any form of arms control that limits its ability to retake Taiwan, even though the growing economic interconnections across the straits have helped build some confidence in a more peaceful solution.

Of greater concern in the long term are the risk of a Sino-Japanese arms race and the lack of any multilateral arms control to forestall that threat. China's acquisition of modern offensive weapons causes increased concern in Japan over Beijing's ability to project power, which will force Japan to follow suit. Japan would take in-flight re-fueling, an aircraft carrier or even Aegis-class ships capable of providing air cover for troop carriers to mean that China wants to control the sea lanes, resources and islands. China might even threaten sanctions against Japanese trading partners or Japan itself. Japan is a major supporter of greater "transparency" in arms control, while China has hidden behind an obfuscating definition of national security. Nevertheless, and somewhat unexpectedly, China did comply with the U.N. Arms Register, a minimal step in building confidence through transparency.

In the absence of any significant arms control, let alone a more ambitious P-5 process for Asia, the prognosis for conventional arms control in Asia is gloomy. A P-5 process for Asia is probably hopeless, and greater transparency will help only at the margins. The problem is not so much China's arms exports, but rather its revanchist and uncooperative behavior. The real issue is how to deal with the political roots of the disputes in which China is involved. Whether it is the Korean conflict or the South China Sea, each has its own dynamic requiring an à la carte approach to arms control.[24] At the root of Chinese policy remains, however, an unsettled debate about how much interdependence it should accept and how much of the irredentist agenda it should abandon. All these issues are part of the basic debate about reform in China and are therefore well beyond the purview of negotiators on conventional arms control.

However, in the second major area of Chinese arms proliferation—the Middle East—there is room for optimism. As noted, Chinese motives for arms transfers are now a mix of some monetary reward and much political symbolism. The political impulses could be satisfied to an important extent by a drawn-out process of negotiations at the P-5 level, where China is a major player. In fact, this is more or less where matters stand today. China decided to boycott the P-5 process after the United States decided to sell F-16s to Taiwan, but it can hardly be argued that the P-5 process was otherwise on the brink of a breakthrough. The United States announced a new round of aircraft sales to the Middle East in between the announcement of the F-16 deal with Taiwan and the Chinese decision to pull out of the P-5 process.[25] Unless the other big four arms exporters challenge China with a firm agreement, it is unlikely China will feel compelled to accept a curb on its arms exports.

Where arms control agreements have been reached, such as the MTCR, China will continue to test the limits of the accord by contesting the relevant details on the range of the weapons. Uncontrolled exports of modern aircraft will encourage China to stretch the terms of the MTCR because it interprets the pact as artificially covering only a limited range of weapons. Nevertheless, China is unlikely to transfer MRBMs as it did to Saudi Arabia, as doing so might undermine its current practice of being a nuisance without becoming defined as the problem.

China is not likely to oppose agreements to increase the transparency of arms transfers in the Middle East, if only because the agreements would not cover its territory. As such, it might be worth beginning serious arms controls in this area, so as to test Chinese sincerity in limiting the proliferation of conventional arms. Now that China has normalized relations with Israel and is more involved in the Arab-Israeli peace process, it is more inclined to be cooperative in the region. Concern over Islamic forces in Central Asia also might lead China to be more cooperative with multilateral arms control in the Middle East.

In sum, China can be pressed to take arms control more seriously, but doing so will be as complex as it was to engage the Soviet Union during the Cold War. The West has enormous leverage with China because of trade, and China has already shown it can bend when the United States exerts sufficient pressure. China is also deeply concerned that technology transfer restrictions will remain in place and even be tightened. There are certainly good arguments for denying China access to certain technologies, even if that move limits Western trade with China and damages the prospects for the economy of Hong Kong.[26] Aid to China can also be restricted, or at least tied to performance indicators, even in the field of human rights. Although China has denied it will respond to such pressure, there is growing evidence that sustained pressure yields incremental results.[27] Support for China's rivals can also influence China, as the sale of F-16s to Taiwan demonstrated. A change in the American and Japanese attitudes to Vietnam might fall into the same category.

Of course, China can—and has—responded to such pressures in ways that harm Western interests. As the F-16 sale to Taiwan illustrates, China has options. Its withdrawal from the P-5 process may be classed as a negative act, but it has the more "positive" option of buying further hardware from Russia. China can spread nuclear weapons technology, block the use of the U.N. Security Council and even foment regional unrest by resuming support and arms for the likes of the Khmer Rouge. In short, China is hard to push around, even though it is essential to make clear to China that its irredentist and independent policies will not always be tolerated. Just as the West's long effort to persuade the Soviet Union that a conflict could not be won, a similar blend of cooperation and pressure will have to be used in relations with China.

The likelihood of constraining China's independent policy, even in the Middle East, is limited. Even if the international system were more stable and amenable to arms control building on an emerging political consensus, China is not ready for such sweeping multilateralism and interdependence. China is a revanchist power in Asia and a dissatisfied one at the global level. This dissatisfaction does not originate so much in communist ideology as in China's nationalism and distrust of those that talk about internationalism when they really mean maintaining the status quo. To be sure, China has gradually become more accommodating to arms control. Further change will come, but only slowly. It may take a generation (or two) before China is ready for meaningful multilateralism and interdependence. In the meantime, its nationalism and sense of great power status can be turned into limited accords on a case-by-case basis.

Appendix

The statistics used in this chapter are drawn from Stockholm International Peace Research Institute (SIPRI), *SIPRI Yearbook 1993: World Armaments and Disarmament,* SIPRI, Oxford: Oxford University Press, 1993, or the 1992 volume, unless otherwise noted. As is well-known, there is no agreement on the statistics for arms transfers. The figures produced by the U.S. Arms Control and Disarmament Agency are too out-of-date for the post-Cold War period. The report prepared by the Congressional Research Service (CRS)—Richard F. Grimmett, *Conventional Arms Transfers to the Third World, 1985–1992,* Washington, D.C.: Congressional Research Service, U.S. Library of Congress, July 1993—is up-to-date but varies greatly from SIPRI's. The CRS data show a drop in the volume of Chinese arms transfers to the developing world between 1990 and 1991, although China's share of the market rose from 4.8 percent to 4.9 percent. SIPRI shows the volume of Chinese transfers increasing and the percentage rising from 5.7 percent to 9.1 percent (see table 9-3). The CRS figures are for all arms, not just major conventional arms, and are in millions of constant 1992 U.S. dollars instead of the millions of constant 1990 U.S. dollars used by SIPRI. SIPRI, which has a long track record in producing such figures, uses a much

Table 9-3. *Chinese Exports to the Third World, 1988–92*

Source of data	1988	1989	1990	1991	1992	1988–92
CRS total[a]	3,359	2,446	1,508	1,123	600	9,036
SIPRI total[b]	2,097	945	1,249	1,705	1,535	7,531
CRS as % of all exports to the third world[a]	7.19	6.61	4.69	5.55	4.72	6.07
SIPRI as % of all exports to the third world[b]	8.85	4.37	7.06	12.88	16.47	8.80

[a]Congressional Research Service (CRS) figures are for all arms, not just conventional weapons, and are in constant 1992 U.S. millions of dollars.

[b]The SIPRI figures are in constant 1990 U.S. millions of dollars.

Source: Richard F. Grimmett, *Conventional Arms Transfers to the Third World, 1984–1991,* Congressional Research Service (CRS), U.S. Library of Congress (Washington, D.C., 1992).

more complex system for valuing arms transfers. Its figures are not intended to represent the real money received by China—something that cannot possibly be known. These discrepancies in figures are not insignificant, as they suggest different trends. As such, the discrepancies cannot be dismissed by arguing that trends are more important than absolute numbers. The CRS figures make it harder to argue that Chinese arms transfers are important, a point that can be made with SIPRI's data. Those who prefer the CRS data ask that the validity of the data, which were gathered from intelligence sources, be trusted without indicating why they believe the SIPRI data are wrong. The track record of manipulation of official American data, whether during the Cold War or the Gulf War, suggests a need for caution in accepting such data on their face. This issue was of particular concern during the Bush Administration, when official policy was to consider China as less of a problem for international security. The author is grateful to Ian Anthony for helping explain the dispute over the data.

Notes

1. For a discussion of the data used in this chapter and the discrepancy between the major sources on Chinese arms sales, see the appendix to this chapter.

2. This is the theme of chapter 2 of Gerald Segal, et al., *Openness and Foreign Policy Reform in Communist States* (London, 1992), 61–110.

3. See also R. Bates Gill, *Fire of the Dragon* (Geneva, 1991), and his "Curbing Beijing's Arms Sales," *Orbis*, XXXVI (1992), 379–397. See also Richard Bitzinger, *Chinese Arms Production and Sales to the Third World*, RAND Note N-3334-USDP (Santa Monica, 1992).

4. Yitzhak Shichor, *A Multiple Hit: China's Missile Sales to Saudi Arabia*, SCPS Papers no. 5 (Kaohsiung, Taiwan, 1991).

5. See chapter 12, "Asia-Pacific," by Andrew Mack, in this volume. See also Gerald Segal, "Managing New Arms Races in Asia/Pacific," *The Washington Quarterly*, XV (1992), 83–102.

6. Tai Ming Cheung, "Ties of Convenience," a paper for the Chinese Council on Advanced Policy Studies (Taipei, 1992); *Far Eastern Economic Review* (3 September 1992), 21.

7. For some discussion of these issues, see *International Herald Tribune* (24 August 1992).

8. For an earlier discussion of Chinese policy on arms transfers, see Anne Gilks and Gerald Segal, *China and the Arms Trade* (London, 1985).

9. John Lewis, Hua Di, and Xue Litai, "Beijing's Defense Establishment: Solving the Arms-Export Enigma," *International Security*, XV (1991), 87–110.

10. Yitzhak Shichor, "Defence Policy Reform," in Gerald Segal (ed.), *Chinese Politics and Foreign Policy Reform* (London, 1990), 77–99.

11. *Asian Defence Journal* (June 1992), 120.

12. Zheng Ruixiang, "Shifting Obstacles in Sino-Indian Relations," *The Pacific Review,* VI (1993), 63–70.

13. *Far Eastern Economic Review* (6 February 1992).

14. Steve Coll, "Secret CIA Escalation in '85 Tipped Afghan Balance," *International Herald Tribune* (21 July 1992).

15. John Bray, "Burma: Resisting the International Community," *The Pacific Review,* VI (1992), 291–296.

16. A Kyodo report on 17 September 1992, in British Broadcasting Corporation (BBC), *Summary of World Broadcasts,* Far East, 1489, i.

17. *International Herald Tribune* (4 April 1992); *Jane's Defence Weekly* (20 June 1992); *Defense News* (16 March 1992).

18. On the chemical weapons issue, see *Defense News* (24 August 1992).

19. Gerald Segal, "China and the Disintegration of the Soviet Union," *Asian Survey,* XXXII (1992), 848–869.

20. For a discussion of some of these issues see Harry Gelber, "China, Strategic Forces and Arms Proliferation," a paper for the CAPS conference (Taipei, 1992).

21. For a general discussion see Robert Scalapino, "Northeast Asia-Prospects for Cooperation," *The Pacific Review,* V (1992), 101–111.

22. Ellis Joffe, "Modernizing Chinese Defence Policy," a paper for the CAPS conference (Taipei, 1992).

23. For some speculation on this process see François Godement, "Policy Dynamics," *Far Eastern Economic Review* (17 September 1992), 26.

24. On à la carte security see Gerald Segal, "Northeast Asia: Common Security or à la Carte," *International Affairs,* LXVII (1991), 755–768; "The Consequences of Proliferation in Asia," a paper for the IISS Annual Conference (Seoul, 1992).

25. *Financial Times* (16 September 1992).

26. On the issue in general see William Schneider, "The Emerging Pattern of Arms Export Controls Affecting Advanced Technology," *Contemporary Southeast Asia,* XIV (1992), 47–59.

27. Simon Long, "The Tree That Wants to Be Still," *The Pacific Review,* V (1992), 156–161.

Part Four

PRINCIPAL PURCHASERS AND RECIPIENT REGIONS, INCLUDING OPPORTUNITIES FOR REGIONAL ARMS CONTROL

TEN

The Middle East and The Persian Gulf: An Israeli Perspective

Gerald Steinberg

FOR THE PAST two decades, the Middle East and Persian Gulf have been the largest markets for conventional arms. In the 1980s, this region accounted for $200 billion in arms sales. Approximately half this total went to two states—Saudi Arabia and Iraq—which, in some years, were the largest two importers in the world.

There are signs this situation may be changing. The radical shifts in the structure of international politics have created some opportunities for conventional arms control in the Middle East and Persian Gulf. Most of the weapons that have been sold or transferred to these regions were produced in the United States and the Soviet Union and were provided within the framework of superpower competition. With the end of the Cold War and the collapse of the Soviet Union, an overall decline in arms sales to the Third World, and particularly to the Middle East and Gulf region, is possible.

Other changes in the region reinforce these trends. The Arab-Israeli peace talks that began in Madrid in October 1991, the Israeli-Palestinian agreements of 1993 and 1995, the 1994 Israel-Jordan Peace Treaty, and the negotiations that took place between Israel and Syria have created a foundation for cooperation and conflict amelioration. In addition, the multilateral working group on Arms Control and Regional Security (ACRS) Working Group, in which some of the major recipients and most supplier state are participating, provides a means of negotiating modalities and details of conventional arms limitations. Although some of the

227

major regional powers, including Syria and Iran, have refused to participate in these negotiations (the government in Teheran is a vociferous opponent of peace talks with Israel, Syria is boycotting the multilateral talks, and Iraq is not included in any of these processes), this forum is a useful foundation for discussing mutual security concerns and threats. For the first time, some of the states in the region have considered confidence- and security-building measures (CSBMs) and arms control as instruments of policy or as potentially useful means of increasing national security.

In addition, the 1991 Gulf War and the belated recognition of the dangers posed by the sale of billions of dollars of arms and military technology to Saddam Hussein, including tanks, aircraft and other conventional weapons, led to reevaluations of arms sales policies among the major suppliers. France, the United Kingdom and the United States spent billions of dollars during the Gulf War to destroy the weapons they sold Iraq. Immediately after the war, the major supplier states met in various frameworks and circulated proposals and initiatives designed to coordinate policy and define criteria and ceilings.

These three factors, in combination, create opportunities for limitations on the sales of conventional arms to the region. Reevaluation on the supplier side and changes in the perspectives of the recipient states have encouraged discussions of policies designed to limit the flow of conventional arms and military technology.

At the same time, a number of factors mitigated against this optimism and narrowed the parameters of any opportunities. The demise of the Soviet Union and its control of Eastern Europe did not end the manufacture and sale of arms from this region. On the contrary, the successor states have used sales of weapons, including T-72 tanks and MiG-29 aircraft, as sources of hard currency and continued industrial employment. They have sold many of these systems to Syria and Iran. Further, the local commanders of military bases in the republics that replaced the Soviet Union sold weapons without involving their central governments. Some of the largest weapons factories in Russia, Ukraine and the former Czechoslovakia continue to export, with the oil-rich Middle East still their primary market. With the end of the Cold War, other producers, including those in North America and Western Europe, are seeking to increase their exports, and, here again, the Middle East is a prime source of earnings. The United States has sold billions of dollars in major weapons to Saudi Arabia and Egypt, and other suppliers can be expected to follow its lead. Syria has acquired and tested Scud-C missiles from the Democratic People's Republic of Korea (North Korea) and is seeking M-9 missiles from China.

Despite the Arab-Israeli peace process, conflicts, tensions and threats of war persist and, with them, the acquisition of weapons. As many analysts have noted, "the norms and mores of deterrence and defense in the region" are not consistent with the view that arms control is integral to national security. Many states are far from status quo powers, and "such Western intellectual concepts as 'deterrence,' 'balance of power,' 'unacceptable damage' and 'wasteful defense spending' are often abstractions with little application."[1]

The Persian Gulf is still an area of instability. Iran is seeking to reassert its status as a regional power (including in Central Asia) following its defeat in the war with Iraq. This goal has led to a massive acquisition program, including tanks, combat aircraft, tactical missiles, electronics and long-range surface-to-surface missiles, and a broad effort to acquire nuclear weapons.[2] This program has in turn triggered increased purchases by Saudi Arabia and other Gulf states. It is not clear how long the international sanctions on Iraq will be maintained and what will happen when they are lifted, but a renewed and massive effort by Iraq to acquire arms can be expected.

Nevertheless, the changes in the international structure and climate have created some possibilities for conventional arms limitations in the Middle East and Persian Gulf. As shown below, the prospects for controls on the transfer of conventional weapons are greater in this region than are limitations on nuclear arms. At the same time, counter-pressures could block any efforts, and the level of conventional weapons acquisitions and deployments may actually increase. Much will depend on the details in the interaction of these factors.

The Conventional Arms Market in The Region

The vast income from sales of oil and the strategic importance of the area have made the Middle East and Persian Gulf the largest market for conventional arms in the world. In the 1980s, the over $200 billion in arms sold to this region constituted two-thirds of the total arms purchased in the Third World.[3] Saudi Arabia and Iraq purchased approximately half these weapons.

Iran, Egypt and Syria were among the seven largest importers in this period.[4] Although Israel's resources are smaller than those of most other states in the region, it spends some 30 percent of its gross national product (GNP) on defense, the bulk of which goes for weapons and military technology.[5]

As a result, the countries of the Middle East have developed massive stockpiles of conventional arms. A Stimson Center report notes that four countries in the region possess more main battle tanks (MBTs) than do the United Kingdom or France.[6] Before the 1991 Gulf War, Iraq had the fifth—and by some measures third—largest army in the world, with 45 divisions, over 6,000 tanks, 5,000 armored personnel carriers (APCs), 4,700 guns and mortars, and 700 combat aircraft. As of 1994, the Syrian army had deployed over 4,800 tanks, including shipments from Czechoslovakia, Poland and the former Soviet Union (FSU), and over 500 combat aircraft. The Israeli military is reported to include 3,800 tanks and over 700 aircraft.[7]

In addition to the quantitative arms races, the Middle East is also characterized by a constant escalation in quality. Saudi Arabia has been able to purchase weapons and platforms, such as airborne warning and control system (AWACS) aircraft, F-15 XP fighter aircraft, upgraded Patriot anti-tactical ballistic missiles (ATBMs), Sidewinder air-to-air missiles, electronic warfare equipment and tactical communications systems that are more sophisticated than those found in most Western European states. In the mid-1980s, Syria received the highly accurate SS-21 ballistic missile and advanced air defense systems from the Soviet Union, whereas most Warsaw Pact members did not have access to similar technology. Similarly, the Soviets sold Su-24 fighter-bombers and MiG-29 combat aircraft to Iran, Iraq, Libya and Syria. The Middle East has also entered the missile age, as demonstrated in both the Iran-Iraq war and when Iraq fired missiles against Israel and Saudi Arabia in the 1991 Gulf War.

The scale of arms purchases in this region results from the interaction of three major factors—the nature of the intense and intertwined conflicts, the strategic importance of this region to outside powers and the revenues from oil production. To understand the nature of the conventional arms races in the region and the prospects for control, each of these factors must be understood.

The Internal Impetus for Acquiring Arms

The demand for arms in the Middle East and Persian Gulf is primarily the result of the intense ethno-national and religious conflicts in the region. The Arab-Israeli conflict has led to a number of major wars in the past five decades, with each war accelerating the arms race among Egypt, Iraq, Israel, Jordan, Saudi Arabia and Syria. Similarly, conflicts between Persian Gulf nations have led to a parallel arms race involving Iran, Iraq, Kuwait, Saudi Arabia and the smaller Gulf states.

Neither of these regions considers war to be "obsolete," and both see the threat of military conflict as an instrument of policy. Wars to expand borders and acquire land, oil or other resources continue: Iraq launched military attacks against Iran and Kuwait in 1979 and 1990, respectively, and Syria and Israel accuse each other of seeking to expand at the expense of the other. The acquisition of conventional weapons is therefore a central element in national security—each state seeks to build up its conventional arsenal, both quantitatively and qualitatively, to strengthen its offensive capabilities, defense and deterrence.

The role of the military in the domestic political systems in many of the states accelerates these conventional arms races. Historically, military juntas deposed the monarchies and post-colonial regimes in many Arab states, including Egypt, Iraq, Libya and Syria, and in most of these cases the military remains the major source of power. President Hosni Mubarak and Prime Minister Hafez Assad were Air Force officers and maintain close ties to the military. In Iraq, President Hussein gained and maintains power through the military. In the remaining monarchies, Saudi Arabia and Jordan, support of the military is necessary for the continued survival of the regime.[8] As a result, the armed forces generally have almost unrestricted access to resources and can use this access to gain the most modern weapons. The purchase of these weapons is a source of prestige as well as military power. These factors were of primary importance in the $2.3 billion arms agreement signed by Washington and Cairo in 1991 and in the arms agreements with Saudi Arabia.

In Israel, a democratic state (the only one in the region), domestic political and economic factors play a role in acquisition policies. The continued threat to security means that defense concerns are dominant, and the military is one of the most powerful institutions in the country, with ready access to resources and decisionmakers. Nevertheless, the share of the government budget allocated to defense has been declining, and in 1992, for the first time, defense spending was not the largest item in the budget.[9]

The External Sources of Arms

From the 1940s until the present, outside powers have had important strategic and economic interests in the region, including access to petroleum, and these interests have contributed significantly to the conventional arms buildup in the region. After World War II and throughout the Cold War period, the United Kingdom, France, the United States and the Soviet Union used the large-scale transfer of weapons to acquire and maintain influence and bases among client states in the region. The United States supplied arms to Iraq and other states in return for their participation in anti-Soviet alliances (the Baghdad Pact), and the British armed Jordan and other clients. Moscow became a major supplier of weapons to Egypt beginning in the mid-1950s and later transferred large-scale supplies of weapons to Algeria, Iraq, Libya and Syria.

In the early 1970s, following the experience of the Vietnam War, the United States increased the emphasis on arms sales to the Middle East as a major policy instrument. Under the Nixon Doctrine, Washington provided weapons to regional allies to strengthen stability without the need to introduce troops. In the Persian Gulf, which became a major focus of American foreign and defense policy, the United States saw Iran and Saudi Arabia as local powers, to which it sold billions of dollars in weapons. American arms sales to Iran increased threefold between 1971 and 1975 and doubled again by 1977. Between 1970 and 1978, Teheran purchased $20 billion in weapons, spare parts and related services from the United States.[10]

The vast increase in arms sales was also significant in qualitative terms. The United States provided many advanced systems and technologies to Iran and Saudi Arabia, including F-15 aircraft, AWACS airborne battle-management systems and air-to-air missiles. It justified these sales on the basis of the need to increase deterrence and security.

The Persian Gulf: Weapons to Recycle Petrodollars

Economic factors are also of central importance in explaining arms sales to the Middle East. This region remains one of the largest arms markets in the world (although on some measures the East Asian market has become larger in the mid-1990s) and, with the end of the Cold War and the resulting large reductions in orders from the defense establishments of the United States and Western Europe, has increased the economic importance of sales to the Middle East. Greater exports provide an alternative to rising unemployment in the defense industries and the demise of some firms. In 1990 alone, the five permanent members of the United Nations (U.N.) Security Council collectively supplied $20 billion in arms to the Middle East.[11]

The massive scale of conventional arms sales in the region would not have been possible without the availability of petroleum revenues to pay for the weapons. For the past two decades, the major suppliers, including the United States and Western Europe, have competed to recycle the petrodollars of the major oil producers, such as Iran, Iraq, Kuwait, Libya and Saudi Arabia.

The acquisition of weapons by these states did not slow significantly after the end of the Iran-Iraq or 1991 Gulf Wars; indeed, it accelerated greatly.[12] Between September 1991 and June 1992, Iran reportedly signed agreements and contracts to purchase weapons worth $8–10 billion over the next five years.[13] Despite disputes regarding repayment of outstanding debts, Russia is still Iran's primary source of weapons. Recent Iranian purchases from Russia are reported to include 400 T-72 tanks, 48 MiG-29s and 24 Su-24 aircraft, 12 long-range heavy bombers, 24 MiG-31s and 24 MiG-27s, as well as early warning and reconnaissance aircraft, Kilo-class attack submarines, 200 self-propelled artillery, SA-6 and SA-5 air defense systems, and 300 Scud-Bs and missiles.[14] Iran has purchased a large number of T-55 tanks from the former Czechoslovakia, and there were reports that Turkmenistan planned to sell much of its air force, including 223 MiG-23s and Su-17s, to Iran.[15]

In the late 1980s, China became Iran's second leading arms supplier. Contracts were signed for $2 billion or, according to some reports, $6 billion in weapons.[16] These transactions include F-6 and F-7 aircraft (equivalent to the MiG-19 and MiG-21), improved T-55 (known as Type-69) tanks, artillery, APCs and anti-aircraft systems.[17] China is also reported to be a major source of unconventional weapons and technology, and Teheran has acquired a number of Scud-C and No Dong surface-to-surface missiles from North Korea and serves as a channel for weapons and technology destined for Syria.[18]

Saudi Arabia has long been the largest arms importer in the world: between 1983 and 1990 it signed agreements to purchase $57.3 billion in arms, military construction and infrastructure (19 percent of the Third World total). The Saudi government signed two framework agreements (known as the Al-Yamamah deal) valued at $30 billion with the United Kingdom in 1985 and 1988 that led to the transfer of Tornado advanced combat aircraft, tactical missiles, and advanced electronics and communications systems.[19] The rate of acquisition accelerated in 1987–90, when the Saudi total was $35.5 billion. In 1990 alone, the Saudis received deliveries of weapons worth $14.5 billion—almost half before Iraq invaded Kuwait.

There has been no let-up in the post-Gulf War period.[20] After the war, 200 M-60A3 tanks, 200 Bradley armored fighting vehicles and 18 F-15C aircraft were delivered. In 1992, U.S. arms sales included 150 M-1 tanks, 200 more Bradleys, 12 Apache AH-64 attack helicopters and many other weapons.[21] The total of American post-Gulf War arms sales to Saudi Arabia (through 1996) will reportedly reach $23 billion, to include an additional 72 F-15 aircraft and 465 M-1 and M-2 tanks.[22] These weapons are all paid for by oil revenues and constitute a major element of petrodollar recycling.

Overlapping Arms Races

Another complicating factor is the degree to which the Arab-Israeli and Persian Gulf conflict zones overlap and interact. This makes it difficult to isolate one of these regions or a small subgrouping of states for the purpose of arms control and limitation agreements. (Iraq, Syria and Turkey constitute a third conflict zone where tensions emerge periodically. This zone has been relatively quiet in the past two decades compared with the other two conflict zones, but imbalances and instabilities could lead to active conflict there.)

Acquisitions by Iraq and Saudi Arabia in particular have a direct impact on the other states in both zones. For example, as a result of the threats posed by Iran and Iraq in the 1980s, Saudi Arabia increased its purchases of advanced weapons systems, including American and European combat aircraft, tanks and missiles. Saudi Arabia also borders on and has maintained a state of war with Israel. The Israelis view arms acquisitions by the Saudis as potentially threatening in the context of an Arab coalition. The 1981 American sales of F-15 aircraft and the AWACS airborne battle management system were perceived in Israel as particularly dangerous, although they were directed toward Iran and perhaps Iraq. In 1992, Israeli Prime Minister Yitzhak Rabin publicly opposed the proposed U.S. sale of seventy-two advanced F-15XP ground-attack combat aircraft.[23] He noted the inherent weakness and instability of the Saudi regime and the possibility that these weapons could fall into the hands of "hostile powers"—Islamic fundamentalists or dictators such as Saddam Hussein.[24] The potential for conflict was illustrated when the Saudis violated agreements with the United States and stationed F-15s at the Tabuk airbase, located close to the Israeli border.

Saudi Arabia is not the only example of overlap in the arms races in the Arab-Israeli and Persian Gulf zones. Iraq has been an active player in both regions since the 1940s. Iraq and Iran have pursued their ancient contest for regional hegemony through massive arms acquisitions and periodic warfare. In the 1970s, as the scope of these acquisitions and deployments increased, and as Saddam Hussein cultivated regional hegemonic ambitions, the Iraqi military threat grew. Following the Iranian revolution in 1979, Iraq invaded Iran, and that conflict continued for eight years and was followed by the Iraqi invasion of Kuwait in 1990.

At the same time, Iraqi forces have been involved in the Arab-Israeli conflict since the 1948 Israeli War of Independence. In May 1967 Iraq deployed forces along the Israeli border and in Jordan as part of the Arab military coalition, and posed a major military threat to Israel until the 1991 Gulf War. The new wide-track highway from Baghdad to Amman and the acquisition of hundreds of German tank transporters allow Iraq to move hundreds of tanks to the Israeli border through Jordan in 72 hours. Weapons purchased by Iraq during the war with Iran (and the American military assistance provided under the pro-Iraqi "tilt") posed a threat to both Jordan and Israel. Prior to the 1991 Gulf War, the Iraqi air force flew in reconnaissance sorties along the Israeli-Jordanian border, and Iraqi ground forces planned maneuvers in Jordan. During the war, Iraqi missiles struck Israeli cities.

In the 1980s, growing military and strategic links between Iran and Syria added another element to the interaction and interconnection between the Arab-Israeli and Persian Gulf zones. During the war with Iraq, Teheran was isolated from most other Middle East and Arab states, while Baghdad received economic assistance and weapons from Egypt, Jordan, Kuwait and Saudi Arabia. Syria, in contrast, provided arms and spare parts to Iran (Iran and Syria are controlled by non-Sunni Moslems, whereas Iraq and most other Middle Eastern Arab regimes are Sunni). After the war ended, these links between Iran and Syria continued, and tanks and North Korean Scud-C missiles enroute to Syria were off-loaded in Iran. In addition, since Lebanon came under Syrian military control in the 1970s, Iran has provided weapons, including short-range missiles, to the Islamic fundamentalist Hezbollah launching raids against Israel.

These interactions, with acquisitions in one zone leading to a chain reaction in the others, makes it difficult to isolate a single zone or conflict area. This factor, in turn, is a major obstacle to conventional arms control in the Middle East.

The Impact of Conventional Arms Transfers

As noted, Middle Eastern states acquire arms to meet their military requirements. For some, including Israel, Jordan, Kuwait and Saudi Arabia, the transfer of weapons has enhanced security, preserved the military balance and, at least to some degree, increased regional stability. These states face real military threats; visible deterrent and defense capabilities are necessary to meet them.[25]

From the perspective of the major powers that are suppliers, the large sales of arms to regional clients and allies failed to achieve their political goals. The Nixon Doctrine in the Persian Gulf collapsed with the overthrow of the Shah of Iran, and the American-made weapons fell into what had become hostile hands. A decade later, in 1990, the Iraqis captured American weapons in Kuwait and used them against U.S. forces. Until August 1990, Egypt, Jordan and Saudi Arabia exchanged military intelligence with Iraq, held joint exercises and transferred American technology and weapons. (Indeed, Jordan continued to serve as a funnel for weapons and spare parts to Iraq even after the invasion of Kuwait and for years after the war ended.)

In recent years, the technological arms race has become destabilizing. In the past, the Israeli qualitative advantage offset the large Arab advantage in numbers and personnel. Now Jerusalem fears its technological lead is eroding, as the United States transfers increasingly advanced weapons to Egypt and Saudi Arabia. At the same time, the Arab states view Israeli qualitative superiority as a potential military threat that has resulted in a situation often described in terms such as a "security dilemma."[26] Each side claims to be responding to the threats posed by the other side, responses that in turn lead to an escalatory spiral. Israel's narrow borders and the mobile nature of warfare that it adopted to "bring the war to the enemy's territory" obscure most efforts to distinguish offensive from defensive weapons, which can reduce the salience of the security dilemma.[27]

The costs of the continuing conventional arms races led Israel to produce a non-conventional deterrent force. The development of nuclear weapons and missiles began in the mid-1950s, when the political leadership feared the resources and access of the Arab states to weapons would eventually provide them with overwhelming superiority and the ability to mount attacks that would endanger Israeli survival. This fear was reinforced after the 1967 and 1973 wars. In 1976, former Defense Minister Moshe Dayan warned that one day the Arabs might decide to "throw against us the thousands of tanks and missiles they are accumulating" in another effort to destroy the Jewish state and that only nuclear weapons could deter that threat.[28] In 1981, Shai Feldman, who advocated an overt nuclear posture, argued, "As Israel moves from [conventional] defense to [nuclear] deterrence, the formal burden imposed by its current posture, as well as the need for enormous quantities of sophisticated conventional weapons, would decrease."[29] A former economic advisor to Prime Minister Yitzhak Shamir advocated similar policies. Even Egypt, which has signed a peace treaty with Israel, continues to acquire billion of dollars in arms, and the Egyptian chief of staff named Israel as the most likely enemy.

Thus, conventional and non-conventional arms are closely linked, and to reduce the threat of a nuclear war in the region, the threat posed by conventional arms must be cut drastically.

New Factors and Opportunities

As noted, some of the major factors contributing to the ever-increasing sale and transfer of conventional weapons to the Middle East have changed. First, the end of the Cold War has removed one of the major sources of the regional arms race, and the demise of the Soviet Union has had a further effect on arms sales. Between 1987 and 1990, the USSR accounted for 31 percent of the sales in the Middle East, compared with 20 percent for the United States.[30] The sudden end to Soviet influence had a particularly strong impact on Syria, which, since the mid-1970s, had become Moscow's major regional client. In the past, Syria could initiate a war with the knowledge that the Soviet Union would replace even massive losses quickly and at little cost (as occurred in 1967, 1973 and 1982). Syria can no longer expect such rapid assistance. In addition, the former Soviet arms factories are now dispersed among a number of countries, and much of the advanced research and development (R&D) capability has been dismantled. It is unlikely that Russia and the successor states will be able to produce new state-of-the-art MBTs, combat aircraft, missiles and other systems. Moreover, the weapons they still produce will not be available in the numbers and under the conditions that existed up to 1990.[31]

Second, in response to the Iraqi invasion of Kuwait and the 1991 Gulf War, the international community, led by the United States, imposed an arms embargo on Iraq. Although there has been some leakage, particularly through Jordan, for the most part this embargo has been effective, and significant purchases of conventional arms purchases by Iraq have essentially stopped. This change has had a major effect on the region, the impact of which will increase as long as the embargo continues. Iraq has been a major threat in both the Arab-Israeli conflict and in the Persian Gulf. The reduction or end of this threat removes one of the major catalysts for arms acquisition in both regions. In addition, with the neutralization of Iraq, it is possible to "decouple," at least to some degree, the two regions. In the realm of conventional weapons, as long as Iraq is unable to acquire new systems, only Saudi Arabia remains an active participant in both regions. At least historically, the role of the Saudi military in the Arab-Israeli conflict has been far smaller than that of Iraq. Thus, the arms embargo and the isolation of Iraq present an opportunity for limiting acquisitions of conventional weapons, particularly in the Arab-Israeli zone.

A third factor is the change in economic conditions in the region. The economies of Egypt, Israel, Jordan and Syria are not based primarily on petroleum exports, and these states have significant difficulty paying for conventional weapons. The purchase of weapons has contributed to large deficits and major economic problems, including unemployment and inflation. Since 1985, Israel has cut its defense budget significantly, in part to fund the absorption of hundreds of thousands of Soviet Jews. U.S. military aid of $1.8 billion annually, which began after the Camp David agreement, has not expanded to match the increased sales to the Arabs, and inflation has steadily eroded purchasing power. In 1988, Prime Minister Shamir acknowledged, "We know the burden of the arms race is devastating to the economies in the region—and it is getting worse."[32] These economic factors have made Israel amenable to proposals for conventional arms control.

There are also some suggestions that major economic difficulties in the Arab and Islamic countries have increased the prospects for arms control. According to Yahya Sadowski,

> Over the past decade, declining oil prices, overpopulation, economic mismanagement, and foreign policy adventurism have combined to wreak havoc on the economies of the Arab states . . . Economically exhausted, many Arabs are searching for cheaper ways of meeting their national security needs.[33]

Jordan's economic crisis, compounded by its response to the pro-Iraq policy during the Gulf War, has effectively prevented the purchase of new weapons for a number of years. In 1988, Jordan defaulted on its debt payments (total external debt was $8 billion, of which $3 billion resulted from weapons purchases).[34] In addition, it cancelled contracts to purchase combat aircraft from France and the United Kingdom.

Chronic economic difficulties and high ratios of foreign debt to GNP also plague Egypt and Syria. Prior to the 1991 Gulf War, Cairo's foreign debt was estimated to have exceeded $50 billion, much of it the result of arms purchases. Syria, as with Egypt, suffers from growing unemployment and a declining standard of living, although the situation improved in the early 1990s, when the value of annual oil exports rose to over $1.5 billion.[35] The increased revenues are, however, insufficient to solve Syria's economic difficulties.

The impact of these economic factors on military spending and arms purchases is compounded by the demise of the Soviet Union, Syria's major supplier, and the demand by Russia that Syria pay in hard currency.[36]

Some analysts, such as Sadowski, argue these factors will limit military budgets and arms acquisitions, particularly by Syria. Others disagree, asserting that the priority given to the military is unlikely to change. The evidence points to the latter conclusion. Between 1990 and 1994, Syrian arms purchases were quite large, including 1,400 T-72 tanks, 48 MiG-29 and 24 Su-24 aircraft and advanced air defense systems from Russia, 250 Bulgarian-made self-propelled artillery and 150 Scud-C missiles from North Korea.[37]

Egypt also continues to purchase advanced weapons, despite the economic and social unrest. In the past ten years, Egypt has replaced its Soviet and Chinese weapons with hundreds of modern American MBTs (M1-A1 and M60-A3) and has acquired over 200 advanced combat aircraft and tactical missiles. Egypt is producing M1-A1 tanks under American license, and has also purchased weapons and military technology from China, France, Germany, Italy, Russia, Turkey, the United Kingdom and others.[38] Since the 1979 Peace Treaty with Israel, Egypt has had no significant external opponents, and this very costly acquisition is seen as a long-term threat by Israel.

The major oil producers, including Iran, Kuwait and Saudi Arabia, are also facing limited resources for arms purchases. Each state has large current account deficits. Iran suffers from a high level of unemployment, over 25 percent based on official figures and estimated to be much higher unofficially, and major problems with its economic infrastructure. The five-year development and post-war reconstruction plan adopted in 1990 called for a hard currency expenditure of $112 billion, which would consume more than the income generated by oil revenues under even the most optimistic projections.[39]

Nevertheless, as noted, military spending and arms purchases among the major oil exporters continue to grow.[40] Teheran, as did Baghdad in the 1980s, has allocated billions of dollars for arms purchases and currently spends one-quarter of its GNP on the military. The evidence demonstrates that Iran will continue to place military acquisitions ahead of economic development. Similarly, after the 1991 Gulf War, Saudi Arabia announced plans to double the size of its armed forces and to acquire the most sophisticated and expensive military equipment available. As Kanovsky concludes, Saudi Arabia, as with Egypt, Iran and Syria, can be expected to continue and to accelerate the rate of spending on weapons and military technology, regardless of its financial situation.[41]

Arms Control, the Peace Process and the Gulf War

Political changes in the region, particularly in the Arab-Israeli arena, provide some basis for discussing arms controls, including conventional limitations. These changes began with the 1978 Camp David accords and the Israel-Egypt peace treaty. Although there is no evidence these agreements contributed to a reduction in the flow of conventional arms (in fact, the treaty led to increased sales of American arms to both Egypt and Israel), the change in political relations, the establishment of direct lines of communication and the application of CSBMs can, at least in theory, facilitate further limitations. The demilitarized zones established in the Sinai between Egypt and Israel and the role of the Multinational Force and Observers (MFO) in verifying compliance have been successful and are a useful model for future agreements.[42]

The Arab-Israeli negotiations that began in Madrid in October 1991, the Israeli-Palestinian Declaration of Principles (DOP) and the 1994 Israeli-Jordanian Peace Treaty also contribute to creating conditions for arms limitation agreements, including conventional weapons. Article 4 of the treaty includes a commitment to the creation of a Middle East version of the European Conference on Security and Co-operation and "the adoption of regional models of security successfully implemented in the post World War era (along the lines of the Helsinki process) culminating in a regional zone of security and stability." In addition, the treaty includes a pledge by the parties

> to refrain from the threat or use of force or weapons conventional non-conventional or of any other kind against each other or of other actions or activities that adversely affect the security of the other Party, . . . to refrain from organising instigating inciting assisting or participating in acts or threats of belligerency hostility subversion or violence, . . . and . . . to ensure that acts or threats of belligerency hostility subversion or violence against the other Party do not originate from and are not committed within through or over their territory. . . .

Section 4 includes a pledge to refrain from "joining or in any way assisting . . . with any coalition organisation or alliance . . . the objectives or activities of which include launching aggression . . ." and a prohibition on "the entry stationing and operating on their territory or through it of military forces . . . of a third party."[43]

In addition, the ACRS Working Group, in which Israel and a number of Arab states (but not Syria) are participating, provides a regional forum for discussion of these issues. The ACRS Working Group has met periodically since 1992, and its members have participated in a number of CSBM exercises in the area of search and rescue. They have also considered a joint declaration of goals and objectives and have agreed to create a regional crisis communications center and three regional risk reduction centers. Although a dispute between Israel and Egypt on the issue of the Israeli nuclear deterrent has delayed the process (with Egypt seeking to block any additional activity until Israel accepts its demands in this area), the forum has established a useful foundation for regional measures such as conventional arms control.[44]

The events surrounding the Iraqi invasion of Kuwait, as noted, provided an additional impetus to arms control in the Middle East, particularly for the supplier states. The costs of the Gulf War are estimated to have been approximately $100 billion, far greater than the profits made in selling weapons to Iraq over the past two decades. The massive conventional force amassed by Saddam Hussein, and the threat this force posed, led to increased discussion of the need for suppliers to limit not only weapons of mass destruction, but also conventional weapons.[45]

Following the war, a number of supplier states announced initiatives and proposals that contained elements of conventional arms limitations. President George Bush's Middle East Arms Control Initiative of May 29, 1991, included a reference to limitations on conventional weapons. Independent French, British and Canadian proposals for guidelines, a registry and advance notification procedures have been discussed by the P-5 and at the G-7 economic summits and U.N. committees.

At the same time, as noted, arms sales to the region have still increased markedly since the end of the 1991 Gulf War. The Bush Administration approved the sale of billions of dollars in advanced weapons to Saudi Arabia, Egypt and the smaller Gulf states. Although the U.S. government claimed these weapons were being transferred to allies to strengthen regional stability, other suppliers viewed the sales as a means of maintaining employment and profits for the American arms industry and led them to seek markets as well.

The history of efforts to limit conventional arms sales in the Middle East demonstrates the difficulties in this process. In August 1949, France, the United Kingdom and the United States announced a coordinated effort to "regulate the flow of arms" to the region. In May 1950, this Tripartite Declaration was formalized with the establishment of the Near East Arms Coordinating Committee (NEACC). However, the extensive interests of all three powers in the region, the competition among them and the failure to involve other major suppliers, including the USSR, weakened the effectiveness of this effort. Further, the language of the declaration included significant loopholes that permitted the acquisition of arms "to maintain a certain level of armed forces to assure their internal security and their legitimate self-defense." The appropriate level for each state was left undefined, a fact that suppliers and recipients exploited.[46] Other efforts following the 1967 and 1973 Arab-Israeli wars, as well as the 1977 Carter initiative, also failed to solve the dilemmas posed by the need to maintain the balance of power and to provide arms to clients.[47]

U.S. conventional arms policy after the Gulf War is reminiscent in many ways of the failed policies of the Tripartite Declaration and the Carter efforts.[48] Writing in *Foreign Affairs,* Alvin Rubenstein stated, "The tentativeness of the arms control proposals unveiled for the region last May are more suggestive of past failures than future promise."[49] Prior to and during the Gulf War, Washington agreed to transfer hundreds of millions of dollars in surplus weapons to Israel. It sold Egypt forty-six F-16C/D fighters and upgraded Cairo's air defenses. In September 1992, as part of his reelection campaign, President Bush announced the sale of seventy-two advanced F-15 XP aircraft to Saudi Arabia, the aim being to provide jobs for American aerospace employees. As a result, U.S. efforts toward conventional arms limitations in the Middle East lost credibility.

Changes in Israeli Policy

From the Israeli perspective, the acquisition of massive conventional arms by states in the region remains a major threat to national security. Israel is a microstate, with no strategic depth and a total population of only 5.5 million. Combined Arab attacks in 1948 and 1973, and the preparations for attack in 1967, threatened the survival of the state, and this scenario continues to be a major concern. As noted, the Peace Treaties with Egypt and Jordan, the Israeli-Palestinian DOP and the disintegration of the Soviet Union have diminished the threat. However, Israel has been concerned and continues to worry about an attack on the Eastern front involving Syria, with potential support from Iraq. Israel is watching the buildup of Syrian forces carefully, and it considers an attack through the Golan Heights as possible at any time.[50]

Prior to the 1991 Gulf War, Iraqi participation in a Syrian offensive was considered likely, and the acquisition of hundreds of German tank transporters increased this capability. At the end of the war, Iraq's capability, although diminished, was still quite formidable. With limited participation by Iraq and Saudi Arabia, the Arabs would have a 2 to 1 advantage in tanks, 3 to 1 in guns and mortars, and 1.6 to 1 in combat aircraft.[51]

Given the political instability in Egypt and its continued acquisition of advanced conventional weapons, this country could rejoin an anti-Israel coalition, a move that would greatly increase the Arab advantage. Were this situation to occur, the Western arms acquired by Egypt would greatly enhance the interoperability of Arab forces. (While Iran has also acquired many tanks and other conventional weapons, it is considered too far away to contribute significantly to a conventional attack on Israel.) The apparent erosion of Israel's technological advantage is a source of growing concern.

As a result of these factors, regional conventional arms limitations would be advantageous to Israeli national security. Israel endorsed the conventional aspects of the May 1991 Bush Arms Control Initiative and has welcomed registries for arms sales and supplier limitations. Former Israeli Defense Minister Moshe Arens supported the concept of a conventional arms freeze, declaring, "The first agreement that needs to be reached is an agreement among the arms exporting countries that they will not export arms into the Middle East."[52] While Israel can be expected to reject efforts to limit its nuclear deterrent, which it continues to view as the ultimate guarantee of national survival, until a full peace agreement is reached and tested for many years, limits on conventional weapons are more likely. If there is to be any progress in Middle East arms control, limitations on conventional weapons will be the first stage.[53]

At the same time, the Israelis are concerned that supplier restraints would not be fully and symmetrically enforced, a situation that would give the Arab states a major advantage. Israel is highly dependent on American arms and technology, and the United States can be expected to honor its agreements. However, the Arab states can use their oil revenues to purchase arms from many other suppliers. China, France, Germany, Italy, North Korea, Russia, the United Kingdom and other suppliers would all have to accept and implement limitations. (The Bush Administration's September 1992 announcement of the sale of F-16 aircraft to Taiwan brought an angry reaction from China. The government in Beijing announced a boycott of the negotiations on arms transfers conducted under the P-5 framework and increased sales to the Middle East and Persian Gulf. The Israeli experience with supplier agreements is not encouraging. The lessons of the Tripartite experience, which allowed the transfer of weapons for "legitimate self-defense," have not been learned. Other supplier agreements, including the Nuclear Non-proliferation Treaty (NPT) and the Missile Technology Control Regime (MTCR) did not prevent Iraq, Libya and Syria from obtaining materials and facilities from European suppliers.[54] Despite these fears and perceived risks, Israel can be expected to take proposals for conventional arms limitations seriously.

A Conventional Arms Freeze?

Despite the political and military changes in the region, arms control in the Middle East and Persian Gulf regions remains a complex and, in most cases, very distant prospect. The number of states involved (over twenty), the overlapping and interacting conflict zones and arms races in the region, the deep and intense nature of the ethno-national and religious conflicts, and the high degree of geographic, demographic, economic and political asymmetry all mitigate against successful arms control. These asymmetries are more pronounced than the situation in Central Europe during the years of negotiations over Mutual and Balanced Force Reductions (MBFR). When considering potential agreements, simple parity arrangements between individual states (Egypt and Israel, and Israel and Syria) are not acceptable because of the possibility of combined attacks.[55] Counting rules involving coalitions based on specific military fronts have been seen as a more promising approach. Such limits on deployments along Israel's eastern front would involve Iraq, Jordan, Saudi Arabia and Syria. Egypt and Saudi Arabia would be central to any limits involving the western front.

The geographic requirements for conventional arms limitations are relatively restricted, as compared with strategic weapons. Tanks, artillery and most tactical fighters are confined to short ranges, in contrast to missile-borne strategic weapons. While limitations on nuclear and chemical weapons or missiles would require the compliance of states from Algeria to Iran, effective conventional arms limitations could be developed within a relatively small group of five or six states in close proximity.

Middle East regional security and arms control has begun with the implementation of modest CSBMs. The next stage would focus on conventional arms involving the Arab-Israeli "confrontation states"—Egypt, Israel, Jordan, Saudi Arabia and Syria. (Militarily, Iraq should be included in this group. As long as outside powers effectively control arms acquisitions and deployments by Iraq, its compliance can be externally guaranteed. If a limitation agreement were negotiated, continued Iraqi compliance would have to be a condition for an end to the international sanctions.) While Iran, Libya and other states in the region have sought to acquire large conventional forces, they are too far removed to have much impact on the Arab-Israeli conventional balance. Thus, as long as Iraq is barred from receiving arms, a freeze on the delivery of conventional platforms to these states would decouple the Arab-Israeli arms race from the Persian Gulf.

A parallel supplier agreement would be a necessary part of any conventional arms limitations for the region. The guidelines and registration proposals discussed to date are small steps in the right direction, but quite limited. In a report published by the Stimson Center, a phased approach is advocated, beginning with a binding international arms registry and an agreement of prior notification for arms sales by the P-5 nations. Proposed transfers that are seen as potentially violating the guidelines would be subject to review in special meetings.[56] The expectation is that this process would lead to public pressure and fewer controversial sales. However, it should be noted there is little prospect for effective public pressure on China's arms sales policy, and public pressure in the United States did not prevent the 1992 sales to Saudi Arabia.

Verification is a necessary and often controversial aspect of any arms limitation agreement that in many cases requires inspectors and intrusive intelligence-gathering. However, limits on major conventional platforms, such as MBTs, artillery, APCs, combat aircraft, helicopter gunships and perhaps naval craft, can be verified with relative ease. These states possess large inventories of these weapons, and a significant change in the balance of power would require the clandestine acquisition of hundreds of tanks and tens of advanced aircraft. The number of potential suppliers of major platforms is relatively small, and an agreement among the P-5 states, which together account for two-thirds of all arms transfers to the Middle East, seems plausible. The participation of second-tier suppliers—Germany, Italy, Poland, and the Czech and Slovak republics—must also be obtained. A limitation agreement that included both suppliers and recipient states could be effectively monitored and verified.

The parties to any arms agreement would have to define the quantitative and qualitative limits. In the past, doing so has posed a number of difficulties with respect to strategic and conventional forces. As noted, the failure to agree on the definition of requirements for a "stable balance of power" and "legitimate self-defense" undermined the effectiveness of the Tripartite Declaration and the Near East Coordinating Committee (1950–55). Similarly, the United States continues to justify multi-billion dollar sales to Saudi Arabia on the basis of "regional balance" and "legitimate self-defense."

To overcome this obstacle, a conventional arms limitation agreement could be based on a freeze on the number of major platforms: MBTs, artillery, APCs, combat aircraft, helicopter gunships and perhaps naval craft. These numbers are relatively well-known, and significant changes could readily be detected. In this way, the complexities of comparing different systems, such as the MiG-29 versus the F-15 or T-80 versus M1-A1, can be avoided. The conventional military balance among the five core states (and the continued prohibitions on Iraqi acquisitions) are considered to be relatively stable, and these states could probably be persuaded to accept a freeze at current levels. Replacements for damaged or destroyed platforms would be allowed, at least in the first stages, whereas significant upgrading, such as exchanging a MiG-21 for a MiG-29, or a T-55 for a T-80 tank, would be prohibited.

A freeze on major platforms would skirt the problems posed by indigenous arms industries. Although Egypt, Iraq, Iran and Israel have local industries, none is capable of independently producing advanced platforms, such as MBTs and combat aircraft. (The Israeli Merkava tank uses a U.S.-made engine, and Israeli-made aircraft, such as the Kfir and the cancelled Lavi, are powered by imported engines.)

In theory, a similar type of arrangement could be applied to the Persian Gulf, with the inclusion of Iran, Iraq, Saudi Arabia and the Gulf states. As in the previous grouping, Iraqi compliance could be externally guaranteed by external supplier agreements. Here, in contrast to the Arab-Israeli conflict zone, no formal negotiating framework exists, although the ACRS Working Group includes most Gulf states and could serve as a nucleus. More important, Iran and Saudi Arabia are among the largest arms purchasers in the world. Competition among the major suppliers continues to be intense, with multi-billion dollar sales from the France, the Soviet Union, the United Kingdom, the United States and, more recently, China. (In the period between 1988 and 1991, over three-quarters of Chinese arms deliveries to the Third World went to the Middle East and Persian Gulf.) Together, these economic forces mitigate against the acceptance of limitations on conventional arms transfers to the Persian Gulf states. The traditional struggle for hegemony between Iran and Iraq, and Saudi fears concerning both sides, also complicate a demand-side conventional arms agreement.

Ballistic Missiles in The Middle East

A number of countries in the region have deployed surface-to-surface missiles in the past decade, and the rate of proliferation is increasing. Egypt, Iran, Iraq, Israel, Libya, Saudi Arabia, Syria and Yemen all have ballistic missiles.[57] Egypt and Syria launched isolated missiles against Israeli targets in the 1973 war, and Iran and Iraq used these weapons against each other's cities. For the six weeks of the 1991 Gulf War, Iraq launched missiles against the civilian populations of Israel and Saudi Arabia.

Although all the missiles fired by Iran and Iraq carried conventional warheads, the major danger missile warfare poses in the Middle East involves non-conventional warheads. U.N. inspectors found that Iraq had produced chemical and biological warheads for its extended range Scud-B missiles, and other states, including Syria, possess chemical warheads.[58]

The Israeli Jericho missile is generally assumed to be part of the strategic nuclear deterrent force, and Iran is actively attempting to develop a nuclear weapons capability. The conventional warheads these missiles carry are relatively small (the thirty-nine Iraqi missiles that were fired at Tel Aviv, Haifa and other Israeli cities caused only one direct fatality), and the capability of combat aircraft and bombers is much greater. However, in cases such as Iran and Iraq, where the ability of the air force to penetrate the air defenses of opposing countries is limited, tactical use of ballistic missiles is an important addition to conventional forces.

Missiles, as with tanks, combat aircraft and other major weapons and platforms are generally procured from outside suppliers. Egypt, Iraq, Libya, Syria and other states received Scud-B missiles from the Soviet Union. These weapons have a short range (200 kilometers) and are highly inaccurate. However, the ranges can and have been extended (in the case of Iraq, with the assistance of technology supplied by Western European and American firms). In the 1980s, Syria acquired more accurate SS-21 missiles from the Soviet Union, and in 1988 Saudi Arabia purchased a number of long-range CSS-2 (Long March) missiles from China. China also sold

Silkworm anti-ship missiles to Iran and signed an agreement involving solid-fuelled M-9 missiles, with a range of 600 kilometers, in the late 1980s.

Efforts to limit the proliferation of missiles to the Middle East have had some success. During the 1980s, Egypt and Iraq collaborated with Argentina on the Condor program. Political changes in Argentina, combined with American pressure and the 1991 Gulf War, led to the suspension of this effort. However, the MTCR failed to prevent the transfer of technology to Iraq, and North Korea is still transferring Scud C and No Dong missiles to Syria and Iran.

Given the strategic importance of ballistic missiles and the link to non-conventional warheads, efforts to limit the proliferation of these weapons and to freeze or reduce their stockpiles are generally considered independent issues. Missiles are linked to non-conventional weapons, and attempts to negotiate limits on both suppliers and recipients are likely to continue to be an independent aspect of arms control.

Conclusions

Despite the successes of the Arab-Israeli peace process, the threat of major war in the Middle East has not ended, and the massive arsenals of tanks, combat aircraft, missiles and other weapons would make a major war very costly for all. Recent changes in the world and in the region have made conventional arms control more plausible, but many obstacles and difficulties remain. Pressures from the defense industries and the massive resources of the larger petroleum producers are the major factors in determining the arms sales policies of the principal supplier states, including the United States. The two conflict zones remain interconnected, with acquisitions by Iran leading to a Saudi response that, in turn, affects Israeli security perceptions and policies. Economic and geographic asymmetries complicate limitation efforts further.

In this region it is clear that conventional arms limitation measures will require that the world's major arms producers and suppliers agree to major and unprecedented restraints in combination with agreements among the recipient states (demand-side limitations). Self-denying supplier-side restraints, beginning with the P-5 and G-7 members, are a necessary, although not sufficient, condition for progress in this area. The negotiation of limitations among the recipient states in the context of the Arab-Israeli peace negotiations would signal recognition of the importance of restraints for regional stability.

As noted, agreements on ceilings for existing weapons and definition of destabilizing new weapons and technologies would provide an important starting point. If a freeze on the levels of major platforms in existing arsenals can be negotiated, further steps toward reductions from these levels can be considered. In addition, once regional verification systems, operated by the states themselves on a mutual basis, have been established to monitor the limits on major platforms, it will be possible to expand the limitations to include tactical missiles, naval systems, and other conventional weapons and technologies.

However, if any of these ideas are to be translated into policy and to lead to agreements, the states involved, both suppliers and recipients, will have to take risks and accept costs. The United States, as the major power in the region and the world and the major arms exporter, must take the lead. Rhetorically, American officials have preached arms control and the benefits of self-denial to the recipients and to the other suppliers.[59] At the same time, the United States has not reduced its exports of conventional arms to the region. In fact, they have increased since the end of the Gulf War. President Bush's announcement of the sale of billions of dollars in highly sophisticated arms to Saudi Arabia in 1992, as part of his election campaign, had a negative effect on any future American efforts to persuade China, France, Russia or the other major producers to accept restraints on their arms sales.[60] Unless there are clear and highly visible examples of supplier restraint, particularly by the United States, conventional arms control in the Middle East will be very difficult to achieve.

Notes

1. Henry L. Stimson Center, "Report of the Study Group on Multilateral Arms Transfer Guidelines for the Middle East" (Washington, D.C., 1992), 16.

2. Sharam Chubin, *Iran's National Security Policy: Capabilities, Intentions and Impact* (Washington, D.C., 1994).

3. Stimson Center, "Report of the Study Group," 18.

4. According to data published by the Stockholm International Peace Research Institute (SIPRI), during the period 1987 to 1991 the rankings were somewhat different. India was the leading importer of conventional weapons in the Third World, with Saudi Arabia, Iraq, Egypt, Syria, and Iran ranking second, third, fifth, tenth, and twelfth, respectively. See "The Trade in Major Conventional Weapons," Ian Anthony, et al., in Stockholm International Peace Research Institute (SIPRI),

SIPRI Yearbook 1992: World Armaments and Disarmament (Oxford, 1992), 271–301.

5. Yair Aharoni, *The Israeli Economy: Dreams and Realities* (London, 1991), 8.

6. Stimson Center, "Report of the Study Group," 19.

7. *The Middle East Military Balance 1993/4,* Jaffee Center for Strategic Studies (Tel Aviv, 1994), 492–493.

8. See J. C. Hurewitz, *Middle East Politics: The Military Dimension* (New York, 1969); Yahya Sadowski, "Sandstorm with a Silver Lining? Prospects for Arms Control in the Arab World," *The Brookings Review,* X (1992), 10.

9. See Gerald M. Steinberg, "Defence Procurement Decision Making in Israel," in Ravi Singh (ed.), *Comparative Study of Defence Procurement Decision Making* (Oxford, forthcoming).

10. Stimson Center, "Report of the Study Group," 22.

11. Richard F. Grimmett, *Conventional Arms Transfers to the Third World, 1984–1991,* Congressional Research Service (CRS), U.S. Library of Congress (Washington, D.C., 1992). See also the discussion of policies in the supplier states in their respective chapters in this volume.

12. Anthony Cordesman, "After the Gulf War: The World Arms Trade and Its Arms Races in the 1990s," unpublished paper cited in Stimson Center, "Report of the Study Group," 18.

13. *Ma'ariv* (Israeli daily) (14 June 1992), citing reports in the Kuwaiti press.

14. *Ha'aretz* (13 and 15 July 1992); *Ma'ariv* (14 June 1992); Grimmett, *Conventional Arms Transfers 1984–1991.*

15. *Flight International* (17–23 June 1992), 14.

16. *Ha'aretz* (9 September 1992), based on reports in *L'Express.*

17. *Ha'aretz* (17 September 1992); *The Washington Times* (4 June 1992), reported that Iran had signed a $1 billion agreement with Russia to purchase 400 T-72 tanks and that the first shipment was delivered in the spring of 1992. According to the Stimson Center ("Report of the Study Group," 20), Iran received or signed contracts to purchase 1,500 tanks in 1991.

18. For a discussion of the role of these sales from the Chinese perspective, see chapter 9, "China," by Gerald Segal, in this volume.

19. Anthony Cordesman, "Saudi F-15 Sale Will Preserve the Balance in the Gulf," *Armed Forces Journal International,* CXXXI (1992), 31–36.

20. Stimson Center, "Report of the Study Group," 18.

21. International Institute for Strategic Studies (IISS), *The Military Balance 1991–92* (London, 1992), 99.

22. *Ha'aretz* (13 and 14 August 1992).

23. The F-15 XP is an "export version" of the F-15E Strike Eagle, considered to be "the most modern version of the most capable fighter the United States has ever built" and a more advanced aircraft than any sold to Israel. The XP version, as with the F-15E, will have ground attack capabilities, and although some F-15 equipment will not be included and the performance of the radar and other components will be downgraded, the platform allows for subsequent improvement with these capabilities. See the analysis in Dr. Natalie J. Goldring, Testimony Before the Subcommittee on Arms Control, International Security and Science, and the Subcommittee on Europe and the Middle East of the House Foreign Affairs Committee (Washington, D.C., 23 September 1992), and Tony Capaccio, "Saudi F-15s to Get Modified Radars, Targeting Gear," *Defense Week* (21 September 1992).

24. *Ha'aretz* (13 and 14 August 1992).

25. See, for example, Cordesman, "Saudi F-15 Sale," 31–36.

26. Robert Jervis, "Cooperation Under the Security Dilemma," *World Politics,* XXX (1978), 167–214.

27. Michael Mandelbaum, *The Fate of Nations: The Search for National Security in the 19th and 20th Centuries* (Cambridge, England, 1988).

28. "Dayan: Atoms, Not Tanks, Should Defend Israel," *Jerusalem Post* (30 November 1976).

29. Shai Feldman, *Israeli Nuclear Deterrence* (New York, 1982), 193.

30. Stimson Center, "Report of the Study Group," 26.

31. See chapter 8, "Russia," by Julian Cooper, in this volume.

32. Address of Prime Minister Yitzchak Shamir to the UN General Assembly, 7 June 1988, as quoted in Stimson Center, "Report of the Study Group," 33.

33. Sadowski, "Sandstorm with a Silver Lining," 7.

34. Eliyahu Kanovsky, "The Economic Consequences of the Persian Gulf War: Accelerating OPEC's Demise," The Washington Institute for Near East Studies (Washington, D.C., 1992), 64–70.

35. Syria's economy was also aided by drug smuggling, which has been estimated to provide up to $5 billion annually. Kanovsky, "The Economic Consequences of the Persian Gulf War," 62.

36. Patrick Clawson, "Unaffordable Ambitions: Syria's Military Build-up and Economic Crisis," Washington Institute for Near East Policy, Policy Paper 17 (Washington, D.C., 1989).

37. Michael Eisenstadt, "Arming for Peace? Syria's Elusive Quest for 'Strategic Parity,'" Policy Paper 31, The Washington Institute for Near East Policy (Washington, D.C., 1992), 37; see also David Butter, "Syria Reaps Rewards of Regional Policies," *Middle East Economic Digest* (27 September 1991), 4; *Ha'aretz* (2 November 1992); Barbara Opal, "Syria to Buy $2 Billion in Soviet Weapons," *Defense News* (8 July 1991), 3, 29, as cited in Eisenstadt above; *Ha'aretz* (3 November 1992).

38. *The Middle East Military Balance,* 264–271.

39. Kanovsky, "The Economic Consequences of the Persian Gulf War," 40–41; Chubin, *Iran's National Security Policy.*

40. Sadowski, "Sandstorm with a Silver Lining," 8.

41. Kanovsky, "The Economic Consequences of the Persian Gulf War," 16.

42. See Geoffrey Kemp, *The Control of the Middle East Arms Race* (Washington D.C., 1992).

43. Gerald M. Steinberg, "Middle East Arms Control and Regional Security in the Middle East," *Survival,* XXXVI (1994), 126–142.

44. Ibid., 126–142.

45. Stimson Center, "Report of the Study Group," 31, citing *The Washington Post* (6 April 1991). The report by the Stimson Center notes that "public opposition in France to the government's arms export policies arose for the first time in response to the fact that French soldiers faced French weapons in the battlefield."

46. Most American analysts view the Tripartite Declaration as a success and blame the collapse of this regime on the 1955 Soviet/Czech arms deal. This analysis ignores the role of the Baghdad Pact and arms sales to Iraq as the trigger that got the USSR involved. See, for example, Nadav Safran, *From War to War: The Arab-Israeli Confrontation 1948–1967* (New York, 1969), 147.

47. After the 1967 war, the United States imposed an embargo on arms sales and proposed a limitation agreement with the Soviet Union. It also called for a U.N.-sponsored agreement to register and limit arms shipments to the area. Moscow rejected the American proposals for joint action and negotiated restraints. In May 1977, President Jimmy Carter announced that arms transfers would be regarded as "an exceptional foreign policy implement." The effects of these new policies in the Middle East were minimal, however, and the rate of arms sales and transfers increased. The United States sold highly advanced weapons, including AWACS and F-14 advanced combat aircraft, to Iran (although they were never delivered), as well as sixty F-15 combat aircraft to Saudi Arabia. Israel was exempted from the overall ceiling imposed initially by Carter. The Carter effort had no visible effect on other suppliers. France sold F-1 Mirage advanced combat aircraft to Iraq, and the Conventional Arms Transfer (CAT) negotiations with the Soviet Union failed. See H. Y. Schandler, R. G. Bell, R. F. Grimmett, and R. D. Shuey, "Implications of President Carter's Conventional Arms Transfer Policy," Foreign Affairs and National Defense Division, Congressional Research Service (CRS), U.S. Library of Congress (Washington, D.C., 1977); Andrew Pierre, *The Global Politics of Arms Sales* (Princeton, 1982).

48. IISS, *The Military Balance, 1991–92,* 99.

49. Alvin Rubenstein, "New World Order or Hollow Victory?" *Foreign Affairs,* LXX (1991), 53–66.

50. Eisenstadt, "Arming for Peace."

51. Joseph Alpher, Zeev Eytan, and Dov Tamari (eds.), *Middle East Military Balance 1989–1990,* Jaffee Center for Strategic Studies (Tel Aviv, 1991).

52. *Arms Control Today* (21 February 1991), 4.

53. The multilateral working group on security and arms control that was convened as part of the Arab-Israeli peace process began by focusing on confidence-building measures, with arms control of any kind to occur at a later stage.

54. In some cases, it can be argued that partial and incomplete limitation efforts are counterproductive as they increase the level of arms and possibilities of conflict. Iraq was a NPT signatory and a member of the Board of Governors, an agreement that did not prevent Saddam Hussein from obtaining nuclear weapons technology. Similarly, while the United States has pressured Israel to accept the limitations of the MTCR, Syria and Iran have increased their arsenals of strategic missiles.

55. Congressional Budget Office, U.S. Congress, *Limiting Conventional Arms Exports to the Middle East* (Washington, D.C., 1992).

56. Stimson Center, "Report of the Study Group," 38.

57. See Martin Navias, "Ballistic Missile Proliferation in the Third World," Adelphi Paper 252, International Institute for Strategic Studies (IISS) (London, 1990).

58. Danny Shoham, *Chemical Weapons in Egypt and Syria: Development, Capability, and Safeguards* (in Hebrew), BESA Center for Strategic Studies, Bar Ilan University (Ramat-Gan, Israel, 1995).

59. See chapter 6, "United States," by Janne E. Nolan, in this volume.

60. *Ha'aretz* (3 December 1992).

ELEVEN

The Middle East and The Persian Gulf: An Arab Perspective

Abdel Monem Said Aly

FOR DECADES, the Arabs and Israelis fought each other to affirm their national identities, territories, and natural resources. For the Israelis, the fight was for a self-recognized sense of nationhood that would gather all the Jews of the world in the holy land of Palestine. For the Arabs, the fight was to rectify the "original sin" of uprooting the Palestinians from their historical homeland and depriving them of their right to self-determination. The conflict between the two sides continued without abatement in international fora and on battlefields for over a half century.

Six wars (in 1948, 1956, 1967, 1969–70, 1973, and 1982) made attempts to resolve the conflict futile. Even when Egypt and Israel, under American sponsorship, signed the 1979 peace treaty, the likelihood of resolving the Palestinian question and achieving peace between the Arab states and Israel defied conclusion. Instead, the two parties remained in a deadly arms race, mobilizing resources and continuously preparing for another, more devastating war. Over time, the conflict extended to a host of increasingly complicated issues, such as the Arab territories occupied in June 1967, the arms races, water supplies, refugees and the economic boycott.

In 1990 and 1991, however, international and regional developments created a new environment conducive to the start of a new peace process and the possibility of a successful conclusion that both parties can accept, even if with considerable resentment. At the same time, the sudden eruption of the Gulf crisis in August 2, 1990 showed the fragility of the stability in the Middle East. The invasion and subsequent annexation of Kuwait by Iraq, followed by Desert Storm and the liberation of Kuwait, left the states of the Middle East searching for a new vision of security and peace.

The region has been always characterized by conflict among races, religions and ideologies. State boundaries, resources, values, uneven economic development, and communal and ethnic tensions have all threatened regional stability. In addition to the local strains and conflicts at the regional level, pressures from the international system have contributed to the tension in the Middle East. In the post-colonial era, the Middle East witnessed continuous and persistent feuds, disputes and conflicts within and among the Arab states, which in a few cases led to military confrontation. Rarely have relations with the non-Arab states in the region been characterized by harmony and cooperation. In very recent history, conflicts have erupted between one or more Arab states and Ethiopia, Iran, Israel and Turkey.

These conflicts have sparked an arms race unmatched in any region except Europe. According to the U.S. Arms Control and Disarmament Agency (ACDA), the Middle East, which accounts for about 3 percent of the world's population, bought on average more than 30 percent of the world's military goods and services in the 1980s. The region devotes more than one-tenth of its output to military expenditures, more than double the ratio for any other region.[1] During the period 1984–87, the Middle East accounted for 61 percent of the total value of all Third World arms transfer agreements and 60.2 percent of arms deliveries. During 1988–91, it accounted for 57.5 percent of all such agreements and 53.4 percent of all deliveries.[2]

The impact of the international system on the security of the Middle East has always been an issue. Tensions or the relaxation of tensions have always been reflected at subregional levels. Next to Europe, the Middle East has been the major area for potential superpower confrontation. On three occasions (1956, 1957 and 1973), this potential came close to reality.[3]

In the last four decades the arms transfer policies of the major powers were an important instrument for intervening directly or indirectly in regional conflicts and building alliances to serve their interests at both the regional and international levels. The result of these policies in the 1980s was unprecedented proliferation of weapons of mass destruction and long-range surface-to-surface missiles in the Middle East. Israel, through its strategic alliance with the United States and the backing of Western Europe, succeeded in building a nuclear arsenal and enhancing its delivery capability so as to provide deep power projection. The Arab countries received conventionally armed Scud missiles from the Soviet Union, the Democratic People's Republic of Korea (North Korea) and China. Some countries in the area acquired the capability to design and manufacture ballistic missiles of substantially greater ranges than those required for battlefield weapons. Iran attributed more than 1,100 civilian deaths to Iraqi missile and air attacks during the Iraq-Iran war (1980–88).[4]

However, the world was changing rapidly. Exhausted by the effort to reach parity in nuclear weapons, military communities, particularly in the West, conceived a new role for conventional power in addressing defense needs. In the 1980s they held a long debate over concepts of war such as "extended battlefield," "deep strike," "follow-on-forces attack (FOFA)" and conventional deterrence.[5] The debate was extended to new force structures based on precision-guided weapons, Stealth platforms, cruise missiles, advanced sensors, and intelligent command, control and communication systems.

Military thinking at the beginning of the decade was motivated by the strategic needs of the European theater. Ironically, the opportunity to implement these new concepts came at the end of the decade in the Middle East. Desert Storm validated and introduced many future doctrinal concepts and combat behaviors. The lessons of the Gulf War were: the importance of the air phase and air assets; the vital role of global reconnaissance and secured communication; continuity of combat day and night; fire power in depth; and electronic warfare.[6] During the war, the U.S.-led forces used more than 7,400 tons of advanced precision-guided munitions. Stealth F-117 strike aircraft used laser-guided bombs (GBU-12 and GBU-24) to hit hard targets. High-speed anti-radiation missiles (HARM) were used to deter surface-to-air missiles (SAMs) and their radar and control centers. F-15 multi-role fighters used low-altitude navigation and targeting infra-red for night (LANTIRN) pods to locate and destroy missile sites and missile launchers.[7]

Certainly, Desert Storm was a turning point in the planning of defense acquisitions for most states in the Middle East. Military planners are now considering a large array of systems that they would have not taken seriously in the past or regarded as particularly important. Further, the end of the Cold War has affected the pattern of arms transfers to the Middle East as a result of the dissolution of the Warsaw Pact and the Soviet Union. Countries in the North Atlantic Treaty Organization (NATO) saw the Middle East as the most likely area to receive the surplus military material resulting from the reductions in conventional forces in Europe.

When in 1990 the Bush Administration declared a new American defense strategy, taking into account the world's changing security environment, it shifted the focus to regional conflicts. Key elements of the new strategy were the forward presence of forces to respond effectively to crises and the development of the Global Protection against Limited Strikes (GPALS) strategy, with systems to protect U.S. forward-deployed forces and allies against limited attacks by ballistic missiles.[8]

On March 6, 1991, President George Bush, speaking to a joint session of the U.S. Congress, defined four key challenges that would have to be met in the Middle East:[9]

—Create shared security arrangements in the region with the help of the United States through American participation in joint exercises involving both air and ground forces and maintenance of a naval presence in the region.

—Control the proliferation of weapons of mass destruction and the missile systems used to deliver them.

—End the Arab-Israeli conflict through a comprehensive peace grounded in United Nations (U.N.) Security Council Resolutions 242 and 338 and the principle of the exchange of territory for peace.

—Foster economic freedom and prosperity for all people in the region.

Two opposite trends seem to be emerging in these new developments in the Middle East. One is a massive buildup of military capabilities by the major parties to the Arab-Israeli conflict as they absorb the lessons of the Gulf War. The other is a serious attempt to settle the long-lived Arab-Israeli conflict. This settlement cannot be achieved unless those involved address the deadly arms race in the area. In fact, the multilateral negotiations on an Arab-Israeli peace accord have arms control as one of five key issues (the other four are water, refugees, economic development and the environment). Recognition of the arms race as a major destabilizing factor in the Middle East crisis is itself a step forward.

Recognition, however, is just the first step. Agreeing on and implementing concrete steps is another, and very difficult, step. Numerous obstacles have to be solved. The security environments of the states in Middle East are different in many respects, and each country has adopted its own defense doctrine. Coming up with a way to compare the weaponry of the various states is an awesome challenge, as all attempts at arms control show. Most important, the Arab-Israeli arms race is highly interconnected with the other arms races in the Middle East, particularly the one in the Gulf region and the one related to inter-Arab rivalries. Any Arab-Israeli arms control arrangement will be very difficult to carry out without Iran's participation.[10]

This chapter presents some ideas for the multilateral negotiations on arms control. Some of the ideas derive from the experience of the Arab-Israeli conflict itself, others from European and world experience. The point of view espoused here is that peace and arms control arrangements can be mutually enhancing for each other. That is, resolution of the political issues in the region will produce a hospitable climate for de-escalating the arms race. In turn, arms control measures will generate mutual confidence and help stabilize a politically very unstable situation. Other arm races in the Middle East, such as the one in the Gulf, will be considered here to the extent they affect arms control in the central Arab-Israeli area.

Arms Races in The Middle East

The proliferation of high-technology weapons in the Middle East in the 1980s is a matter of concern to all who care about peace in this troubled area. For more than four decades the conflicting parties have been searching for high-quality weapons to the maximum extent possible. Israel has rationalized its acquisitions on the basis of the need to counter the general superiority of the Arab states in quantities of weapons.[11] In 1979, Abba Eban, former Israeli foreign minister, said in front of the Israeli Parliament that Israel's existence hangs in the balance between "the Arab quantity and the Jewish quality."[12] Almost ten years later, when Yitzhaq Shamir, then the prime minister of Israel, was asked whether Israel's launching of the Ofeq-1 satellite could affect the arms race in the region, he answered, "it has no connection with the arms race. If we are speaking about races, then it is more a race over technological and scientific capabilities."[13]

No Arab will disagree. For example, as one study of the regional conflict concludes,

the balance of power between the two conflicting parties indicates—with few exceptions—a clear Israeli superiority over the Arab side. The Israeli superiority is necessarily qualitative and has been capable of constraining the Arab quantitative superiority in population, area, and military and economic resources.[14]

Regarding the military balance, Arab estimates[15] are close to those of the Israelis.[16] All numerical evaluations show that the Arab states, when considered together, have a clear strategic advantage over Israel. Taken individually, however, only three Arab states—Egypt, Iraq (before the second Gulf War) and Syria—have rough parity with Israel when "defense" and "offense" calculations are made. Even so, when Israel has been on the attack, typically it has gone far beyond its borders (as established by the U.N. 1947 partition resolution). It swallowed up all of Palestine in 1967, as well as Sinai and the Syrian Golan Heights. Israel is able to reach from Baghdad to Tunis and from the fertile crescent to the heart of Africa.

As noted, the Arab-Israeli conflict is not the only arms race in the Middle East. The Iran-Iraq war and the second Gulf War contributed immensely to this reality. Ironically, the quantitative advantage of the Arab world has proven less effective in the Gulf theater than in the conflict with Israel. Iran has more population than the Arab Gulf states put together. During the Iran-Iraq War, Iran had an estimated 6.2 million men eligible for military service out of a population of 45.2 million, whereas Iraq, the largest Arab Gulf state, had only 2.03 million men eligible for military service out of a population of 15.5 million.[17] In terms of mobilized first-line ground forces, Iran had a 4.8 to 1 edge over Iraq, in total mobilized ground forces 3 to 1, and in manpower reserves 13 to 1.[18]

When Iraq's somewhat unfavorable geostrategic position is added to its quantitative lag in manpower, its search for technologically advanced weapons becomes understandable. However, the large arsenal Iraq amassed during its war with Iran persuaded Baghdad to launch its invasion of Kuwait and thereby militarily to blackmail other Gulf states with less population and capabilities. Thus, the recent Gulf War led to a further spiraling in the race for high-technology weapons in the Gulf states aimed at counterbalancing not only Iran but also Iraq.

When the technological capabilities of nations in the Middle East are measured, Israel emerges as the only country able to design, manufacture, test and market military technologies and equipment.[19] Egypt, Iran and Iraq are trying to develop a capability to adapt foreign technology to local needs, maintain the equipment and improve it. Other Arab countries have continued to rely almost completely on foreign companies and consulting firms. In terms of arms output, only Egypt among the Arab countries is a major producer. As compared with Israel, however, Egypt's arms production capability is substantially less.[20] Israel has significant *technological* superiority over the Arab world in the areas of conventional, nuclear and space weapons. This technological edge gives Israel three strategic advantages:

—Although small, Israel is far more able than other states in the region to deal as a partner in the international weapons market and with its allies, particularly the United States.

—Israel has more military options for dealing with different situations because of its advanced technologies.

—Israel is better able to stay abreast of technological changes and ahead of its adversaries.

This technological imbalance has threatened the security of the Arab states. In response, they have looked to, and sometimes debated, four strategies: (1) modernization of their armed forces; (2) substitution of their imports with indigenous arms; (3) long technological leaps forward when developing weapons systems; and (4) the qualitative use of their quantitative superiority.[21]

The Arab world has pursued modernization by importing modern weapons systems. Between 1976 and 1985 it spent more than $380 billion, or about eight times Israeli military expenditures, to modernize its armed forces.[22] Between 1984 and 1991, Egypt, Iraq, Libya, Saudi Arabia and Syria spent more than $130 billion on arms purchases.[23] Since 1988, however, the debt crisis, the decline in oil prices and the limited capabilities of some countries to absorb and man new weapons systems have constrained Arab ability to purchase more arms.[24] This trend was reversed after the Gulf War.

The Arab world looked to import-substitution to deal with the technological gap. Egypt, Qatar, Saudi Arabia and the United Arab Emirates (UAE) established the Arab Organization for Military Industries for that purpose in 1975. The last three states withdrew from the organization after the decision was made at the Baghdad summit in 1978 to boycott Egypt. It was the last collective Arab effort to manufacture arms. As noted, Egypt is the only Arab country to have developed a respectable capability to produce weapons for local consumption and export. Between 1983 and 1987, Egypt exported weapons valued at $862 million.[25]

The strategy of making long leaps forward to assimilate new technologies first appeared in research and scientific institutions in Egypt. The idea was to utilize the existing scientific base, particularly in Egypt, to build a pan-Arab scientific research and development (R&D) capability in nuclear, space, electronics, information and energy technologies. "[T]echnological leakage from the west through reverse engineering, add-on engineering, imitation and scientific theft" would enhance the effort.[26]

As to qualitative use of the Arab quantitative superiority, it was posited that the Arab states did not need more expensive platforms, particularly aircraft, but rather less expensive and more effective delivery systems such as missiles. Given this strategy, the Arab focus on chemical, missile and force multiplier technologies is understandable.[27]

The Arab states used the four strategies simultaneously. In search of quality and security, they purchased highly advanced arms (for example, F-15 and Tornado aircraft); they substituted arms imports with local production (for example, the Al-Hussain, Al-Abbas and Al-Faw-1 missiles); they sought to build advanced technological bases (such as the Condor III missile);[28] and they took advantage of their quantitative advantage.

National Defense Programs

The Gulf War led to a reappraisal of national security requirements by the countries of the Middle East and sparked a new round of the arms race there. This is reflected in the current weapons acquisition programs of the major countries that compose the Arab-Israeli balance.

ISRAEL

Conventional weaponry is undergoing a revolution as a result of the developments in microelectronics, sensors, high-energy lasers, very high speed computers, precise navigation systems, artificial intelligence, fiber optics, stealth technology, and advanced materials and composites.[29] Israel intends to use these technologies to support a new type of military force, dominated by a few, high-technology precision weapons systems, all-weather combat systems, electronic warfare, effective intelligence and accurate targeting.[30]

Israel seeks to employ new technologies to ensure its security and survival by:

—Concentrating on civil and military R&D to achieve a degree of self-reliance.[31]

—Entering into agreements for R&D with other countries, especially the United States, aimed at developing new weapons and making better use of its arsenal.[32]

—Engaging in scientific theft, particularly from the United States, to upgrade its technological capabilities.[33]

In these ways Israel has been able to produce a wide range of advanced weapon systems, at the same time that it has been importing highly sophisticated ones.[34] The U.S. and Western armies are in fact using several tactics and technologies that have evolved out of the U.S.-Israeli defense relationship. For example, U.S. use of electronic warfare during preemptive assaults has roots in the tactics Israel applied in the 1967 Arab-Israeli war. Through Israel the United States has been able to test its weaponry in combat and has gotten innovative technologies applicable to U.S. military items, while Israel has received enough equipment, money and know-how to develop its own advanced fighter, Lavi, its own modern main battle tank, the Merkava, and a host of advanced missiles, including the Barak anti-missile missile and the upcoming Arrow anti-ballistic missile (ABM) system. Israel has cooperative relationships with several leading U.S. companies, including Martin Marietta, General Dynamics and Raytheon.[35]

In addition to technological cooperation, Israel has managed to obtain from the United States all the key weapons of its conventional arsenal. Israel's air assets now include the F-16, the F-15 Eagle, the F-4 Phantom II, the E2c Hawkeye, the C-130 Hercules, the Boeing 707, the AH-1 and Cobra helicopters. It will soon receive around eighteen AH-64 Apaches and twenty Blackhawks. Israel is considering adding the F/A-18 to its multi-role inventory because it might have to expand the scope of its tactical requirements—according to Israeli air force planners, Israel might be called on to deploy multi-role aircraft against targets demanding deeper penetration and higher altitudes, for which the F/A-18 is uniquely suited. Israeli fighter pilots have made clear they want to replace their McDonnell-Douglas F-15 Eagles with the forthcoming Lockheed F-22 advanced tactical fighter. However, Israel will not be able to afford the $100 million F-22, unless the United States drastically boosts its military aid.[36]

Israel is planning to deploy around five Patriot batteries before the end of the 1990s. However, by the year 2000 the Israeli Arrow missile system could provide roughly three times the range of the Patriot as well as offering a higher operating altitude. U.S. and Israel technicians are conducting experiments in the Arrow program that will help establish the larger ABM technology base necessary to build regional and theater ABM networks.[37]

The U.S. air-to-ground AGM-144 Have Nap missile is essentially the same as the Israeli Popeye missile (a stand-off precision-guided air-to-ground missile). Israel has a reputation for the development and production of unmanned aerial vehicles (UAV) and already has contracts with the U.S. Navy, Marine Corps and Army for several systems.[38]

Israel is now developing and producing, in addition to the Popeye, the Python III missile (short-range air-to-air), the Barak missile and the ADAMS point defense missile interceptors, as well as a wide range of advanced electronic warfare systems, reactive armor suits (Blazer) and other items. Israel is one of a few countries producing a first-rate tank, the

Merkava. It was developed to ensure that Israel would have a tank regardless of changes in world politics.[39]

As for Israel's navy, it is growing by deploying more capable high performance submarines and sea-based long-range ballistic and cruise missiles supported by a satellite surveillance network. Israel has increased its ship point defense capabilities against aircraft and missile attacks, including sea skimmers, with the Barak missile system. The unmanned helicopter Hellstar developed by Israeli Armaments Industry will soon be delivered to the Israeli navy, so that Israel will be the first country to operate such a vehicle.[40] In the plans are two newly built Dolphin-class attack submarines and three SAAR 5 missile corvettes. The Dolphin-class submarines will be far superior to what the Arab navies have. They are expected to carry long-range missiles, so that Israel will be able to extend its theater of action and targeting capabilities.[41]

In addition to Israel's qualitative edge in conventional weapons, it has two areas of absolute superiority. It began a new era in the technology race in the Middle East—this time in space—when, on September 19, 1988, it launched the Ofeq-1 satellite, followed on April 2, 1990 by Ofeq-2 and on April 6, 1995, by Ofeq-3. The perception of the Arab countries is that Israel is developing space assets to enhance its military capabilities through spying, jamming, reconnaissance, command and control, and battle management.[42] There is no comparable Arab program.

The nuclear race is almost equally one-sided. Most experts on the subject agree that Israel possesses not only a nuclear capability but nuclear warheads and delivery system.[43] The destruction of Iraq's nuclear program as a result of the Gulf War leaves the Arab world with virtually no nuclear capability. This reality is at odds with predictions in the 1970s that Iran and Libya would have a nuclear bomb by 1985 and Egypt, Iraq, Kuwait and Syria by 1990, and that Egypt, Iraq and Libya would be small nuclear powers before the end of the twentieth century.[44] With the exception of small research reactors in Egypt, Iraq and Libya, the Arab world has no nuclear capability.[45]

EGYPT

Egypt can produce all its small arms and 95 percent of the ammunition, in addition to high explosive and anti-tank bombs. It has also entered the more complicated fields of armored vehicles, aircraft, cannons and missiles. Egyptian arms production has gone in four directions:[46]

—Design and production of less complicated weapons.

—Development and modification of Soviet weapons by reverse engineering.

—Production of arms licensed by different countries (including the United States, the United Kingdom, France and Brazil).

—Cooperation with other countries to develop and produce new weapon systems.

As a result, Egypt has finally assimilated the more advanced technologies of optoelectronics, remotely piloted vehicles and missiles.[47]

Egypt is now converting from old Soviet equipment to new military systems from the West. With the delivery of 680 ex-U.S. Army M-60A1 MBTs, it has been able to abandon at least part of the fleet of T-55 tanks with L7 105mm guns. Egypt has begun local assembly of a series of 555 M1-A1 Abrams MBTs with U.S. technical assistance. By 1996, Egypt had 147 of them. Negotiations on the procurement of artillery multiple-launch rocket systems (MLRSs) are underway. On order are one Grumman E-2C, six C-130s, fifteen CH47s, eight DMN F-7s and twenty-four AH-64A Apaches. Negotiations for the purchase of a further forty-six Turkish-built F-16C/D are being completed.[48]

Egypt's defense procurement policy is geared around U.S. dollars for security assistance. Until 1988, Egypt had for many years been among the top four Middle East buyers of military equipment, with around 43 percent of it from the United States, including more than 170 F-16 aircraft, several E2-C Hawkeye surveillance planes, numerous M-113 armored personnel carriers (APCs), 40 Maverick missiles, a variety of glide and cluster bombs, and Stinger missiles.[49]

SAUDI ARABIA

Desert Storm was a turning point for Saudi defense planning just as it was for other countries in the Middle East. Subsequently, Saudi planners considered a large array of U.S. systems they would not have taken seriously or regarded as important prior to the Gulf War. Saudi Arabia announced a very ambitious, although tentative, plan to expand its defense forces to about 250,000 men. The aim was to bring Saudi forces up to the size and effectiveness of all the coalition forces deployed during the Gulf War.

Riyadh has pursued a prudent diversification policy, procuring from non-U.S. sources items the United States could not or would not deliver (that is, the equivalents of the Tornado strike aircraft or the Chinese DF-SA intermediate-range ballistic missile [IRBM]). The Saudis often grow weary of U.S. caution over technology transfer and fear of upsetting the regional balance. However, the Saudis now have five airborne warning and control system (AWACS) aircraft and more than fifty F-15s. The ground forces will have up to 465 M1-A2 Abrams MBTs and some 600 M-2 Bradley mechanized infantry combat vehicles (MICVs) before the year 2000; more than twelve multiple-launch rocket systems (MLRSs) are already deployed. Saudi forces also use the Stinger missile and the Bell 406 Combat Scout and AH-64 Apache helicopters. They have 116 TOW launchers with 2,000 anti-tank guided missiles. Some requests for purchases included over 150 Hellfire anti-tank missiles and more than 2,000 Maverick air-to-surface missiles.[50] The Al-Yamamah 1 expansion program called, most notably, for seventy-two Tornado and thirty Hawk trainers; delivery of these items is being completed. AlYamamah 2 called for a further forty-eight Tornados, sixty Hawks and forty WS-70A Blackhawk helicopters. Six Patriot batteries with 384 missiles were ordered; a further fourteen batteries were requested.[51]

Finally, Saudi Arabia concluded a $5 billion sale for seventy-two F-15E/H aircraft with the Bush Administration and has opened discussions for the purchase of four additional AWACS for an estimated $1.5-$2 billion. It has also been reported that Saudi Arabia is considering becoming a full equity member of the European Fighter Aircraft (EFA).[52]

SYRIA Syria still intends to pursue a military option to liberate its occupied territories if the peace talks fail. The Syrian Army has 1,400 T-72 MBTs, acquired from the former Soviet Union and Czechoslovakia. It currently fields four armored divisions.[53] Until recently, Syria had only a few short-range Scud-B missiles, as well as some SS-21s and FROG-7s. Purchases of Scud-Cs from North Korea and M9 medium-range tactical ballistic missiles from China have been reported. The Syrian Air Force is well-equipped with MiG-29 Fulcrums and SU-24 aircraft, and an additional forty-eight Fulcrums are to be delivered.[54]

IRAN Iran is making major efforts to reorganize and modernize its armed forces, which suffered severe losses during the long war with Iraq. At the same time, it has expanded its arms industry to support a growing military machine. Between 1984 and 1991, Iran signed arms transfer agreements valued at $19.8 billion, of which arms worth $16.1 billion have been delivered.[55] The value of covert U.S. deliveries in 1985–86 is not included in these estimates, nor are black market agreements and deliveries. One estimate is that arms deliveries to Iran in the period 1983–90 came to $39.5 billion.[56]

The West's active military presence in the Gulf, its policy of a permanent prepositioning of heavy defense equipment in the area and the plans for large-scale arms purchases by Gulf states have created a new security environment for Iran. Furthermore, it still considers Iraq to be an important threat and will do so as long as Saddam Hussein remains in power.

Iran is a fundamentalist state that strongly believes it should extend its Shi'ite Moslem ideas to other countries. It is actively involved in Lebanon and is a growing factor in Sudan. It has wide-ranging strategic interests with the numerous Shi'ite populations in the Gulf region. Future oil and gas disputes with Saudi Arabia and Qatar could result in Iranians trying to impose their wishes by force. Vital oil resources and installations in Saudi Arabia are within 150 km of the Iranian coast, a situation that lends itself to blackmailing Riyadh with the threat of a tactical missile attack. Such a threat would pose tremendous risks to the disembarkation ports in the Persian Gulf and intervention forces intending to use those ports. Iran made a very important move when it expelled all Arab nationals from the island of Abu Musa, a step that caused considerable concern in the area.[57]

Tehran's most intense political effort is aimed at the former Soviet Moslem republics in Central Asia, especially those bordering Iran. Azerbaijan, Tajikistan and Turkemenistan are all targets for Iranian intervention. Given the situation in the ex-Soviet Moslem republics and the economic difficulties facing ex-Soviet officers and officials, the fear is that Iran could acquire nuclear weapons from them. Iran is also making a significant effort to establish R&D and production facilities that eventually could lead to a nuclear capability.[58] Former Director of U.S. Central Intelligence Robert M. Gates testified to Congress that Iran was seeking a nuclear bomb and could have one by the year 2000. U.S. authorities have sought to block deals between Iran and other nations such as Argentina and China to obtain equipment that would have helped Iran manufacture nuclear weapons.[59]

Russia agreed to supply Iran with 400 of the latest version T-72 MBTs and some 134 MiG combat aircraft. These acquisitions were intended to allow Iran to restore its depleted air force in a remarkably short time.[60] Of the 115 Iraqi Air Force combat aircraft that escaped to Iranian airports during the Gulf War, a large number are in serviceable condition, and at least some could be kept in operation with the assistance of Russia and/or China.[61] Iran negotiated with Italy for the sale of CH-47 Chinook medium-lift helicopters that it could use for civilian and military purposes.[62]

The latest figures on Iranian arms purchases indicate orders from Brazil, China, North Korea (which is rapidly becoming one of Iran's best suppliers, with a contract for almost 200 Scud-Bs and Scud-Cs) and the states of the former Soviet Union. Bulgaria delivered over 10,000 rockets and SAM launchers from its stocks.

Iran is also trying to rebuild its naval capability, a move that could pose a significant threat to oil-shipping routes through the Straits of Hormuz. Recently Iran bought three ex-Soviet Navy Kilo-class submarines, which could be used with long-range air patrols and shore-based Silkworm anti-ship missiles to support Iranian military operations in the Gulf.[63]

OTHER GULF STATES

The other Gulf states face a tremendous military challenge from Iran and Iraq. To deal with that threat, they need outside help from either the United States or other Arab states. The armed forces of *Kuwait* were severely beaten in the early stages of the Iraqi invasion, and most of their equipment was destroyed or transferred to Baghdad. After liberation, Kuwait was determined to rebuild its armed forces with the best equipment. It is expected to spend over $9 billion on arms purchases and to make other defense-related expenditures.[64] Among the major weapon systems it is procuring are forty F/A-18 Hornet fighters, currently being delivered under a $1.9 billion contract. A deal worth $2.5 billion covers the purchase of several Patriot and Improved Hawk air defense missile batteries.[65] After negotiating with both the United Kingdom and United States regarding the selection of a new MBT to rebuild its armed forces, Kuwait decided to buy 236 U.S. M1-A2 tanks, to be delivered over a two-year period starting in 1994.[66] It was also reported that Kuwait may order French naval equipment, including Simonneau Marine fast patrol boats, La Combattante 4 fast missile corvettes, Aerospatiale MM-40 Exocet anti-ship missiles, Matra Mistral air-defense missiles and possibly Eridan-Class Minehunters.[67]

Bahrain is going to purchase the AH-64 Apache attack helicopter. The United States sold it some MLRS systems, in addition to one squadron of F-16 fighters and two dozen M60-A3 tanks that were part of a fifty-four tank deal.[68]

The *UAE* is currently receiving 137 Mirage 2000s and 39 Hawk combat aircraft. It has expressed a need for a second batch of modern fighters. The competition for this follow-on order is largely between additional Mirages and American F/A-18C/Ds. The UAE signaled its willingness to diversify its traditional suppliers by signing a deal for 500 Russian BMP-series MICVs.

The *Sultanate of Oman* has signed a $225 million contract to buy two missile corvettes equipped with an advanced combat system.

The Navy of *Qatar* is planning to build four VITA-type large missile craft for $187.5 million.[69]

Strategic Asymmetries

Although the Arab countries have a quantitative advantage in manpower, weapons and territory over Israel, the latter has a clear qualitative strategic advantage over them. The Arab states have been making a relentless effort to close this qualitative gap for many years. Their programs to acquire missiles, chemical weapons and even nuclear weapons—at least by Iraq—should be understood in this context. For its part, Israel's fear of the Arab quantitative edge has ignited its own acquisition of high-quality weapons, including those of mass destruction. This Israeli effort in turn has sparked an Arab effort to match Israel. The asymmetry in the quantitative and qualitative capabilities of the two sides is the dynamic of the arms race in the Middle East and creates a very unstable situation.

The Iran-Iraq War and the Gulf War ignited another arms race in the region. Since Iraq, Saudi Arabia and Iran are also prospective military adversaries of Israel, the two arm races have become highly interconnected, a situation that further contributes to regional instability.

Arms Transfer Policies After The Gulf War

After the Gulf War there was much talk by Western defense companies about controlling international exports of military equipment. The reality, however, has been feverish activity in the Arab states, Israel and elsewhere, backed by advertising campaigns claiming "combat-" or "battle-proven" weaponry.

Interestingly, the top five arms exporters to the region are the permanent members of the U.N. Security Council.[70] Since the Gulf War, the United States has ranked first in arms transfer agreements with Third World countries, with a share of the total market of 44.8 percent in 1990, up from 23.6 percent in 1989 (constant U.S. dollars). In 1990 and 1991, the total value of U.S. arms deliveries worldwide was $9.6 and $13.5 billion, of which the Middle East's share was $3.1 and $3.2 billion, respectively.[71] The Middle East and North Africa experienced a 21 percent surge in weapons acquisitions in the aftermath of the Gulf War, with big increases registered by Iran, Saudi Arabia, Syria and the UAE. The Middle Eastern purchases focused on the high technology and precision-guided weapons characteristic of the Gulf War.[72]

The United States has a clear advantage among arms exporters because of the scale of its role in the Gulf conflict. Its arms transfer policy in the region before the war seemed to have worked extremely well. Its ability to move a half million soldiers and over 1,000 aircraft, and to establish a joint command with Saudi Arabia, was substantially facilitated by the compatibility of Saudi and American equipment and infrastructure.

The American arms transfer policy involves such issues as national security, regional arms balances, the U.S. defense industrial base, future competitiveness, jobs, ethnic politics and morality. The U.S. military industry, for example, employs roughly 1.3 million people, including the largest concentration of scientists and engineers of any industrial sector in the country. In 1990 it exported products valued at $39 billion and produced a positive trade balance of $27 billion, the largest of any industrial sector.[73]

The U.S. weapons systems used in the Gulf War are based on 1970s technology, which in the mid-1990s is being phased out of U.S. Department of Defense procurement. The systems include the AWACS, F-14, F-15, F-16, Abrams tank, Bradley infantry fighting vehicle, MLRS, Apache attack helicopter and Patriot ground-to-air missile. Their replacements, which will be based on technologies of the 1980s, will not move into the production phase until the late 1990s. In the interim, exports to friendly countries can help keep trained labor forces busy and production lines warm.[74]

In 1990 the United States was the top exporter of arms, displacing the USSR for the first time since 1985. With 45 percent of the total Third World market, it also displaced the USSR there as the largest arms supplier, a position the USSR had held since 1983.[75]

Ex-Soviet and Eastern European countries urgently need hard currency from weapons sales and will be hard-pressed to give large discounts and grants for arms purchases. They also use their arms sale policy to pressure the West for economic aid and international recognition. Examples are: the Czechoslovak government's decision to sell 200 T-72 MBTs to Syria, despite very strong pressure from the United States, and to continue negotiating with Iran over the transfer of as many as 1,500 T-54/55 MBTs that were to be withdrawn under the Conventional Forces in Europe (CFE) treaty.[76] Russia was also trying to sell BTR-80 APCs to Oman and BMP-3 vehicles to the UAE.[77] Recently, and despite U.S. pressure, Russia decided to deliver three new submarines to Iran.

From 1987 to 1990 China ranked fourth among all suppliers in the value of its arms transfer agreements with the Third World, but in 1990 moved up to third, with nearly $2,600 million in agreements, or 6 percent of all such business (in constant U.S. dollars). Of particular interest is China's ability and willingness to sell various missiles to Third World states. In the latter half of the 1980s, it sold and delivered CSS-2 IRBMs to Saudi Arabia, Silkworm anti-ship missiles to Iran, and anti-tank and other surface-to-surface missiles to various purchasers.[78]

It is clear that the end of the Cold War, followed by the CFE Treaty in Europe, combined with the political consequences of the Gulf War, will lead to increasing competition among arms suppliers to the Middle East. At the same time, just a few countries dominate the arms market there: the five permanent members of the U.N. Security Council—the United States, Russia, France, the United Kingdom and China—account for 86 percent of the weapons sold to the Middle East. Further, they are the only ones capable of selling major weapon systems and sustaining their supplies for an extended period.[79]

Proliferation of Ballistic Missiles and High-Technology Weapons

The Middle East is expected to face major security problems in the 1990s because of the global proliferation of ballistic missiles, long-range manned and unmanned vehicles, and precision-guided weapons. Long-range surface-to-surface missiles hold special technical glamour following their use in the Iran-Iraq "war of cities" and the Gulf War. Equally important are the proliferation of long-range offensive aircraft, which often exceed reasonable defense requirements, and the acquisition of precision-guided weapons.

The basic character of conventional arms proliferation is changing radically in the 1990s. Precision-guided weapons are providing a growing number of countries with the ability to strike at critical economic targets such as key factories, refineries, desalinization plants and power facilities. Because the acquisition of such weapons produces imbalances in military capabilities between the haves and have-nots, the latter have an incentive to look to weapons of mass destruction to rectify the imbalance. That is, the acquisition of precision-guided weapons spurs the proliferation of weapons of mass destruction.

A number of Middle Eastern states, including Egypt, Israel, Iran, Libya, Saudi Arabia, Syria and Yemen, already possess ballistic missiles capable of carrying weapons of mass destruction. Ballistic missiles are the most visible symbol of a proliferation that thrives on fear and competition. There is also an economic incentive to acquiring ballistic missiles in that they are often less expensive than acquiring, sustaining and modernizing large conventional forces.

Israel's approach is to acquire and combine all the key elements of an integrated deep strike capability. Its strategic missile force consists of the Jericho-1 missile, which can throw a 750 kilogram warhead over ranges of

450 km. An improved version, Jericho-2, should have a range of 750 km, and an even more improved model, probably similar to the Shavit space launch vehicle, will reach a range of 1,450 km.[80] It can safely be argued that these missiles, whether equipped with conventional or unconventional pay-loads, offer a real strategic counter-force capability. This has been material-ly aided by the growing Israeli-U.S. technological connection in the fields of advanced electronics and precision navigation.

Moreover, Israel has a fleet of the most advanced strike aircraft, equipped with refuelling capabilities that can double and triple their strike range. Israel also has several models of short-, medium- and long-range UAVs for battle support and strike purposes. Stand-off precision-guided air-to-ground, air-to-air and anti-ship weapons are key elements in the Israeli inventory.[81]

There is a growing fear that the proliferation of such weapons will alter the balance of power between regional and superpower states. William H. Webster, a former director of the U.S. Central Intelligence Agency, pre-dicted in 1989 that by the year 2000 at least fifteen nations would be producing their own ballistic missiles. Non-proliferation experts at the Pentagon consider the proliferation of missiles as complicating U.S. security relationships with the Middle East and other Third World regions. To address the concern, the United States is supporting two policies: first, the development of anti-tactical ballistic missile (ATBM) systems for re-gional use that can intercept and destroy ballistic missiles in flight; and, second, enhancement of the Missile Technology Control Regime (MTCR) by encouraging other countries to join it, with special efforts to get China, Russia and other exporters of ballistic missiles to the Middle East to participate.[82]

The application of this policy in the Middle East for the most part has favored Israel. Through strategic cooperation agreements with the United States, participation in the Strategic Defense Initiative program and joint venture projects, Israel was able to acquire the restricted technologies most needed for its strategic projects. Co-development of the advanced Arrow ATBM, in addition to deployment of the Patriot batteries in Israel, will likely jeopardize the limited Arab deterrence capabilities in the foreseeable future. Egypt, Syria and the rest of the Arab world in general are facing a window of vulnerability that they will have trouble accepting over the long term. Despite all possible guarantees by the United States or the internation-al community, suspicion feeds suspicion. Even in peacetime, the perception that a rival nation has the ability rapidly to project nuclear and advanced conventional power on a large scale can be a compelling incentive to undertake a counter-effort to acquire similar weapons.[83]

The most notable success of the MTCR regime has been to stop development of the Condor II project by Argentina, Egypt and Iraq.[84] The destruction of Iraq's ballistic missile launchers and missile production facilities during the Gulf War and later by the U.N. inspection teams have been in some ways an even more direct and intrusive application of U.S. policy. In the meantime, the United States continues to pressure China and North Korea to stop delivering ballistic missiles to Iran and Syria.[85]

The West needs to make its policy regarding the proliferation of advanced military technologies more balanced and to stop distinguishing between "good" and "bad" proliferation, depending on the nation concerned. It should formulate a convincing and reliable policy that treats proliferation as a long-term problem involving cumulative risk, and not as a case-by-case matter. A comprehensive structure of policy and legislation that addresses nuclear, biological and chemical weapons, missiles and other relevant delivery systems is needed. Until the major powers, particularly the United States, are willing to be part of a process in which they will often take positions on individual issues and cases at odds with the desires of friends and allies, their suggested policies and actions will not be effective.[86]

The Prospects for Conventional Arms Control

The situation in the Middle East is intolerable. Motivated by the Arab-Israeli and other conflicts in the Gulf region, countries in the Middle East are embarking on a new phase of the arms race. This trend runs counter to the mood of conflict resolution, particularly in the Arab-Israeli arena. The United States and other Western countries, because of the size and importance of the military industry in their respective economies, are not hindered in selling arms by the search for arms control arrangements in the Middle East. Ex-Soviet bloc countries and China, because they are starving for hard cash, are looking for markets for their military equipment. Both sides have found that market in the Middle East. For differing reasons, the suppliers and recipients of arms in the Middle East have raised the level of high-technology weapons in the region, particularly after the Gulf War.

What are the prospects for arms control in the Middle East? In looking at future possibilities, it is important to review past experience with this issue, as well as the current multilateral negotiations.

The Lessons of the Past

Throughout the Arab-Israeli conflict, Egypt, Israel, Jordan and Syria have accepted arms control measures. The ceasefire and armistice agreements of 1948 and 1949 contained provisions for neutral and demilitarized zones on Israel's borders with Egypt and Syria and within Jerusalem. The U.N. agreements that ended the Suez War in 1956 included provisions for a U.N. peacekeeping force on both sides of the Egyptian-Israeli border. While Israel declined to accept the force on its side of the border, Egypt accepted and thereby endorsed unilateral control of the movement of its forces in certain parts of Sinai. In fact, between 1957 and 1967 Egypt unilaterally declined to deploy the main elements of its armed forces in Sinai, a clear signal it intended no aggression to Israel.[87]

While these measures were intended to de-escalate existing conflicts, after the 1973 war arms control measures were used to enhance the unfolding peace process. In the first disengagement agreement in 1974, Egypt not only accepted limitations on its armed forces in certain areas east of the Suez Canal, but it also accepted the presence of U.N. forces and curtailment of its air defenses to the west of the Suez Canal. In the second disengagement agreement in 1975, Egypt accepted, in addition to the demilitarized zones, certain confidence-building measures (CBMs)[88] such as early warning systems, electronic censoring and notification of military movements to the U.N. force in Sinai.[89]

The most ambitious of these arrangements came in 1979 in the Egyptian-Israeli peace treaty. It divided Sinai into three zones. Zone A was not to have more than 1 mechanized infantry division, with up to a total of 230 tanks and 22,000 personnel. Zone B was not to have more than 4 battalions of border units, equipped only with light weapons, and up to a total of 4,000 personnel. Zone C would have only Egyptian civil police. On the Israeli side of the border, Zone D was not to have more than four infantry battalions with up to a total of 4,000 personnel and up to 180 APCs and their military installations and field fortifications.[90] These zones were to be supervised by early warning systems and multi-national forces.

The provisions of the peace treaty have dramatically reduced the possibility of a surprise attack for both Egypt and Israel. They have led to a process of military cooperation to monitor implementation of the agreement. Furthermore, they established the precedent of an asymmetrical balance of forces and territory in such a way as to address Israeli insecurities. Finally, they reduced to a minimum the possibility of accidental clashes on land or sea or in the air.

Egypt and the other Arab states perceive Israel's nuclear capability not as a deterrent weapon, but as one for compellence. It is a destabilizing factor in the Middle East in that it is a constant incentive for other countries to acquire weapons of mass destruction. In 1974, at the 29th session of the United Nations, Egypt and Iran introduced a resolution to establish a nuclear weapons-free zone (NWFZ) in the Middle East. A majority of the 138 members the U.N. General Assembly adopted Resolution 3263, with only Israel and Myanmar abstaining. At subsequent General Assembly meetings the resolution was reconfirmed, and starting in 1980 there was no opposition to it. Several Arab countries—Bahrain, Jordan, Kuwait, Mauritania, Sudan and Tunisia—participated in introducing these resolutions beginning in 1975.[91]

Egypt has gone as far as taking unilateral steps toward arms control. It saw the Egyptian-Israeli peace treaty as a way to curtail, if not eliminate, the Israeli nuclear arsenal. Although this hope did not materialize, Egypt ratified the Nuclear Non-Proliferation Treaty (NPT) in 1982, and in 1986 froze all its nuclear programs.[92] Second, through different international fora, Egypt pursued the idea of establishing a NWFZ in the Middle East. Third, during the Paris Conference on Chemical Weapons in January 1989, Egypt supported multilateral efforts to impose a total ban on chemical weapons (CWs) and asked that the Chemical Weapons Convention include effective security guarantees for its members not only against the use or threat of use of CWs, but also against the use or threat of use of any weapons of mass destruction. The countries that possess nuclear weapons refused this link.[93] Egypt based its position on a plan proposed by President Hosni Mubarak that would make the Middle East free of all weapons of mass destruction.

Attempts throughout the Arab-Israeli conflict to control arms exports to the conflicting parties have proved unsuccessful. The 1950 Tripartite Declaration of France, the United Kingdom and the United States, which were then the major Middle East arms suppliers, opposed any forceful revision of the 1947 armistice lines and committed the three countries to ration the shipment of offensive arms to the region. The Kennedy and Carter Administrations made other attempts.[94] However, because of the lack of progress in settling the conflict, the Arab side saw these arms control attempts as intended to make the Arabs accept an unacceptable status quo. In addition, the Cold War climate provided the rationale for the parties, along with other powers in the area, to participate in a costly arms race in the Middle East.

The Multilateral Negotiations of the 1990s

In light of developments in the early 1990s, arms control and regional security are, not surprisingly, of paramount importance in the Arab-Israeli peace process. When the multilateral peace negotiations got underway after the Madrid Conference, a committee was established to address these issues. In addition, regional security and arms issues have been a focus of the various bilateral talks involving Israel and Jordan, Lebanon, the Palestinians and Syria. However, the scope and aims of these bilateral negotiations are limited.

The principal venue for discussions on regional security and arms control in the Middle East is the Committee on Arms Control and Regional Security (ACRS). It has only been able to address these issues in a general way, however, because of sharp differences among the main parties concerned, particularly between Egypt and Israel. These differences fall into two main areas:

(1) One is whether to approach the peace process from a political or security standpoint. This issue emerged in the discussion of the relationship between the bilateral tracks and the multilateral negotiations, particularly in the security and arms control areas. The Arab side called for a strong linkage between the bilateral talks and the multilateral negotiations such that there would be balanced progress across the two. The underlying premise was that peace would never prevail in the region unless the problems of security and arms control between the Arabs and Israelis were resolved, particularly as regards nuclear arms. The Israelis, in contrast, insisted on postponing the security and arms control arrangements to a later stage, following conclusion of a comprehensive settlement of all the political issues between the two sides, or even until there was an apparent guarantee the settlement would last. Israel's position thus delayed negotiations on security and arms control issues until a political settlement was reached.

(2) The second area involves priorities to be established within the arms control agenda of the multilateral negotiations. From the beginning the Arabs insisted on discussing the arms issues between the Arabs and Israelis up front, including Israel's nuclear arms. They believed that the Israeli arsenal was not justified in a country that is preparing for peaceful relations. In contrast, the Israelis only called for small steps, such as CBMs.

After seven rounds of discussion, the two sides began to shift their positions, albeit uncertainly. The impetus was Israel's beginning to accept the idea of opening its military arsenal to inspection and negotiations, although not until there was peace with the Arabs. For their part, the Arabs paid more attention to confidence-building at the regional level. In that framework the parties discussed many confidence-building ideas, such as establishing centers for conflict prevention, networks for regional communication, information banks for issues relating to regional control and regional security centers.

The two sides agreed to establish centers in certain countries. There was to be a Regional Center for Security in the Middle East (in Oman) and a Regional Net for Communication (in Egypt), among others. A number of such centers were to be set up temporarily outside the Middle East, to be moved into the region after a peace settlement. To date, however, little real work has been done on implementing the centers.

Thus, after several years of negotiations on arms control and regional security between the Arabs and Israelis, there is little to show. The parties have not even been able to resolve issues relating to the organization and arrangements of the negotiations. Moreover, the negotiations have not established any common ground for addressing arms control, particularly nuclear arms. As a result, the Arab countries, particularly Egypt, have been looking for other fora and frameworks for discussing Israel's nuclear capability, such as the Geneva Disarmament Conference. The evidence proves, to a large extent, that multilateral negotiations are not a sufficient mechanism for discussing and resolving the disagreement over arms control between the Arabs and Israelis.

Arms Control and Resolution of the Arab-Israeli Conflict

The record shows that there are important precedents for any future Arab-Israeli arms control, and the prospects are not as bleak as might be thought. In fact, the experience of the Egyptian-Israeli arms control measures is expected to help in bilateral negotiations between Israel on the one hand and Jordan, Lebanon, Syria and the Palestinians on the other. At the same time, the lessons of this experience fall short of providing guidance on how to address the major dynamics of the arms race in the Middle East more forcefully, particularly Israeli fears of the Arab quantitative advantage and Arab fears of the Israeli qualitative advantage.

Obviously, the Egyptian and Arab call for a NWFZ in the Middle East is an attempt to reduce Israel's nuclear strategic advantage. It has to be mentioned here that the Arab world does not perceive Israeli nuclear capability as a deterrent but rather as a compellence weapon to make the

Arabs accept the unacceptable status quo. However, the Egyptian proposal also responds to a concern that the Israeli nuclear arsenal is igniting another arms race in other weapons of mass destruction and therefore is destabilizing not only the Arab-Israeli arena but also the rest of the Middle East. The proliferation of advanced missiles, which poses a considerable threat to most countries in the area, compounds the threat posed by the acquisition of chemical and mass destruction weapons.

The last two Gulf crises demonstrated that the problems in the Middle East are much larger than the Arab-Israeli conflict. Nevertheless, this conflict remains a major one as far as arms control is concerned. Fortunately, the Gulf War has created a general consensus in the United States, Europe and the major powers that peace and security in the Middle East cannot prevail without limiting the acquisition of arms, particularly weapons of mass destruction. U.N. Security Council Resolution 687, which calls for a permanent ceasefire in the Gulf, demands the elimination of all Iraqi chemical and biological weapons, the dismantling of its nuclear facilities and limitations on the range of its ballistic missiles. The resolution is important not just because it sets precedents in the Middle East context, but also because it states explicitly that these measures are taken as steps toward the creation of a zone in the region free of nuclear weapons of mass destruction and their delivery systems (Article 14).

The success of this effort depends on having the concerned parties developing ideas and proposals that will enhance the prospects for peace and security in the Middle East.[95] In October 1990 a group of experts presented the U.N. Secretary-General with a study on effective and verifiable measures that would facilitate the establishment of a NWFZ in the Middle East. The study suggested practical measures to cap Israeli nuclear capabilities by putting the Dimona reactor under the safeguards of the International Atomic Energy Agency. This step would keep the Israeli nuclear "deterrent" intact until further political steps are taken that could lead Israel toward accepting a NWFZ. The study introduced a host of other ideas applicable to the Middle East, drawn from the European experience with arms control and confidence-building. The study is interesting because it is not confined to the nuclear field only, but also puts forward ideas on limiting other mass destruction and conventional weapons, including missiles.

What the study lacks is the necessary linkage between the establishment of the NWFZ in the Middle East and the overall settlement of conflicts, particularly the Arab-Israeli one. Also lacking is a timeframe for bringing

Israel into a NWFZ in the Middle East. Although certain asymmetries might be acceptable to facilitate agreements, symmetrical and reciprocal arrangements should be the target. Thus, if Israel were to keep its nuclear weapons while safeguarding the Dimona facility, these weapons should be phased out over time. They should be reduced in number as a part of the CBMs. Others could be eliminated in the context of new international guarantees or "traded" for peace treaties with Arab countries. The remaining Israeli nuclear weapons should be eliminated once full normalization of relations and different types of economic and functional cooperation are initiated. The same process could be applied to chemical weapons for both sides of the conflict.

The idea here has two aims. The first is to link arms control measures with a political timetable for an overall settlement. The second is to eliminate the most devastating weapons from the area. These steps cannot be achieved without transparency of information about the weapons of mass destruction in the inventories of both sides. The arms control talks in Europe could not have resulted in anything without prior agreement on which arms the talks were intended to control. Transparency should be the first step in multilateral arms control negotiations in the Middle East.

Transparency is also important in negotiations on conventional weapons. Both sides should provide information not only about the inventories of weapons they hold, but also about weapons under development. A moratorium on the acquisition and development of high-technology weapons should be implemented during the negotiating process. An alternative is to establish a moratorium on the deployment of these weapons. This step would be particularly important in the case of long-range (more than 150 km) ballistic missiles and ATBMs such as the Israeli Arrow. The arms-exporting countries should work out a ban on exports of cruise and long-range ballistic missiles. Israel should halt its plans to expand its sea projection capabilities, particularly sea-launched long-range conventional and nuclear missiles. The acquisition of advanced submarines should be stopped during the Arab-Israeli negotiations. These steps would prevent the triggering of a new naval race that would make arms control measures difficult in the future. CBMs such as notification of naval movements and cooperative sea operations against drug smuggling or terrorist actions by regional powers could enhance both the possibilities of arms control and the mutual trust necessary for peace in the Middle East. Other CBMs such as arrangements for dealing with incidents at sea, notification of exercises and agreements to prevent dangerous military activities can contribute to achievement of this goal. These measures would be aimed at capping the existing level of arms held by both sides. It would also be important to reduce the levels of certain categories of weapons such as tanks and artillery. To address Israeli fears of the Arab quantitative advantage, the reductions should be asymmetrical. The Arab side could restructure and redeploy its armies in such a way as to reduce Israeli apprehension. A shift from standing armies to greater reliance on reserves should be considered. Such an approach is one way to address the quantity versus quality issue in the Arab-Israeli arms race.

A major obstacle to application of these ideas is Iran, which is not part of the multilateral arms control negotiations. Iran is important to the military balance in the Gulf and to the entire Middle East. It also is a revolutionary power that is not satisfied with the status quo. At the same time, it is still exhausted from its long war with Iraq, and it faces enormous internal domestic difficulties. At this point, the balance is not in Iran's favor. Consequently, capping the Iranian arms buildup cannot be accomplished without the cooperation of the supplier states, particularly the P-5, which need to limit their sales to the region.

Success in restraining arms transfers to the Middle East will not be easy, given the economic difficulties in the West, the former Soviet Union countries and China. They all will continue looking to exports of arms to decrease their deficits, create jobs and generate hard cash. Restructuring their economies and converting military into civilian industries will take a long time. In the interim, the Middle East will remain the largest possible arms market.

At the same time, conditions could not be better for actions by the major suppliers to restrain their transfers of weapons. The Cold War has ended, a more or less stable balance exists in the Middle East, the United States is playing a dominant role in the area, and recipient countries are having to address serious economic difficulties. All these factors are conducive to new attempts to restrain the supply of arms to the region, particularly those that may lead to a new wave of the arms race or to destabilization of the existing balance.

There are also a number of precedents for action by the supplier states. The MTCR, established in 1987 by a U.S.-led coalition of major industrial powers, marked the first step in restricting exports of missiles, their components and the technology that can be used in their manufacture. Another coalition of major industrial countries formed the Australia Group to restrict trade in CWs and precursor chemicals. More needs to be done. These efforts have had limited impact on the Middle East because they did not include all the major suppliers and did not take a balanced approach to the region. The only achievements of the MTCR regime have been to stop the Condor II program and to pressure China to reduce its sales of surface-to-surface missiles to Syria and Saudi Arabia.[96] None of these initiatives has addressed Israel's nuclear or space program, although both have significant implications for the arms race in the Middle East.

Following the Gulf War, on May 29, 1991, President Bush announced an initiative to have the P-5, the largest arms exporters, discuss guidelines for restraining transfers of destabilizing conventional arms and weapons of mass destruction. The initiative also called for the eventual elimination of surface-to-surface missiles. In a meeting in London in October 1991, the P-5 agreed to exercise restraints in their arms exports and to avoid transfers that would aggravate conflicts and increase tensions in the Middle East.[97]

This initiative has been replaced by the Wassenaar Arrangement.[98] With respect to the Middle East, the Wassenaar countries might seek such measures as a ban on the transfer of weapons with new technologies that are not currently in the region. Another possibility might be for the major suppliers to agree to make public every arms sale to the Middle East. A third, and probably the most difficult one, would be to tax all arms sales to the Middle East. The revenues generated would be used for economic cooperation among Middle East countries, particularly those involved in the Arab-Israeli conflict.

Experience shows, however, that bright ideas are not enough to solve conflicts. What creates peace is a common political understanding. The principal suppliers should establish not only balanced and fair guidelines for controlling arms transfers, but also political measures to reduce the tensions that continue to spark the arms race. Forceful backing of the peace process should go hand in hand with limitations on arms transfers to the area.

Notes

1. U.S. Congressional Budget Office (CBO), *Limiting Conventional Arms Exports to the Middle East* (Washington, D.C., 1992), 2–3.
2. Richard F. Grimmett, *Conventional Arms Transfers to the Third World 1984–1991,* Congressional Research Service (CRS), U.S. Library of Congress (Washington, D.C., 1992), 22, 38.
3. Abdel Monem Said Aly, "The Superpowers and Regional Security in the Middle East," *Regional Security in the Third World* (London, 1986), 208.

4. Shahram Chubin, "Iran and the Lessons of the War with Iran: Implications for Future Defence Policies," in Shelley A. Stahl and Geoffrey Kemp (eds.), *Arms Control and Weapons Proliferation in the Middle East and South Asia* (New York, 1992), 95–112.

5. Andrew J. Pierre, "Enhancing Conventional Defense: A Question of Priorities," in Andrew J. Pierre (ed.), *The Conventional Defense in Europe* (New York, 1986), 9–39.

6. "The Eight Lessons of Success," interview with General Forray, *Military Technology,* XV (1991), 26–27.

7. "US Air Force Performance in Desert Storm," *Military Technology,* XV (1991), 146–157.

8. Richard Cheney, "U.S. Defense Strategy for an Era of Uncertainty," *International Defense Review,* XXIV (1991), 7–9.

9. Rosemarie Hollis, "Security in the Gulf: No Panaceas," *Military Technology,* XV (1991), 52.

10. Henry L. Stimson Center, "Report of the Study Group on Multilateral Arms Transfer Guidelines for the Middle East" (Washington, D.C., 1992), 15–16.

11. A. S. Aly, "Quality vs. Quantity: The Arab Perspective of the Arms Race in the Middle East," 61–74, in Shelley A. Stahl and Geoffrey Kemp (eds.), *Arms Control and Weapons Proliferation in the Middle East and South Asia* (New York, 1986), 61–74.

12. As quoted in Osama A. Al-Kholi, "Technological Education and Its Contribution to the Arab-Israeli Conflict," *Al-Mustaqbal Al-Arabi* (March 1986), 98 (in Arabic).

13. Jerusalem Domestic Service, September 19, 1988 (in Hebrew), reproduced in U.S. Department of State, Foreign Broadcast Information Service (FBIS), FBIS-NES (20 September 1988), 22.

14. Osama Al-Ghazali Harb, *The Future of the Arab-Israeli Conflict* (Beirut, 1987), 19 (in Arabic).

15. See various issues of *The Arab Strategic Report* (Cairo) (in Arabic).

16. See various issues of *The Middle East Military Balance,* Jaffee Center for Strategic Studies (Tel-Aviv, various years).

17. David Segal, "The Iran-Iraq War: A Military Analysis," *Foreign Affairs,* LXVII (1988), 946.

18. Segal, "The Iran-Iraq War," 954.

19. See the interviews with Prime Minister Yitzhak Shamir and Director of the Israeli Space Agency, Yuval Neeman, by the Jerusalem Domestic Service (in Hebrew), as transcribed in FBIS-NES (19 September 1988), 22–23; Helena Cobban, "Israel's Nuclear Game: The US Stake," *World Policy Journal,* V (1988), 419–427; Moshe Arens, "The Lavi and the Future of High-Tech in Israel," *IDF Journal,* IV (1987), 10. See also Ron Bouskela and Tamar Karavan, "High Tech Supplement: Israeli Company Profiles," *IDF Journal,* IV (1987), 63–68.

20. Stockholm International Peace Research Institute (SIPRI), *World Armaments and Disarmament: SIPRI Yearbook 1985* (London, 1985), 240.

21. Aly, "Quality vs. Quantity," 66–69.

22. M. E. Said, "The Political Economy of Defence in the Arab World," a paper presented at the First Arab Strategic Conference (Amman, 1987), 9 (in Arabic).

23. Grimmett, *Conventional Arms Transfers,* 59.

24. Abdel Monem Said Aly, "The Arms Trade and Regional Conflict: Supplier's Policies and Behavior and Their Consequences," a paper presented at the 39th Pugwash Conference on Science and World Affairs (Cambridge, Mass., 23–28 July 1989), 6–7.

25. Aaron Karp, "The Trade in Conventional Weapons," *SIPRI Yearbook 1988: World Armaments and Disarmament* (Oxford, 1988), 176.

26. M. E. Said, "Star Wars Between Israel and America," *Qadava Fikria* (October 1988), 132–133, and M. Kadry Said, "The Future of Conventional Deterrence in the Arab-Israeli Confrontation," *Al-Siassa Al-Dawlva* (April 1988), 255 (in Arabic).

27. For more information about Arab missiles and chemical capabilities see Lewis Dunn, "Chemical Weapons Arms Control," *Survival*, XXXI (1989), 209–224; "Chemical Addiction," *Defence and Foreign Affairs* (April 1989).

28. For the story of the Condor III missile, see *Spiegel* (1 May 1989), 153.

29. Gerald Green, "Approaching 2000—Technology and Defence," *National Defence*, LXX (1985), 17.

30. "Position for the Future," an interview with Major General Moshe Peled, *Military Technology*, XV (1991), 50.

31. *Statistical Abstract of Israel, 1985* (Jerusalem, 1985), 659; Wolfgang Flume, "Focus on Israel: Israeli Defence Industry—Peacetime Link in the Economic Chain," *Military Technology*, XI (1987), 93–96.

32. Steven M. Shaker and Howard B. Shaker, "Israeli Weapons Technology and the US Military," *National Defense*, LXX (1986), 38.

33. Claudia Wright, *Israel's Special Relationship with the United States* (Washington, D.C., 1986), 27–31.

34. See Arens, "The Lavi and the Future," and Flume, "Focus on Israel: Israel Defense Industry."

35. "The Lessons of Combat: Israel's R&D Center," *Military Technology*, XV (1991), 51–53.

36. Tim Ripley, "Israel's Pilots Took to the Future," *International Defence Review* (March 1922), 261.

37. Marvin Leibstone, "US-Middle East Defence Cooperation," *Military Technology*, XV (1991), 67.

38. Leibstone, "US-Middle East Defence Cooperation."

39. "The Lessons of Combat," 53.

40. Philip L. Bolte, "The Shrinking Tank Market," *Military Technology*, XV (1991), 36.

41. Israel Leshem, "Current Israel Naval Program: A Status Report," *Military Technology*, XIV (1990), 79–98.

42. Talaat Musalaam, "The Strategic Causes for the Israeli Participation in Star Wars," *Drasat* (September 1987), 14–15 (in Arabic); *The Jerusalem Post* (20 September 1988).

43. On the subject see: Paul Jabber, *Israel and Nuclear Weapons* (London, 1972); Paul Jabber, *A Nuclear Middle East: Infrastructure. Likely Posture and Prospects for Strategic Stability* (Los Angeles, 1977), particularly 24–27; Yair Evron, "Israel and the Atom: The Uses and Misuses of Ambiguity, 1957–1967," *Orbis*, XVII (1974), 1326–1343; Shlomo Aronson, *Israel Nuclear Option* (Los Angeles, 1977); Leonard S. Spector, *Going Nuclear* (Cambridge, Mass., 1987); Louis R. Beres (ed.), *Security of Armageddon: Israel's Nuclear Strategy* (Lex-

ington, Mass., 1986); Mark Gaggney, "Prisoners of Fear: A Retrospective of the Israeli Nuclear Program," *American Arab Affairs* (Fall 1987), 75–96; and "Revealed: The Secrets of Israel's Nuclear Arsenal," *Sunday Times* (London) (5 October 1986). See also Yazid Sayigh's interview with Frank Bambay in *Al-Mustaobal Al-Arabi* (June 1988), 122–125 (in Arabic); Rodney W. Jones, *Small Nuclear Forces* (Washington, D.C., 1984), 24–27; Rodney W. Jones, *The Proliferation of Small Nuclear Forces,* The Washington Papers (Washington, D.C., 1984), 24–27.

44. Rodney W. Jones, *Proliferation of Small Nuclear Forces,* The Washington Papers (Washington, D.C., 1984), 24.

45. Adnan Mustafa, *The Arab Nuclear Energy* (Beirut, 1983), 63–64 (in Arabic).

46. See *The Arab Strategic Report,* 1985, 1986, 1987, and 1988.

47. See *The Arab Strategic Report,* 1985, 1987, and 1988.

48. "World Defense Almanac 1991–1992," *Military Technology,* XVI (1992), 160.

49. Leibstone, "US-Middle East Defence Cooperation," 68.

50. Ibid., 67.

51. "World Defense Almanac 1991–1992," *Military Technology,* XVI (1992), 183; David Isby, "The International Market for Combat Aircraft," *Military Technology,* XVI (1992), 10–24.

52. Barbara Opal, "Saudi's Explore Additional Buys of AWACS Planes," *Defense News* (7–13 September 1992); Isby, "The International Market for Combat Aircraft."

53. "World Defense Almanac 1991–1992," *Military Technology,* XVI (1992), 184.

54. David Eshel, "What If the Peace Talks Fail?" *Military Technology,* XVI (1992), 239.

55. Grimmett, *Conventional Arms Transfers,* 57.

56. Richard F. Grimmett, "Arms Trade with the Third World: General Trends 1983–1990," *International Defense Review,* XXV (1992), 59.

57. David Eshel, "Arms Race in the Gulf," *Military Technology,* XVI (1992), 67.

58. Eshel, "Arms Race in the Gulf," 63.

59. *International Herald Tribune* (17 November and 1 December 1992).

60. Eshel, "Arms Race in the Gulf," 64.

61. "World Defense Almanac 1991–1992," 163.

62. "Italy May Sell Iran Boeing CH-47 Copters," *Defense News* (24–30 August 1992).

63. Eshel, "Arms Race in the Gulf," 65.

64. Ibid., 67.

65. Ibid., 67.

66. "Abrams Victory in Desert Tank Duel," *Jane's Defence Weekly* (17 October 1992), 5.

67. "Kuwait May Buy French Naval Gear," *Defense News* (24–30 August 1992).

68. Eshel, "Arms Race in the Gulf," 67.

69. Ibid., 67.

70. Bob Hutchinson, "Leaner Years for Arms Industries," *International Defense Review,* XXV (1992), 12.

71. CBO, *Limiting Conventional Arms Exports,* 15.

72. Grimmett, "Arms Trade with the Third World," 55–60.

73. Joel L. Johnson, "In Search of a Sensible U.S. Arms Transfer Policy," *Military Technology,* XV (1991), 12.

74. Johnson, "In Search of a Sensible," 14.

75. Ibid., 17.

76. Bolte, "The Shrinking Tank Market," 45.

77. Philip Finnegan, "Russia Extends Mid East Arms Sale Hunt," *Defence News* (15–21 June 1992), 1.

78. Grimmett, "Arms Trade with the Third World," 57.

79. CBO, *Limiting Conventional Arms Exports,* 10–11.

80. Tamir Eshel, "Iraq Thrust Towards Strategic Weapons: A Mini Superpower in Formation," *Military Technology,* XIV (1990), 99.

81. Liebstone, "U.S.-Middle East Defense Cooperation," 66–67.

82. Barbara Starr, "Ballistic Missile Proliferation: A Basis for Control," *International Defense Review,* XXIII (1990), 265.

83. Seth Carus, "Missiles in the Middle East: A New Threat to Stability," *Policy Focus* (Washington Institute for Near East Policy) (June 1988), 9–10.

84. M. Kadry Said, "The Future of Conventional Deterrence in the Arab-Israeli Confrontation," *Al-Siassa Al-Dawlva* (April 1988), 249–256 (in Arabic).

85. Starr, "Ballistic Missile Proliferation," 265.

86. John S. McCain, III, "Proliferation in the 1990s: Implications for U.S. Policy and Force Planning," *Military Technology,* XIV (1990), 266.

87. Mustafa Alwi, "Disarmament and the Settlement of the Arab Israel Conflict," *Al-Siassa Al-Dawlva* (July 1978), 63–65 (in Arabic).

88. Alwi, "Disarmament and the Settlement," 65; M. A. Al-Gamassi, *The Al-Gamassi Memoire. The October 1973 War* (Paris, 1990), 480–483 (in Arabic).

89. Alwi, "Disarmament and the Settlement," 65.

90. "Text of the Egyptian-Israeli Peace Treaty," 392–409, in Kamal Hassan Ali, *Warriors and Negotiators* (Cairo, 1986) (in Arabic).

91. Wahid Abdel Majed, "The Declaration of Establishing Nuclear Weapon Free Zone in the Middle East," *Al-Siassa Al-Dawlva* (July 1987), 48 (in Arabic).

92. Nadia Mustafa, "Egyptian Politics and the Nuclear Option: Vision, Behavior, and Constraints," *Al-Siassa Al-Dawlva* (July 1989), 24–59 (in Arabic).

93. Esmat A. Ezz, "The Chemical Weapons Convention: Particular Concerns of Developing Countries," *UNDIR Newsletter* (March 1989), 7.

94. Stimson Center, "Report of the Study Group," 8–15.

95. Mustafa, "Egyptian Politics and the Nuclear Option."

96. CBO, *Limiting Conventional Arms Exports,* 27.

97. Stimson Center, "Report of the Study Group," 38.

98. The Wassenaar Arrangement is discussed in chapter 15, "Toward an International Regime for Conventional Arms Sales," by Andrew J. Pierre, in this volume.

TWELVE

Asia-Pacific

Andrew Mack

IN THE AFTERMATH of the Cold War, defense budgets in most of the Third World and on both sides of what was the Iron Curtain are declining. By contrast, defense spending throughout most of the Asia-Pacific region is increasing—although less rapidly than growth in growth domestic product (GDP).[1] As an arms-importing region, Asia-Pacific still ranks below the Middle East, but far above the rest of the developing world.

East Asia's share of global arms imports and related licensed production of major conventional weapons rose from 12.4 percent in 1984 to 21.1 percent in 1993.[2] While the Middle East is expected to be the world's largest arms recipient until at least 2000, accounting for about 30 percent of all international arms transfers, East Asia will remain the second biggest buyer, with Japan, the Republic of Korea (South Korea) and Taiwan the three largest arms importers.

However, while East Asia's *share* of global arms imports rose, the *absolute* level of arms acquisitions and licensed production fell between 1988 and 1993. According to the Stockholm International Peace Research Institute (SIPRI) (which measures the value of armed *transfers* each year, not new *agreements*), the value of arms transferred and produced under license in 1988 was just under US$7 billion and in 1993 just over $4.5 billion.[3] The major cause of the decline has been the end of the Cold War and the consequent improvement in the regional security environment.

The ongoing import-led military buildup in East Asia is transforming regional military capabilities in ways that could be destabilizing should political relationships deteriorate seriously in the future. States throughout the region are increasing their power projection capabilities with the acquisition of new combat, surveillance and early-warning radar (AEW) aircraft, sophisticated missile systems, air-to-air refuelling capabilities, naval surface combatants and submarines.

The types of weapons systems being acquired reflect the fact that the security focus of most regional states is increasingly outward-looking—the domestic insurgencies that characterized much of the region in the 1960s, 1970s and, in some cases the 1980s, have either disappeared or are waning.[4]

Weapons acquisitions also reflect the facts of strategic geography. In Asia-Pacific, unlike Europe, the Middle East or South Asia, the key states are either islands or are located on peninsulas or archipelagoes, and security planning necessarily focuses on the maritime realm. In addition, the introduction of 200-mile Exclusive Economic Zones (EEZs) created new missions for regional maritime forces. These new missions determine the nature of many of the weapons platforms and surveillance systems that are being purchased. Only on the Korean peninsula is there a confrontation across a land frontier comparable with that on what used to be the central front in Europe.[5]

The Regional Arms Trade

Nearly three-quarters of all arms transfers to Asia-Pacific flow to the heavily armed states of Northeast Asia. Most states in the region still rely on imports to modernize their defense capabilities, although, as the industrial base of the more advanced regional states becomes more sophisticated, the trend toward domestic arms production, mostly under some form of licensing agreement, may increase. China produces a full range of conventional weapons, albeit of outmoded design. Japan, too, produces most of its own weapons—although the more sophisticated systems are produced under license. The Democratic People's Republic of Korea (North Korea) and South Korea produce a wide range of weapons systems, as does Taiwan. The only other countries that have significant levels of domestic arms production are Indonesia and Singapore.

Licensed production increased rapidly in the 1960s, 1970s and 1980s. The number of major conventional weapons systems produced under license in Australia, Indonesia, Singapore and Taiwan, for example, went from one in 1967 to twenty-four in 1988.[6] The greater the share of a nation's weapons systems that are produced domestically, the less effective regulation of arms transfers will become as a means of controlling military arsenals. Recent evidence suggests, however, that indigenous weapons manufacturing by so-called "third tier"-producing countries—which in-

clude Indonesia, North and South Korea, Singapore and Taiwan—has "sta-
bilized and even declined."[7]

Supply and Demand

Although the absolute level of arms imports into the region has fallen,
East Asia remains the second most important arms market in the world after
the Middle East. Both supply-side and demand-side pressures are likely to
ensure that it will remain so. Supply- and demand-side pressures determine
the volume and rate of arms transfers to the region. On the *demand* side of
the equation these pressures include:

—The particular security concerns of regional states.

—The need to modernize outdated equipment. Much of the region's
military equipment is obsolete or obsolescent. In 1992, for example, some
84 percent of the region's combat aircraft were based on pre-1966 designs,
while South Korea's destroyers and most of Taiwan's major surface combatants were World War II vintage.[8] State-of-the-art weapons systems may
sometimes be sought for reasons of prestige rather than national security.
Especially in states where the armed forces play an important role in
politics—Indonesia, Myanmar and Thailand are obvious examples—
military prestige "wish-lists" can have a significant impact on defense
procurements, even when they have little relevance to national security
needs.

—The continued rapid rates of economic growth tend to drag arms
expenditures upwards. Absolute increases in defense expenditures, however, do not necessarily mean that the *share* of defense expenditure in
national income will increase. In much of the region, defense expenditure
has risen absolutely while falling as a percentage of GDP. Recent research
indicates the single best indicator for increased defense expenditure is not,
as conventional strategic wisdom might suggest, an increase in perceived
external threats, but rather the rate of increase in GDP.[9] If GDP rises,
defense expenditures—and arms imports—will tend to rise *regardless* of
perceived threats in the external environment. This pattern helps explain
such apparently anomalous situations as that in Thailand, where, even
though the major perceived threats—from China and Soviet-backed Vietnam—had disappeared, defense expenditures continued to rise. The somewhat depressing implication of this finding is that national economic
decline may be one of the most effective means of controlling rising
defense budgets and hence arms imports.

—Corruption is a further factor. In many regional states, powerful individuals or groups within government and the military seek to purchase weapons systems primarily for the pay-offs that accompany them. In Thailand, for example, a 1992 report claimed that "commissions" from arms sellers to senior Thai military officials averaged "15–20 percent of any deal."[10] The clear implication here is that individual greed rather than strategic need was a major factor in determining arms purchases.

Supply-side factors include:

—The desire of the major weapons-exporting states and corporations to replace markets lost as a consequence of the end of the Cold War.

—Concern over preventing the decline or collapse of domestic defense industries for domestic political reasons. This factor was clearly the motivation behind the Bush Administration's decision in the run-up to the 1992 election to permit a $6 billion sale of F-16s to Taiwan—a breach of a 1982 agreement with China.

—The concern of major powers, primarily the United States, to support allies and friends.

—The willingness of particular weapons-supplying states and corporations to provide various corrupt inducements to buyers.

The most important factor underpinning this decline has clearly been the end of the Cold War, and not *simply* that the resulting improvement in the East Asian security environment reduced the demand for weapons. The economic crisis in Russia has forced changes in Moscow's arms transfer policies that have also had a dramatic effect on the region. North Korea and Vietnam, once at the top of the regional arms-importing league, now import virtually nothing from Russia. The reason is not that the Russian government refuses in principle to sell weapons to Leninist regimes, nor that North Korea and Vietnam have no desire to modernize their defense forces. Rather it is that, since 1991, an increasingly impoverished Moscow has demanded that its former allies pay for all their imports, including arms, in hard currency, which they do not have. Moscow's hard-nosed policy plunged the North Korean economy deep into crisis; Vietnam's economy was less affected. Both countries lacked the hard currency reserves to pay for weapons, and neither could have borrowed from abroad for this purpose even had they wanted to.

To compensate for the loss of traditional markets, in the early 1990s Moscow began to push hard for sales to non-traditional customers in the region. Its most notable success to date has been the sale of eighteen MiG-29 Fulcrums to Malaysia. Other countries have expressed interest, but no more. One reason is that the stunning successes of Western arms relative to Russian arms in the Gulf War have made the latter systems considerably less desirable than the West's.

Russia's most important market in the Asia-Pacific region, indeed the world is China, and the relative importance of the Chinese market has grown considerably since North Korea and Vietnam effectively ceased to be Russian customers. The Chinese have bought over 440 T-72M tanks and 26 Su-27 fighters, the first relatively modern combat aircraft in China's inventory. (Whereas the Su-27 is a 1970s design, most of China's combat aircraft are of 1950s and 1960s design.) Reports of further orders of Su-27s and Su-24s and MiG-29s and MiG-31s have appeared from time to time in the press, but thus far no deliveries have been made, nor have any licensed production facilities been established. The Russians reportedly even offered to sell their supersonic, long-range Tu-22 Backfire bomber[11] and 3,000-km range AS-15 missile[12] to China. Any such deals would generate great concern in the region. Other Russian items on China's military wish-list include rocket engines, improved radars and missile guidance systems and, possibly, an aircraft carrier.

The combination of a booming economy, the perceived need for military modernization and a desire not to be too dependent on Western suppliers means that China is likely to continue to be Moscow's most important arms market for the foreseeable future.

While the United States is increasingly being challenged by its European arms-producing rivals in East Asian markets, it retains a number of advantages over its competitors, particularly in the case of combat aircraft. The ubiquitous U.S. F-16, for example, is still the single most popular fighter/strike aircraft in the region. The F-16s continuing popularity arises in part because so many of them were purchased during the Cold War and regional air forces are familiar with their operation, and in part because of their advantages in cost and performance. Air and ground crew familiarity, commonality of spare parts, availability of upgrades, and inter-operability with U.S. and other regional forces are also factors. Indonesia, Singapore, South Korea, Taiwan and Thailand either have, or are in the process of acquiring, F-16s. Australia, Japan and Malaysia have ignored the regional preference for F-16s, acquiring F-15s, FA-18s, and MiG-29s and FA-18s respectively. Taiwan is acquiring 60 Mirage 2000s as well as 150 F-16s, and both Japan and Taiwan have indigenous fighter programs—the FSX and the IDF respectively.

China's arms sales to the Third World have declined dramatically, from $5.9 billion in 1987 to $500 million in 1994.[13] The decline was primarily the result of the collapse of China's markets in the Middle East. China's major East Asian customers have been Myanmar and Thailand; low price was a major selling point, although corruption is also alleged to have been a factor in sales to both countries.[14] Further large orders from Thailand seem unlikely since the Thai military is unhappy with the quality of the arms, including tanks and naval surface combatants. Myanmar reportedly spent $1.2 billion in 1990 on a range of Chinese weapons systems, including tanks and some twenty-four F6 and/or F7 fighters (Chinese versions of the MiG-19 and MiG-21).[15]

A Buyer's Market

The end of the Cold War and the resulting reduction in demand for weapons in Europe and the United States mean that worldwide there is now a buyer's market for weapons systems. In Asia, unlike in other regions, demand remains buoyant.

Considerations of cost and a desire to acquire particular weapons technologies are now greater determinants of procurement decisions in regional states than ideology or alliance is (Japan may be an exception to this rule). One consequence of this trend is that for a number of regional states in Asia-Pacific, the United States is no longer automatically the supplier of choice. The highly competitive nature of the current arms market also means that regional states can, and do, demand sophisticated state-of-the-art weapons systems that supplier states would once have been reluctant to sell. There are, of course, limits to what can be supplied—the United States is not about to permit the transfer of Stealth fighters to the region. However, regional states recognize that the nature of the global arms market gives them new bargaining leverage and that advanced weapons systems that one country refuses to sell another country probably will.

Taiwan is a case in point. To avoid offending China, the United States had long denied Taipei the F-16s it sought, and the Taiwanese turned to the French firm Dassault and negotiated to buy sixty Mirage 2000s, much to China's fury.[16] Taipei had considered buying MiG-29s, Israeli Kfir C-7s and the Italian AMX, as well as the Mirage, and is currently building 140 of its largely indigenous IDF fighters.

U.S. concern not to offend China has been a real restraint on American arms sales to Taiwan. The Taiwanese, however, have often been able to find other suppliers. The Mirage deal is not the only example. When Taipei was denied access to U.S. Harpoon anti-ship missiles (ASMs), it negotiated a licensing deal with Israel to produce the Gabriel ASM. Then, in mid-1992 the Bush Administration, confronted with the very real possibility of defeat in the election and anxious to avoid further factory closures in the pre-election period, decided to endure the predictable Chinese outrage and sell Taiwan the F-16s after all. The package deal for 150 aircraft was worth nearly $6 billion. Taiwan's huge foreign reserves, and its determination to build up its defense forces, make it an extraordinarily attractive market for arms corporations. As noted, governmental concerns in the arms-exporting state that it not offend China is a constraint on sales.

The United States also denied itself sales opportunities in Indonesia when it refused to sell arms to the Suharto government until it improved its human rights record. The French and British have no such scruples, however, and have been competing vigorously for Indonesia arms contracts. While politics—as in the cases of Indonesia and Taiwan—may have caused the United States to lose some arms sales, it is by no means the only reason. According to James Blackwell, "many U.S. defense officials and military officers have not adjusted to the notion that they must now compete on an even basis with European companies which often enjoy government backing."[17] Sellers are more aggressive than ever—as France's elevation to first place in world arms sales in 1994 indicates.

With some 3,000 new fighters and strike aircraft reportedly being procured by Asia-Pacific states during the next decade—and an equal number of existing aircraft being upgraded—the market opportunities for extra-regional producers are considerable.[18] Regional states have been actively shopping around in their search for the right weapons systems at the right price. Seeking new fighters, both the Philippines and Taiwan looked at the Israeli Kfir fighter, which is a third the price of the F-16. The Thai military sought (ultimately unsuccessfully) to buy AMX fighters from Italy in the early 1990s;[19] Taiwan also expressed interest in the AMX. During the same period, a number of states, including Malaysia, South Korea and Thailand, expressed interest in the European multi-role Tornado, while China, Malaysia, South Korea and Thailand considered the MiG-29.

The high cost of the most sophisticated modern fighters—the F-15, F/A-18, Tornado, Mirage 2000, and even the F-16—restricts the ability of Asia-Pacific states, particularly in Southeast Asia, to buy them in large numbers. High prices are one reason the trend in imports of military aircraft into the region has been downward since the mid-1980s. Relatively cheap, light fighter/trainers can, however, complement the more sophisticated combat platforms, and using the extra numbers to provide more comprehensive area coverage makes good strategic sense. The Europeans have more to offer here than the United States. British Aerospace has had great success selling its light multi-role (including light strike) Hawk throughout the region. Brunei, Indonesia, Malaysia and South Korea have bought or ordered Hawks, and other countries are considering them.

European arms manufacturers have also been highly successful in selling naval platforms to the region, although often the purchasing country demands a high degree of offsetting. Demand is likely to continue to be strong: "some 200 new major surface combatants are programmed for procurement [in the 1990s], and about 50 more are under serious consideration."[20] Taiwan ordered six Lafayette frigates from France in 1992 to complement the eight U.S. Perry class frigates it is building under license. Indonesia has bought thirty-nine former East German Navy ships.[21] Brunei is getting three missile attack boats from the United Kingdom; the Philippines is acquiring three similar boats from Spain—and three more from Australia; and Malaysia is buying two sophisticated frigates from the United Kingdom and has signalled its intention to buy eighteen offshore patrol vessels. Myanmar has bought three coastal patrol boats from Yugoslavia in addition to the six it acquired from China.[22] Australia is building eight "ANZAC class" German Meko light frigates for its Navy, with a further two being produced for New Zealand, and Singapore is buying a landing ship and four mine counter-measures ships from Sweden.

The fact that the United States does not produce conventional submarines means this important market is the exclusive preserve of Europe and Russia. East Asian states are likely to acquire some thirty-six submarines in the 1990s.[23] Australia is building under license six highly capable Swedish-designed Collins Class submarines, for which air-independent operation is a future option. Japan is acquiring 155 new submarines, to be built in-country; Taiwan is seeking six to ten new boats. The French reportedly offered Rubis class nuclear attack submarines to Taipei.[24] South Korea may be seeking to acquire as many as sixteen German Type 209/3s.[25] In Southeast Asia, Indonesia, Malaysia, Singapore and Thailand are all contemplating buying submarines, although the numbers will likely be small. Indonesia is the only Southeast Asian state with submarines, with two old German-built ones, although Australia has six Oberon class boats that it will replace in the 1990s with the even more capable Collins class. Russia is doing less well than Europe in submarine sales. Currently, only China, which is buying Kilo class submarines, is a Russian customer. North Korea has twenty-two obsolete Soviet submarines, but will not be able to acquire any more. Pyongyang not only lacks the finance to buy major weapons systems; in addition, Moscow has undertaken not to sell it any new arms.

While no regional navies currently deploy aircraft carriers, since 1992 there have been persistent reports that China is seeking to buy a carrier from Russia. These reports have caused some concern in the region, since such an acquisition would considerably increase China's ability to project air power over the contested waters of the South China Sea. In October 1992, Chinese Foreign Minister Qian Qichen announced that China had abandoned plans to buy the carrier.[26] However, this and other denials have met with skepticism among regional security analysts, who believe that the reason is a temporary financial constraint, rather than a change in heart.[27]

Security planners in Japan are known to be interested in acquiring a small aircraft carrier, while Thailand has ordered a light carrier from Spain. Even small carriers are capable of operating very short take-off and landing (VSTOL) combat aircraft, such as the Harrier.

Within the region, the desire for increased self-reliance in defense, national pride and a need to generate domestic employment continue to impel some states toward indigenous weapons production—even where it may not be cost-effective from a purely economic point of view. Japan's FS-X fighter program and Taiwan's IDF program are obvious examples. Insofar as the trend continues—and it is very evident in the shipbuilding area—traditional suppliers from the United States and Europe will sell relatively fewer off-the-shelf weapons systems, and relatively more of the high-technology components for those systems that regional states are unable to produce themselves.

The Strategic Implications of The Buildup

It is sometimes claimed that the flow of modern weapons systems to the Asia-Pacific region is of no great consequence because regional states are not seeking to acquire major power projection capabilities. It is true that no states in the region appear to be seeking the sorts of forces necessary to mount successful invasions against neighbors, although the North Koreans have long had massive offensive forces configured to "seize and hold territory," and China's buildup and provocative posturing in and around the Taiwan Strait have caused great concern in Taipei and elsewhere in the region. "Power projection," however, is an idea that embraces more than the capability to invade other states. It also includes the capability to strike distant military targets—at sea as well as on land. In this latter sense power projection capabilities in the region *are* growing. While no country in Northeast or Southeast Asia has long-range strike aircraft comparable to Australia's F-111s, the modern combat aircraft they have acquired, or are acquiring, are formidable power projection platforms. The acquisition of air-to-air refuelling capabilities and some form of AEW, both of which are potent force multipliers, enhance the strike range and capability of these aircraft. Long-range maritime patrol aircraft, such as the Orion P-3, can also deliver ASMs over long distances.

Acquisition of over-the-horizon ASMs—mostly Harpoon and Exocet—by almost all navies in the region represents a further transformation in regional naval strike capabilities. A small fast attack craft, aircraft or submarine armed with Harpoon missiles can, for example, strike over a greater range and with far more accuracy than a salvo from a World War II battleship. Fired from over the horizon, a Harpoon can blow a frigate in half.[28] Quiet, modern submarines armed with ASMs are a particularly lethal combination.

Some regional states are acquiring—and in the case of China and North Korea exporting—a variety of short- and medium-range ballistic missiles. China, North and South Korea, and Taiwan all have largely indigenous missile programs based on knowledge gained by reverse-engineering missiles imported from Israel, the United States or the USSR many years ago. Japan, South Korea and Taiwan also have space-launch programs of varying degrees of sophistication that could rapidly be converted into missile programs, and Indonesia has an embryonic space-launch program. Apart from China, no countries in the region have long-range cruise missile programs, although the current Liberal-National Coalition government in Australia has expressed interest in acquiring conventionally armed Tomahawk cruise missiles as an eventual replacement for the aging F-111s. Any such move could be destabilizing—it could lead other countries to seek similar offensive capabilities.

The ballistic missiles currently deployed in the Asia-Pacific region do not have much power projection capability when armed with conventional warheads. Against countries with limited air defense capabilities they are far less cost-effective than strike aircraft.[29] The real concern is that they may be matched with nuclear, chemical or biological warheads. The Pentagon has claimed that China, Myanmar, North Korea, Taiwan and Vietnam have offensive chemical weapons capabilities.[30] North Korea and Taiwan are suspected of having biological warfare capabilities.[31]

The gravest threat, of course, would be missiles armed with nuclear warheads. There is *some* risk this threat could be realized in the medium and long term. The October 1994 U.S./DPRK Agreed Framework is supposed to have put a halt to North Korea's nuclear weapons program, but the North may have acquired smuggled fissile material from Russia (there were unconfirmed reports to this effect in 1992), or have created a clandestine underground nuclear program. Either option could have provided North Korea with enough plutonium for a number of nuclear weapons. It is also possible the North may have diverted enough plutonium from the research reactor whose operations it has declared to build one or possibly two nuclear weapons. South Korea and Taiwan have both sought, over some twenty years, to acquire nuclear weapons technologies, although without success. Japan's commitment to plutonium production is generating considerable regional concern, and not just in North Korea. Japan has enough weapons-usable plutonium for hundreds of nuclear weapons.

Were the United States to withdraw its security commitment to the region, that step would considerably increase the risk that regional states would seek to go nuclear. Many regional states are concerned that this possibility is real. It is true there are supply-side regimes that seek to control the transfer of nuclear and missile technology. East Asian states, however, already have the technical capability to make nuclear weapons and build missiles. Moreover, U.S. leverage over the possible proliferator states (Japan, South Korea and Taiwan) has been declining since the end of the Cold War, as the nuclear industries of the regional states have become increasingly self-reliant, and as the sources of supply of nuclear materials and technologies have diversified. The implication is that supply-side regimes are likely to be of declining efficacy.

The Potential Effectiveness of Supply-Side Restraints

In the aftermath of the Gulf War there were many demands for restraints on the global arms trade. In the United States, for example, a March 1991 poll found that 82 percent of Americans wanted a multilateral agreement to limit arms transfers to the Middle East.[32] In October 1991, in response to these concerns, the China, France, Russia, the United Kingdom and the United States—the P-5 members of the U.N. Security Council—negotiated an initiative to prevent "destabilizing arms transfers." While the P-5 initiative was intended primarily to restrain the arms buildup in the Middle East, it was hoped it could serve as a model for restraining arms transfers to other regions of the world.

The P-5 initiative did not, however, call for *reductions* in arms transfers. It in no way proscribed arms exports for the "legitimate right to self-defense," as enshrined in the U.N. Charter. Herein lay the first problem: what constitutes "legitimate self-defense" is frequently contested. Nation states invariably claim their weapons purchases are for "legitimate self-defense." Only very rarely is their aggression as blatant as Saddam Hussein's. Indeed, the language of the P-5 document was so ambiguous and potentially contradictory it seemed pretty much designed to fail. Almost every clause offered loopholes. For example:

—Arms transfers that "prolong and aggravate an existing armed conflict" were proscribed, although the import of arms for "legitimate self-defense," which was permissible under the guidelines, might well "prolong . . . an existing conflict."

—Arms transfers intended solely to meet the "needs of legitimate self-defense" were to be permitted, but arms transfers that "increase tension in a region" were proscribed. Clearly the former could lead to the latter.

—Arms transfers that could create "destabilizing military capabilities to a region" were proscribed, yet a weapons system that one state believed was "destabilizing" another might see as enhancing "legitimate self-defense" and thus consider to be permissible.

—Arms transfers that seriously undermined the recipient state's economy were proscribed, even though arms transfers for "legitimate self-defense" could have precisely this consequence.

Even if these hopelessly contradictory guidelines could have been implemented, it is unlikely they would have precluded *any* of the arms transfers to the Asia-Pacific region over the past decade. According to William Hartung, "State Department officials involved in the Big Five discussions have already indicated that they cannot conceive of any [military] sale the United States would be prevented from making under the guidelines."[33] Such questions quickly became academic, however, following the U.S. sale of F-16s to Taiwan. In response, Beijing announced it would no longer be bound by the ambiguous guidelines. Whatever hopes the international community may have entertained about negotiated multilateral restraints on arms transfers were dashed.

In February 1995, the Clinton Administration announced a new conventional arms transfer policy. It views such transfers as a legitimate instrument of U.S. foreign policy when they "enable the United States to help friends and allies deter aggression, promote regional security and increase interoperability of U.S. forces and allied forces."[34] This policy essentially *encourages* arms sales, not discourages them.

It is clear the Clinton Administration is not interested in curtailing arms transfers to East Asia, except, of course, to states that have incurred deep U.S. political disapproval. As Richard F. Grimmett of the Congressional Research Service writes,

> Although the Administration has emphasized that its decisions on arms transfers will not be driven by commercial considerations, but primarily by national security, the Clinton arms transfer policy holds that supporting a strong, sustainable American defense-industrial base is a key national security concern, rather than a purely commercial matter. In so doing, the Clinton policy publicly elevates the significance of domestic economic considerations in the arms transfer decision-making process to a higher degree than has been the case in previous administrations.[35]

There is *some* support in the U.S. Congress and the arms control community for greater restraint on arms transfers. On February 1, 1995, Senate Appropriations Committee Chairman Mark Hatfield (R-Ore) and House International Relations Committee member Cynthia McKinney (D-Ga) introduced the Code of Conduct on Arms Transfers Act of 1995. The bill attracted 109 House and Senate co-sponsors, which were not enough for passage. It has also been fiercely denounced by states such as Malaysia. U.S. officials indicated the administration did not support the legislation. The Code of Conduct would prohibit arms exports to any government that does not meet the criteria set out in the Code, unless the president exempts a country and Congress affirms the exemption. The conditions a country must meet to be eligible for U.S. weapons as stipulated by the Code are:

—Have a democratic form of government.

—Respect the basic human rights of citizens.

—Pursue non-aggression (against other states).

—Participate fully in the U.N. Register of Conventional Arms.

As a recent report by the Federation of American Scientists notes, "the Code's criteria are all primary foreign policy tenets of past and present U.S. administrations. Nevertheless, 90 percent of the record $14.8 billion in U.S. arms sales to the Third World in 1993 went to states which do not meet the Code's criteria."[36]

Without U.S. leadership, or at the very least support, there is no chance any global restraint regime will be successfully negotiated. Currently Washington does not appear to be interested in either leading or supporting moves to create such a regime. Nor, it must be said, is there interest among any East Asian states. Geoffrey Kemp speaks for many security planners when he argues that preventing war "may mean providing additional weapons to friendly countries rather than seeking to restrict their inventories."[37] This view is clearly common in Washington and is based on the assumption that tilting the military balance in favor of friends and allies enhances deterrence and thus reduces the risk of aggression. The possible dangers of this approach are discussed later.

A Walk on The Demand Side

"Far-reaching arms control agreements among . . . countries will depend on progress to resolve regional conflicts,"[38] an observation Geoffrey Kemp made about the situation in the Middle East but that applies with equal force to Asia-Pacific. The focus of demand-side approaches to controlling global arms transfers is, as the term suggests, not on controlling the supply of weapons and weapons technologies to particular regions, but rather on the security (and other) concerns that give rise to demands for the arms in the first place. If conflicts can be resolved, the demand for arms will be reduced, and arms flows will slow without any supply-side regimes being instituted. Arms control regimes, as is often noted, are most difficult to negotiate when most needed, and least difficult when least needed.

The assumption that states may become more secure by acquiring more arms—"if you want peace, prepare for war"—has a venerable military history. It is also the precept that legitimized arms transfers under the P-5 guidelines. As the United Kingdom's Lord Caithness put it at the October 1991 P-5 meeting, "the right of self-defence . . . is meaningless if states cannot also acquire the means to defend themselves."[39] However, while "Peace through strength" policies may be appropriate means of enhancing security in some contexts, CBMs would not have been a sensible response to Adolf Hitler in 1939—and may be just as inappropriate in others.

Where the central security problem a state confronts is the threat of unprovoked aggression, deterrence is an appropriate response, and arms transfers will enhance the security of the threatened state by increasing its deterrent and war-fighting capabilities. Where the threats are less tangible or imaginary—the adversarial states have mutual suspicions but no aggressive intentions—arms transfers may exacerbate the "security dilemma" by increasing fear, suspicion and hostility. "Security dilemmas" arise when one state takes the defensive preparations of another as evidence of an offensive intent. An action-reaction "conflict spiral" may result, leading to crises and culminating in a violent military confrontation that neither side originally sought. In other words, arms transfers intended to enhance deterrence and reduce the risk of aggression may, perversely, increase the possibility of unintended war.

Avoidance of the risks inherent in conflict spirals and security dilemmas requires greater stress on demand-side strategies of reassurance and less on offensive deterrence. Striking an appropriate balance between deterrence and reassurance is not, however, easy. Too much emphasis on deterrence can exacerbate security dilemmas and may trigger self-defeating arms spirals. Too much reassurance, on the other hand, may appear as appeasement and undermine deterrence. In areas of tension in the Asia-Pacific region, the deterrence/reassurance balance appears tilted too far in the direction of deterrence—only Japan pursues a self-conscious policy of seeking to reassure its neighbors, not the least by eschewing acquisition of offensive weapons. Confronting strategic uncertainty, the traditional response of regional security planners has been to increase defense preparedness, which usually means increasing arms imports.

It is precisely because the strategic future is so uncertain in East Asia, and because many states view each other with muted suspicion, that the case for greater emphasis on reassurance strategies is compelling. Such strategies may include engagement in bilateral and multilateral security dialogue, military transparency, confidence- and security-building measures (CSBMs) and a shift toward defense-dominant force structures and strategies. The "instability" risks associated with arms races do not exist when the force structures being strengthened are purely defensive.[40] One reason Japan has been able to rise to third place in defense spending globally without causing undue regional concern is that its force structure is defensive—it does not consist of long-range bombers, missiles, aircraft carriers and the like. Japan simply lacks the physical capability to invade and subjugate its neighbors, a fact that has reassured them enormously about its intentions.

Regional interest in reassurance strategies has been growing steadily in the 1990s. The value of pursuing security dialogues and CSBM regimes is no longer contentious in the mainstream security discourse in the region, a radical change from the 1980s. A region-wide security dialogue has been institutionalized in the annual high-level Association of South-East Asian Nations (ASEAN) Regional Forum (ARF) meetings. ARF has met three times—in 1994, 1995 and 1996. Working groups of officials have been established to pursue issues that ARF member states designate as important and bring recommendations back to the annual meeting. Issues that are too sensitive to handle at the official level will be dealt with in so-called "Track II" fora, whose members include academics, experts from think-tanks and officials who participate under the polite fiction that they are there in a "private capacity." Reports and recommendations from Track II meetings may inform the deliberations of ARF meetings.

East Asian approaches to confidence-building are very different from those of the West. Coming from lawyer cultures, Westerners tend to see dialogue as leading to negotiation on substantive security issues—with the point of the negotiation to reach verifiable legal agreements. The ASEAN approach, in contrast, sees dialogue primarily as a means of building relationships. With good relationships, both the risk of conflict and the need for formal agreements are reduced. ASEAN's Western "dialogue partners" have frequently found the modus operandi of ARF somewhat frustrating, since there is little agreement on matters of substance.

Within ARF there has been no discussion of limiting arms transfers into the region. On the contrary, defense modernization, which requires arms transfers, is widely supported. Even the most modest steps toward increasing transparency in arms transfers are often resisted. There has been some discussion of creating a Regional Arms Transfer Register—an idea that originated with Malaysia—but it has little support within the region. Even if there were greater transparency in arms transfers, it is not clear it would promote moves toward control.

Demand-side strategies ultimately have to go beyond the pursuit of modest transparency measures and other low-level CSBMs and seek to manage and ultimately resolve regional conflicts. Arms restraint, as Ernest Graves has noted, "depends very much on the outcome of efforts to resolve longstanding political conflicts."[41] There is growing recognition in the region that such approaches are necessary, and preventive diplomacy and conflict management are on the ARF agenda.

Even successful conflict resolution strategies would not address all the factors that determine the demand for arms acquisitions. They would, for example, be irrelevant if the demand for arms imports arose from corrupt inducements offered by sellers or commissions sought by buyers. Conflict resolution is irrelevant where states import arms because they perceive them to be legitimately necessary to replace outdated equipment. States frequently pursue defense modernization in the absence of external threats. Conflict resolution is also irrelevant if, as seems to be the case, economic growth rates and not external threats are the major determinant of military expenditure and arms imports.

Finally, conflict resolution has little direct relevance to perhaps the most important *strategic* factor underpinning the military buildup in Asia-Pacific—the fear the United States may withdraw from the region. Notwithstanding constant reassurances from U.S. officials, regional security planners are concerned that growing isolationist sentiments at home sooner or later will impel the United States to withdraw from the region. The original rationale for the U.S. presence in Asia—the containment of Soviet communism—disappeared with the break-up of the USSR. In the mid-1990s East Asian states are militarily more self-reliant, and nationalism and anti-Americanism are increasingly evident among America's allies, while trading conflicts are a source of constant tension. Regional states doubt the political and economic ability of the United States to sustain its commitment in the long term. Even if the United States stays formally engaged in the region, there are fears the "Somalia Syndrome" would force a withdrawal if American forces were drawn into a regional conflict and sustained large numbers of casualties. Whether such regional perceptions are correct or not is of little consequence: they exist, and they have a real impact on regional defense policies. Insofar as fear of a U.S. withdrawal is a factor driving arms acquisitions in the region, it is difficult to see what can be done to reassure regional states and thus reduce the demand for weapons transfers. Verbal assurances are inadequate.

Conclusions

The probability that arms transfers to the region will be controlled by supply-side means is not great. The most intractable problem is that neither buyers nor sellers see any need to curb arms transfers to the region—quite the contrary. Since the motivations for arms acquisition are many and various, no one strategy will lead to restraint. The demand-side approaches that are receiving increasing attention in the region may help create a security environment where the demand for weapons begins to fall. Ultimately, however, changes in the region's economic relationships may provide the greatest disincentive for war and thus decrease the demand for arms procurement.

Market-driven trade and foreign direct investment strategies are increasing regional economic integration. As regional economies become more and more enmeshed, the costs of war rise and the benefits decrease. When access to land and raw materials were the keys to national wealth, war was economically rational. That criterion no longer applies—states get rich by expanding trade and increasing domestic productivity.

Market-driven economic growth also strengthens civil society vis-à-vis the state and encourages the emergence of pluralist political institutions—South Korea and Taiwan are obvious regional examples. One of the best established findings of international relations research is that democracies do not fight each other. The implication is that the continued spread of democracy in East Asia may well be a force for regional peace.

Interstate war is, in any event, extremely rare in the modern world. Of the thirty-four major armed conflicts in 1993 (a "major armed conflict" is one in which there have been at least 1,000 battle-related deaths), not one was an interstate war. Statesmen have yet to adjust to the idea that interstate war, at least between the relatively developed states, is becoming increasingly rare. Security planners will naturally be skeptical. If, however, the extraordinary decline in interstate war is not accidental, *if,* as suggested, it is caused by profound changes in the structure of the global economy and polity, and *if* an increasingly strong global norm against resort to war in international disputes reinforces these disincentives, then the prospects for reducing arms levels may be quite good.

What the above suggests is that the encouragement of foreign trade, foreign direct investment and the growth of democratic institutions in East Asia may be a better strategy for reducing arms acquisitions than either the traditional supply-side strategies of the arms control community or the most ambitious confidence-building proposals on the agenda of ARF.

Notes

1. The author thanks Michael Shaik for his research assistance on this chapter.

2. Bates Gill, "Arms Acquisitions in East Asia," in Stockholm International Peace Research Institute (SIPRI), *SIPRI Yearbook 1994: World Armaments and Disarmament* (Oxford, 1994), 552.

3. Gill, "Arms Acquisitions in East Asia," 552.

4. Only in the Philippines is the military *primarily* concerned with counter-insurgency. Indonesia confronts minor insurgencies in East Timor and Irian Jaya. The Burmese military confronts residual resistance from ethnic insurgent groups.

5. The Sino-Soviet, Sino-Vietnamese, and Sino-Indian borders have all been sites of military confrontation in the past. China, however, has moved to resolve the border tensions with most of its neighbors, and European-style confidence-building measures have been implemented on the Sino-Indian and Sino-Russia borders.

6. Office of Technology Assessment (OTA), U.S. Congress, *Global Arms Trade: Commerce in Advanced Military Technology and Weapons* (Washington, D.C., 1991), 166.

7. Gill, "Arms Acquisition in East Asia," 556.

8. Ibid., 556.

9. Daniel P. Hewitt, "What Determines Military Expenditures," *Finance and Development,* XXVIII (1991), 22.

10. Tai Ming Cheung, "Officers' Commission: Arms Procurement Driven by Profit Rather Than Need," *Far Eastern Economic Review* (2 July 1992), 13. See also Kenneth Stier and Bao Anyou, "Khaki Commerce," *Asia, Inc.* (October 1992).

11. Cheung, "Officers' Commission," 20.

12. See Antony Spellman, "US, French Fighter Sales to Taiwan Nudge Mainland China Closer to Russia," *Armed Forces Journal* (January 1993), 16.

13. Richard F. Grimmett, *Conventional Arms Transfers to Developing Nations, 1987–94,* Congressional Research Service, U.S. Library of Congress (Washington, D.C., 1995), 29.

14. Stier and Anyou, "Khaki Commerce."

15. John Badgley, "A Clear Gathering of Miraculous Success," *Peace and Security* (Spring 1992), 11.

16. On November 18, 1992, the Taiwanese Air Force signed a Memorandum of Understanding with Dassault Aviation covering the sale of sixty Mirage 2000–5 multi-role fighters. The $2.6 billion deal included an additional forty fighters. Spellman, "US, French Fighter Sales," 16.

17. James A. Blackwell, "Defense Industrial Cooperation in Asia: U.S. View," a paper presented at the 5th KIDA/CSIS International Defense Conference (Seoul, 13–14 October 1992), 23.

18. Panitan Wattanaygorn and Desmond Ball, "A Regional Arms Race," *The Journal of Strategic Studies,* XVIII (1995), 149.

19. "Thailand," *Asia-Pacific Defense Reporter* (February-March 1992), 33.

20. Wattanaygorn and Ball, "A Regional Arms Race," 151.

21. The thirty-nine ships include sixteen corvettes and nine minesweepers. See "German Ships for Indonesia," *Miltech* (September 1992), 111.

22. "New Base Is Boost to Naval Power," *Jane's Defence Weekly* (12 September 1992), 32.

23. Wattanaygorn and Ball, "A Regional Arms Race," 152.

24. Ibid., 152.

25. Spellman, "US, French Fighter Sales," 16.

26. "China Reportedly Won't Buy Carrier," *Japan Times* (9 October 1992).

27. Notwithstanding the foreign minister's denial, China's President Yang Shangkun said in November 1992 that the government *had* decided to buy a carrier. See "Chinese Leader Urges Purchase of Carrier," *The Washington Times* (30 November 1992).

28. However, few countries in the region have the target-acquisition platforms necessary for effective over-the-horizon targeting of ASMs such as Harpoon and Exocet.

29. The Chinese make this point constantly—and with some logic—when responding to U.S. criticism of China's missile exports. U.S. officials have never satisfactorily explained why exporting relatively short-range and inaccurate missiles with modest payloads is destabilizing, while the export of Western hit aircraft that can strike at distant targets with great accuracy and much greater payloads is not.

30. Rear Admiral Thomas A. Brooks, in testimony before the Seapower, Strategic and Critical Raw Materials Subcommittee of the House Armed Services Committee, U.S. Congress, March 9, 1991, 57–58.

31. Rear Admiral William O. Studeman, in testimony before the Seapower, Strategic and Critical Materials Subcommittee of the House Armed Services Committee, U.S. Congress, March 14, 1990, 54.

32. Cited in William D. Hartung, "Curbing the Arms Race," *World Policy Journal,* IX (1992), 22.

33. Hartung, "Curbing the Arms Race," 23.

34. Grimmett, *Conventional Arms Transfers 1987–94,* 8.

35. Ibid., 18.

36. Federation of American Scientists, http://www.fas.org/pub/ gen/atwg/conduct/s326.html.

37. Geoffrey Kemp, *The Control of the Middle East Arms Race* (Washington, D.C., 1991), 10.

38. Ibid., 10.

39. Statement by Lord Caithness, U.K. minister of state for Foreign and Commonwealth Affairs (London, 17 October 1991).

40. The idea of a "defensive" weapons system per se makes little sense, since such systems can be force multipliers for *offensive* systems in an offense-dominant force structure. It is possible, however, to talk about defense-dominant *force structures—* those that have powerful forces for defense but very weak forces for offensive operations. For an evaluation of the pros and cons of defensive strategies and force structures in a regional context, see Andrew Mack, "The Theory of Non-Provocative Defense," *Korean Journal of Defense Analysis,* III (1991), 241–260.

41. Ernest Graves, "The Future of U.S. Security Assistance and Arms Sales," *Washington Quarterly,* XIV (1991), 91.

THIRTEEN

South Asia

Rodney W. Jones

NUCLEAR WAR has a greater chance of breaking out in the next decade in South Asia than in any other developing region.[1] Next to the Middle East, South Asia is one of the most chronically volatile and proliferation-prone regions. India may be preparing to test nuclear weapons, according to recent reports, a step bound to provoke an open nuclear arms race in the subcontinent.[2] The unresolved territorial dispute between India and Pakistan over Kashmir has figured in three wars and brought the two nations to the brink of yet another in 1990 and 1993. Numerous other external and internal sources of conflict in the region aggravate the risk that India and Pakistan will assemble nuclear weapons and launch them in a catastrophic exchange. India's long-standing conflict with China is marked by disagreements over Tibet's aspirations for autonomy and the demarcation of the India-China boundary in the Himalayas. From India's standpoint, the antagonism stems from China's 1962 invasion and humiliating defeat of Indian forces in the northeast, China's nuclear threat—first demonstrated in 1964—and China's political and military assistance to rival Pakistan. More recently, the Soviet collapse and emergence of independent states in Central Asia has opened a dynamic new regional arena south of Russia to competing Chinese, Pakistani and Indian interests.

During the 1980s, Pakistan bore the brunt of pressure in the region as a result of the Soviet invasion of Afghanistan: over three million Afghan refugees entered Pakistan, reaching nearly 3 percent of Pakistan's population. The Soviet withdrawal brought only partial relief, since warfare continued with a struggle for power among armed Afghan groups. Countless automatic and other weapons intended for the Afghan *mujahideen* (resistance fighters) were diverted into Pakistan. They fueled the expansion of local banditry in the tribal provinces and along the trunk highway connecting Punjab and Sindh with Karachi, giving a new edge to chronic ethnolinguistic and regional tensions in Pakistani society and politics.

Illicit traffic in small arms through porous borders has similarly aggravated the otherwise home-grown Indian problem with Sikh extremists in Punjab. After the Soviet withdrawal in 1989, the diffusion of Afghan arms also enabled a younger generation of Kashmiri militants to resort to guerrilla warfare tactics to support their political demands for independence. Fortunately, the "Kalashnikov culture" has not permeated deeper into India's interior, where Hindu-Muslim tensions erupt periodically, as they did in December 1992 when Hindus razed an ancient mosque in Ayodhya.

The internal conflicts in South Asia, particularly those adjacent to common borders, those that threaten confidence in national viability and those that feed Hindu-Muslim communal animosities, could inflame popular feelings and provoke crisis and military confrontation between India and Pakistan.

A critical question is whether policies to restrain conventional arms transfers can help alleviate these risks in South Asia and the political causes of insecurity. The chapter posits that such policies can have a beneficial impact, provided they are designed with due understanding of legitimate, local requirements for military security and emphasize control of the most serious armaments proliferation problem in the region—nuclear weapons.

Nuclear weapons are the most compelling reason for conventional arms restraint in South Asia. Given the region's relatively weak purchasing power for advanced conventional arms, it is hard to justify making novel conventional arms control measures a high priority there, unless it can be shown they will help stem the proliferation of nuclear weapons. Fortunately, there is reason to believe that restraint of conventional arms transfers can be designed with nuclear proliferation controls as a key objective. The issues of conventional and nuclear arms tend to be intimately related in a proliferation-prone region, as a few examples suggest.

First, even though the fissile materials and other technical requirements for making and securing nuclear weapons are distinctive, the military infrastructure and logistics supporting the deployment, early warning, delivery and employment of these weapons can overlap considerably with those for modern conventional weapons. Regulating the transfer of dual-capable conventional arms and related equipment can benefit nuclear non-proliferation objectives.

Second, denial of conventional arms to a threatened state can, in the absence of other meaningful provisions for its security, increase that state's interest in nuclear alternatives, particularly if its adversary can threaten its survival with conventional means alone. Sufficient conventional defenses can alleviate this nuclear proliferation incentive, unless the adversary is already nuclear-armed.

Third, provisions for regional security in conventional arms control regimes can be important buffers against political pressures to go nuclear. Confidence-building measures (CBMs) and arms limitation agreements that reduce the chances of a conventional war between two nuclear-capable adversaries could reduce pressure for the acquisition, threatened use or use of nuclear weapons.

The urgency of checking nuclear proliferation in South Asia enhances the importance of prudently managed conventional arms control measures.

Trends in Defense Expenditures

Defense budgets and imports of arms both grew rapidly in South Asia in the 1980s.[3] At the end of the decade, however, that growth slowed.

India's Defense Spending

India's *official* defense expenditures climbed through the 1980s to an all-time high in 1990, but fell in the early 1990s, while Pakistan's continued to rise marginally (see figure 13–1).[4] India's defense payout more than doubled from $3.72 billion in 1979 to a peak of $10.0 billion in 1990, but dropped sharply between 1991 and 1993 to a level of about $7 billion.[5] This drop was the result mainly of two factors: the involuntary interruption of Soviet arms transfers to India after the Soviet Union's demise; and the requirements set by the International Monetary Fund for supporting India's recovery from the 1990 foreign exchange reserves crisis. In 1994 and 1995, however, the Indian defense budget resumed real increases, reflecting the cost of heavy Army deployments in troubled Kashmir.[6]

The rapid growth in Indian defense expenditures in the mid-1980s coincided with the acquisition of the Jaguar and Mirage 2000 aircraft that Prime Minister Indira Gandhi had contracted for in the early 1980s to modernize the Air Force.[7] It also reflected the modernization of naval and ground forces sponsored by active service chiefs such as Admiral Nanda for the Navy in the 1970s and General Sundarji for the Army in the 1980s.[8]

As a burden on its overall economy, India has kept its official defense cost to under 4 percent of gross domestic product (GDP) (see figure 13–2),

Figure 13-1. *Total Defense Expenditures, India and Pakistan, 1970–95*

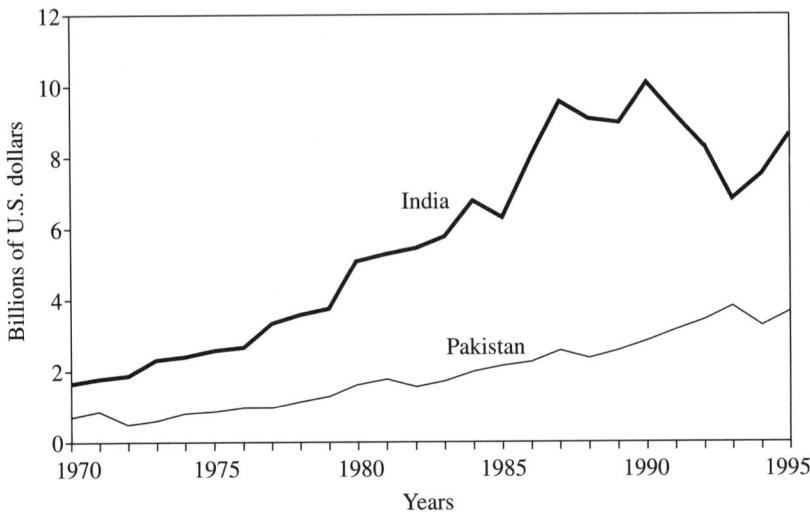

Source: Data from country pages, year-by-year prices, and exchange rates in International Institute of Strategic Studies (IISS), *The Military Balance* (London, various years).

Figure 13-2. *Defense Expenditures as a Percent of Gross Domestic Product, India and Pakistan, 1970–94*

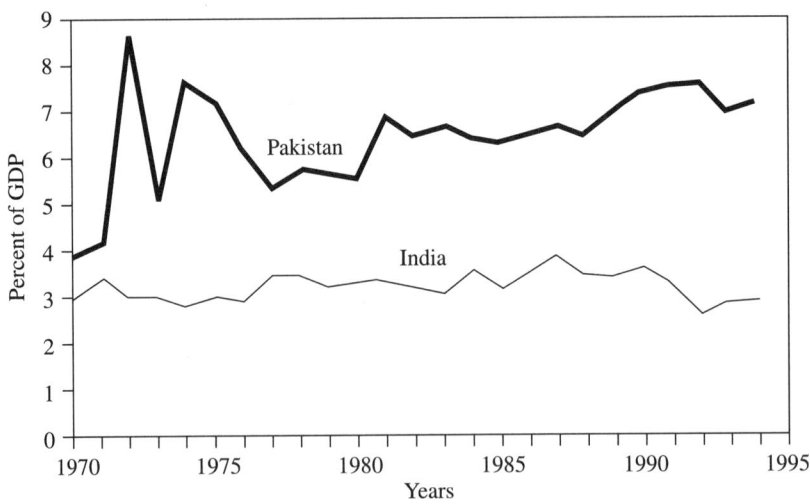

Source: Computed from U.S. dollar equivalents in International Institute of Strategic Studies (IISS), *The Military Balance* (London, various years).

with an average of about 3.5 percent since the early 1970s. By 1992, Indian defense expenditures had dropped below 3 percent of GDP and represented only 2.8 percent of GDP in 1993 and 1994.

Published figures on India's defense expenditures do not reveal the actual cost of India's arms imports from the Soviet Union, which involved a complex barter and "rupee-ruble" trading relationship.[9] Under these arrangements, India reportedly repaid in rupees long-term Soviet loans entered into to finance imports of Soviet military hardware and spare parts. The Soviet Union in turn used the rupees to pay for large quantities of tea, pharmaceuticals and other consumer products imported from India. The collapse of the Soviet Union disrupted these arrangements. The successor states, primarily Russia and Ukraine, and individual defense firms have been trying to get payment for their arms exports in hard currencies.[10]

Pakistan's Defense Spending

Pakistan's official defense expenditures doubled from $1.18 billion in 1979 to $2.58 billion in 1989, hovering in those years at levels that averaged one-third of India's (figure 13–1). In the 1990s, however, Pakistan's defense expenditures continued to climb, reaching $3.7 billion, about half India's level in 1995. As a burden on the economy, Pakistan's defense spending has averaged 6.5 percent of GDP, more than twice India's relative share (figure 13–2). In the early 1990s, Pakistan's defense burden once again reached 7.5 percent of GDP, before settling at about 7 percent of GDP in 1993–94, about two and a half times as big a bite of GDP as India's defense burden was.[11]

China has provided some military support to Pakistan since the 1965 Indo-Pakistan war and has been perceived by India as a threat that has figured in its defense preparations since the early 1960s. A meaningful comparison of trends in China's defense expenditure with India's would therefore be of some interest. Bear in mind that China's principal defense concerns come from the north, the Pacific and Southeast Asia, rather than just from the Indian subcontinent, so that there is no reason to expect a one-to-one comparison of India's and China's defense expenditures and force size.

Comparison with Chinese Defense Spending

Regrettably, there are still no reliable data on Chinese defense expenditures. China's official data greatly understate real expenditures and the

Figure 13-3. *Defense Expenditures, China and India, 1980–94*

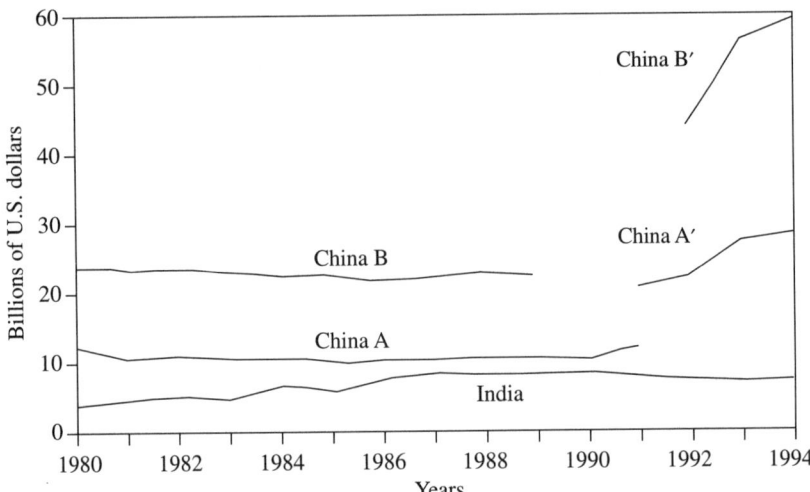

Note: China A figures are at 1985 exchange rates, and China B figures are in constant 1989 U.S. dollars. China A′ and B′ figures are based in part on purchasing power parity (PPP) calculations, using new estimation methods. Source for China A and India—tables in International Institute of Strategic Studies (IISS), *The Military Balance* (London, various years); for China B—U.S. Arms Control and Disarmament Agency (ACDA), *World Military Expenditure and Arms Transfers* (Washington, D.C., 1990); and for China A′ and B′, see International Institute of Strategic Studies (IISS), *The Military Balance 1995/6* (London, 1995), chapter on "China's Military Expenditure," 270–275.

actual burden on the economy,[12] and reputable sources provide sharply different estimates. Figure 13–3 plots two earlier contrasting estimates for China—the lower China A-line is derived from "conservative" estimates in the 1980s by the International Institute for Strategic Studies (IISS),[13] while the higher B-line is based on U.S. government estimates in the 1980s from the U.S. Arms Control and Disarmament Agency (ACDA).[14]

The IISS figures (China A-line) suggest that the real decline in China's defense expenditures beginning in the late 1970s resulted in the 1980s in a substantial narrowing of the wide gap between China and India. The ACDA figures (China B-line), which are probably closer to China's real expenditures and defense burden during those years, show a similar trend, including a narrowing of the gap, but suggest that China's annual defense expenditures remained well over twice the value of India's in 1989.

These opposite trends during the last decade of a decreasing defense outlay by China and an increasing one by India are related to developments in India's force structure that gave it the capability to counter the once formidable threat of a conventional military invasion by China, as happened in 1962. India modernized and China cut back the respective ground and air forces that each could project and sustain in the Himalayan theater, a shift in the local conventional balance in India's favor. This fact made it easier for India to repair its relations with China.

In the 1990s, however, China's defense expenditures reversed direction and increased sharply—a reflection of the cost of modernization of conventional forces and of fresh imports of modern Soviet defense systems after a long hiatus.[15] The IISS had estimated the dollar value of China's 1991 defense expenditures to be $12.03 billion (using 1985 prices and exchange rates), a 13 percent increase over its estimate of $10.617 billion for 1990.[16]

More recently, Western analysts of Chinese defense expenditures have produced revised estimates drastically upgrading the relative value of Chinese defense expenditures and indicating that they are actually more than three times the levels official Chinese figures suggest.[17] The China A' and B' lines in figure 13–3 indicate the sharp jumps in IISS and ACDA 1994 estimates to $28.5 billion and $56 billion, respectively. Not only are China's defense expenditures far larger than India's therefore, they are also pulling ahead at a much faster rate.

Economic Reform and Defense Cutbacks

The slowing of defense growth in Pakistan and its reversal in India in the early 1990s are encouraging signs that reform policies are taking hold in both countries. This shift allows economic development, export competition and integration into the world economy to receive priority attention. Events could, however, suddenly alter this hopeful pattern. For example, an overt nuclear arms race, rather than the shadowy nuclear competition between Delhi and Islamabad of recent years, could uncap the restraint on defense expenditures and drive those expenditures and imports of arms up sharply. If the gap continues to widen at the present rate, the surge in Chinese defense budgets in the 1990s could also put new pressure on India.

Smaller South Asian Countries

The trends in budgets and imports in the smaller states—Bangladesh, Sri Lanka, Myanmar and Nepal—may follow different paths from those of India and Pakistan. However, the paths of the smaller states typically had slight effect on the basic regional picture through 1990.[18] In 1990 India and Pakistan's combined defense expenditures came to about $11.3 billion, accounting for some 92 percent of the region's total of $12.3 billion in 1990.[19] In that year, India's defense costs, which were three times the size of Pakistan's, accounted for 69 percent of the regional total. The slowdown in India's defense expenditures and the acceleration of those of Bangladesh, Myanmar and Sri Lanka in the 1990s altered the profile somewhat. In 1994, for instance, India and Pakistan's combined defense expenditures of $10.7 billion accounted for 88 percent of the regional total of $12.2 billion, of which India's share declined from 69 percent in 1990 to 60 percent in 1994.[20]

Arms Transfers to The Region

As an arms-importing region, South Asia ranks far below the Middle East and typically falls below the Asia-Pacific region, where the number of importing states is larger and purchasing power for conventional weapons has been rising dramatically.[21] Slow economic growth in the past, increased indebtedness, Pakistan's loss of U.S. military assistance, disruption of former Soviet (now Russian) arms supplies to India, and the imperative of conserving resources to restructure their economies have placed heavy new constraints on Indian and Pakistani options for their defense budgets and arms imports for the 1990s.

Import Surge of the 1980s

In the 1980s, South Asia's defense imports, particularly India's, achieved unprecedented prominence. According to the SIPRI estimates depicted in figure 13–4,[22] India imported $21.6 billion of major weapons during the eight-year period 1983–90, or an average of $2.7 billion a year, with a peak in 1987 of over $4.5 billion. India was the number one arms importer in the Third World during the latter part of this period, 1986–90, exceeding the cumulative totals attributed each to Iraq and Saudi Arabia.[23] The financing of these imports may help account for the surge in India's defense expenditures during roughly the same period (figure 13–1).

Since 1990, India's ranking as a Third World arms importer has fallen to fourth, with arms deliveries worth $800 million in 1991, following Saudi Arabia ($7.1 billion), Afghanistan ($1.9 billion) and Iran ($1.5 billion).[24] There were signs in 1994–95 that India was resuming a major arms procurement relationship with Russia, and India's defense budget rose by 6–7 percent in real terms in 1995.[25]

Figure 13-4. *Value of Transfers of Major Weapons to India, Pakistan and Iran, 1983–90*

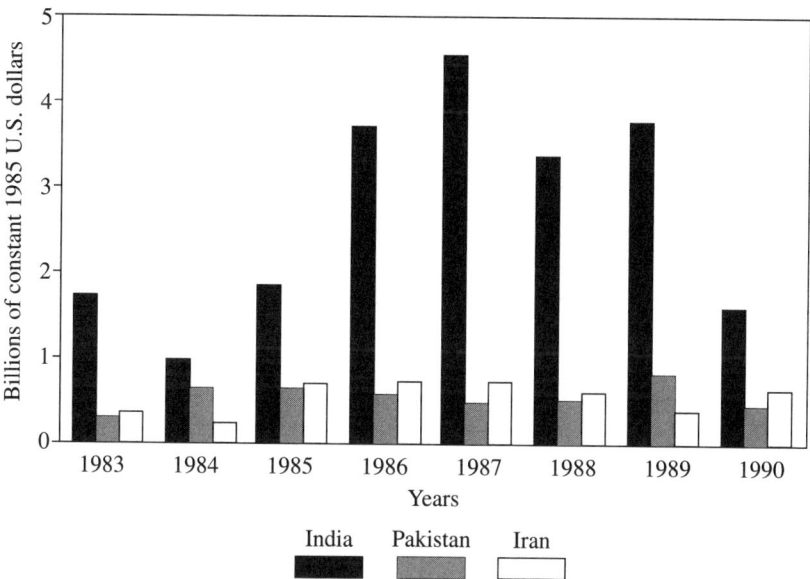

Source: Stockholm International Peace Research Institute (SIPRI), *SIPRI Yearbook: World Armaments and Disarmament* (Oxford, 1988–91).

The total of Pakistan's arms imports in 1983–90 (figure 13–4) was $4.343 billion, or an average of $543 million a year—an impressive amount, but only about one-fifth of India's figure. Pakistan's imports have dropped from this level since 1990, partly as a result of the suspension of U.S. security assistance in late 1990 as mandated by the Pressler Amendment.[26]

Comparison with Chinese Arms Imports

ACDA has estimated the average of China's arms imports during 1983–89 to be $439 million annually.[27] China's arms imports appear to have been somewhat less than Pakistan's in those years, and thus much lower than India's.

China was able to keep imports at this suppressed level because it had a domestic defense production base adequate to meet the requirements of its armed forces—albeit with essentially 1960s-vintage equipment and at the expense of all but a gradual modernization of conventional arms.

If reports of Chinese agreements with Russia and Ukraine to buy SU-27 aircraft,[28] an aircraft carrier and other modern equipment prove true, they could boost the value of China's arms imports in the 1990s to levels similar to India's in the 1980s.[29] The IISS estimated China's external (hard currency) military procurement to be worth about $1.5 billion in 1994.[30]

Arms Exports from South Asia

South Asia has not been a major source of arms exports. In contrast to Brazil and China, India has not made a major push to market its indigenously produced arms in the international market. Its own forces absorb most of its production, and retired equipment is uncompetitive or has negligible export value. Nevertheless, in the late 1980s India announced new arms export policies that removed the normative inhibitions and opened the way for public sector firms to promote arms sales. In addition, India has discussed sales with countries such as Kuwait and Afghanistan and has sold small quantities of arms to neighboring states, such as Sri Lanka. India's licensed production of MiG-21/23 and now MiG-29 aircraft gives it the facilities and experience to bid for maintenance contracts for aircraft sold by Russia to Southeast Asian states such as Malaysia. Pakistan has had an even smaller export role but has, for example, resold Chinese-origin combat aircraft to Bangladesh.

Domestic Arms Production and Assembly

India and Pakistan have relied heavily on imports to modernize their armed forces for the last two decades. The smaller states in the region, which are still in the early stages of building basic defense services, depend totally on imports for their major equipment.

India has established an extensive defense industry base in the public sector[31] that is largely self-sufficient in the production of small arms, ordnance, light armored and motorized vehicles, light aircraft and certain aircraft avionic components. Under foreign-licensed assembly and co-production arrangements, India's defense firms have also produced (or perennially claim to be almost ready to produce) main battle tanks (MBTs), armored personnel carriers (APCs), supersonic combat aircraft, helicopters, several types of guided missiles (anti-tank, air-to-air and surface-to-air), naval frigates and diesel submarines.

India's R&D activities in defense extend to: radar, communications, avionics, optics and other military electronic applications; airframe construction, special materials and jet aircraft engine development; assembly of digital processors; and development of space, missile and satellite systems. The defense R&D organizations provide system integration capabilities for upgrading the communications, surveillance and combat capabilities of imported aircraft and naval vessels, as well as independent knowledge useful for evaluating, pricing and procuring foreign arms and components.

The scientific, technical and capital requirements for successful original design, development and production of weapon systems that meet world standards are extraordinarily difficult to achieve. India's decade-long efforts to produce, for example, an indigenous MBT (the Arjun) and high performance fighter plane (light combat aircraft) have consumed significant resources and been inconclusive. So far, imports of the Soviet-design T-72 MBT and MiG-29 fighter aircraft have replaced planned procurement from these indigenous programs.

Significant advances in India's domestic defense development programs will require foreign collaboration, greater investment, guaranteed procurement of indigenous products, privatization of engineering and production and, possibly, a determined export program to spread cost and pay for imported tooling and components. Once unthinkable, such steps could become practical by the turn of the century if India continues its economic liberalization and shifts allocations from the public to the private sector. India's long-postponed diplomatic recognition of Israel in late 1992, a major foreign policy departure, could open the way to sophisticated Israeli military expertise and collaboration on a variety of weapon systems.

Pakistan also aspires to an independent defense production base. It has established, partly with Chinese assistance, a significant range of facilities for the production of basic ordnance and for the service, maintenance and rebuilding of basic defense equipment, from armored vehicles to fighter aircraft and jet engines. Its defense R&D establishment and its engineering, metallurgical and chemical industries are insufficient, however, to support a diversified, modern defense industry or actual production of major weapon systems. At the same time, they may enable Pakistan to maintain some imported equipment and augment some of its ground force capabilities in modest ways.

Modernization and Buildup of Conventional Forces in South Asia

Figures 13–5, 13–6 and 13–7 show important elements of the military buildup in South Asia in the 1980s.[32] These figures compare changes in the

Figure 13-5. *Composition of India's and Pakistan's Air Forces, 1979–95*

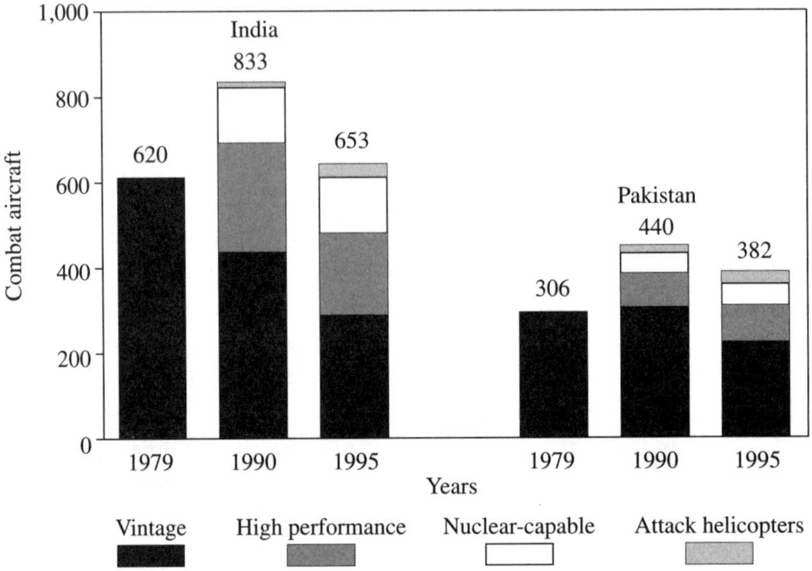

Source: International Institute of Strategic Studies (IISS), *The Military Balance 1979–80, 1990–91, 1995/96* (London, 1980, 1991 and 1995).

numbers and categories of major deployed Indian and Pakistani combat aircraft, armored vehicles and naval systems at the beginning and end of the decade. They highlight the relative size and degree of modernization of the respective Indian and Pakistani military services over time, accomplished through upgrades in mobility, firepower, range and nuclear-capable delivery systems.

Air Power

With respect to combat aircraft, figure 13–5 shows the extent to which both sides have replaced "vintage" combat aircraft with modern versions and have deployed attack helicopters. The newer combat aircraft typically have greater ranges and payload than their predecessors. A significant fraction of these high performance aircraft (for example, India's Jaguars and Mirage 2000, and Pakistan's F-16s) can be considered generically nuclear-capable.[33]

Aspects of the modernization of the air force not displayed in the figure include the acquisition of more capable air-to-air and air-to-ground missiles

Figure 13-6. *Composition of India's and Pakistan's Ground Forces—Armor and Artillery, 1979–92*

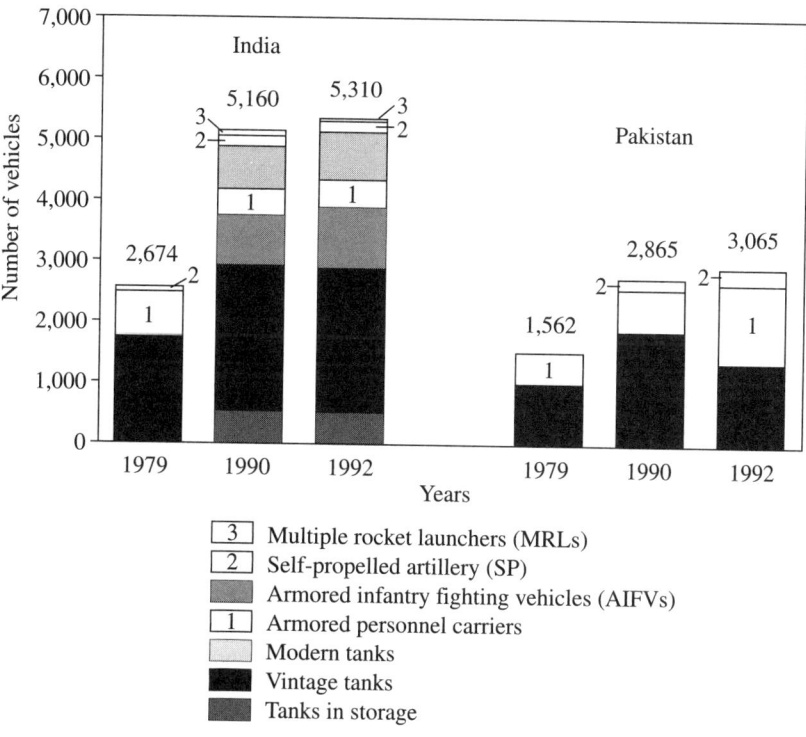

Source: International Institute of Strategic Studies (IISS), *The Military Balance* (London, various years).

on both sides, and India's growing military air transport fleet and advantages in reconnaissance and surveillance aircraft—all important features of the air balance.

Ground Force Firepower and Mobility

There has been a steady buildup, as well as modernization, of armored vehicles (figure 13–6). Both sides have increased their inventories of vintage, but locally effective, MBTs (for example, India's Vijayanta, based on the British Chieftain, and Soviet-origin T-55s, Pakistan's Chinese Type-59s and U.S.-built M-48A5s). However, while Pakistan ended the decade without acquiring an up-to-date tank, India had equipped its ground forces with 900 modern Soviet-design T-72s.[34]

Figure 13-7. *India's and Pakistan's Blue-Water Capabilities, 1979–92*

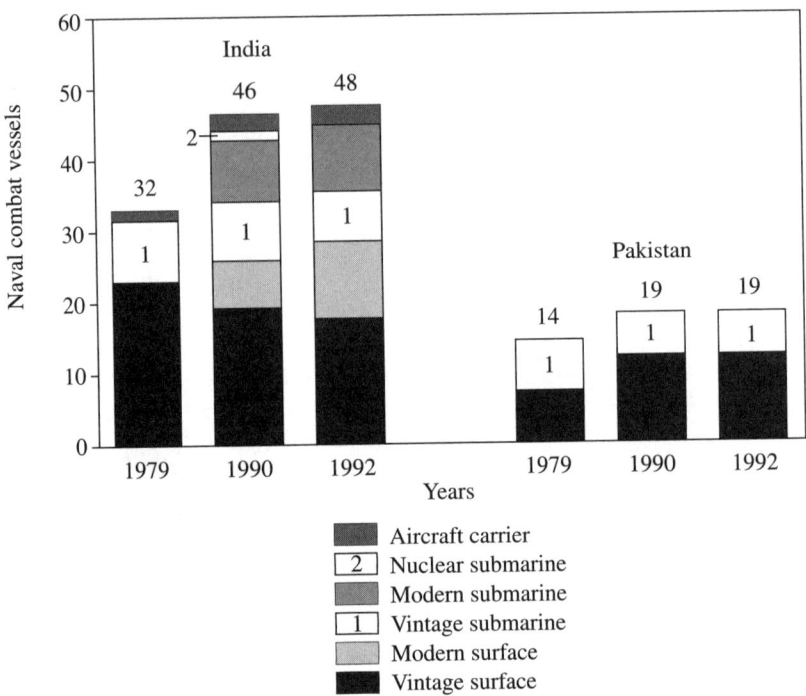

Source: International Institute of Strategic Studies (IISS), *The Military Balance* (London, various years).

A greater mobility of ground forces on both sides is reflected in the increases in APCs and addition of self-propelled heavy artillery. By 1991 India had also acquired from Soviet sources 80 multiple-rocket launcher vehicles and 800 armored infantry fighting vehicles (BMP-1/2)—augmenting the rapid offensive capability of its ground forces in open terrain.

Naval Power

India's acquisition of a two-fleet navy—consisting of two old but functioning aircraft carriers, modern diesel submarines and modern surface ships (destroyers and frigates)—is shown in figure 13–7.[35] India has equipped, as has Pakistan, its surface ships and submarines with potent anti-ship missiles (India, for example, uses the Exocet, Pakistan the Harpoon). In addition to its carrier-based combat aircraft (the Sea Harrier, a very short take-off and landing fighter airplane), India has formed a dedicated land-based squadron of maritime attack aircraft (Jaguars) and has deployed eight long-range maritime surveillance platforms (the Soviet Tu-142 Bear-F), for which Pakistan essentially has no counterparts.

Conventional Force Balances

During the modernization and buildup of the 1980s, India made greater gains and widened the gap with Pakistan in terms of modern equipment. In the category of *high performance combat aircraft,* for example, India increased its ratio of superiority during that time from 2:1 to 3:1. In the category of *armored vehicles,* the aggregate ratio of tanks and armored infantry vehicles for the two countries between 1979 and 1992 remained virtually the same at about 1.6 to 1. As of 1992, however, 35 percent of India's tanks were modern, while none of Pakistan's were, and 64 percent of India's APCs were modern, heavily armed infantry fighting vehicles, whereas Pakistan still had only traditional, lightly armed APCs.

India's *surface navy* became much more powerful than Pakistan's over the decade, having acquired its own air strike capability and the means to blockade Karachi and cut off Pakistan's maritime links with the outside world. India's combined naval, submarine, air strike and air transport forces also provide a modest maritime power projection and intervention capability in the Indian Ocean. India used this capability, with the Sri Lankan government's agreement, to insert a large Indian "peacekeeping force" during Sri Lanka's internal conflict in July 1987 (the force was withdrawn in 1989 after its impact proved inconclusive) and to suppress an insurrection in the Maldives in November 1988.

The Nuclear Dimension

Buildup of Nuclear Capabilities

India first demonstrated its nuclear weapons capability in May 1974 in a so-called "peaceful nuclear explosive" experiment. The test occurred almost ten years after China had detonated its first nuclear weapon in October 1964, and six years after India rejected signing the Nuclear Non-Proliferation Treaty (NPT). The possibility that China would blackmail India with its nuclear threat was a factor in the debate within India in the 1960s over whether to sign the NPT or, instead, to develop or maintain the option to build nuclear weapons.[36]

Since the 1974 explosion, India has maintained a policy of ambiguity, denying it has a nuclear weapons program or plans for one. It hedged on this point increasingly under the tenure of Prime Minister Rajiv Gandhi in the late 1980s, as evidence accumulated that Pakistan had made technical breakthroughs with uranium enrichment and was achieving its own nuclear weapons capability.

In the mid-1980s India began testing military prototype, nuclear-capable missiles based on booster technologies that had been under development in its civilian space program. The significance of the test of the longer range Agni missile prototype, described by R&D officials as a "technology demonstrator," was that India had an intermediate-range ballistic missile (IRBM) capability under development with the range to target urban areas in China.

The implication was that if Pakistan went nuclear, and India decided to become a nuclear power, India would have to be able to deter China as well, rather than just Pakistan. This implication continues to apply.

Although neither India nor Pakistan has declared that it has nuclear weapons, or intends to deploy them, unveiling their concealed nuclear programs or testing nuclear devices would ignite an open nuclear arms race. Pakistani officials publicly stated as early as 1992 that Pakistan has the means to construct nuclear explosive devices,[37] and plausible hints of Pakistan's breakthrough date back to the Brass Tacks crisis of 1986–87.[38] It is now widely believed that those Pakistani assertions are credible and that highly-enriched uranium devices need not be tested to be confident they work.

The active internal conflict in Kashmir and the still smoldering Khalistani (Sikh) separatist movement in the Indian state of Punjab represent flashpoints for war between the two countries. Both India and Pakistan claim the Indian-held state of Jammu and Kashmir, the Kashmiri part of which has been the setting of armed resistance to Indian rule since late 1989. Punjab lies on India's border with Pakistan. The danger of the outbreak of a conventional war over Kashmir and the Sikh unrest in Punjab climbed to a crisis peak in early 1990 and has oscillated at high levels since. India has flooded Kashmir with 600,000 troops since 1993 in its effort to suppress the militant resistance. The risk that such a conflict could escalate to the nuclear level remains deadly serious.[39]

Slow Pace of Chinese Nuclear Modernization

While the extent of the nuclear weapon capabilities in Pakistan and India remains unclear, it is worth noting that nuclear forces deployed by China

Figure 13-8. *China's Nuclear Arms—Deployed Systems, 1980–95*

A. Intercontinental range

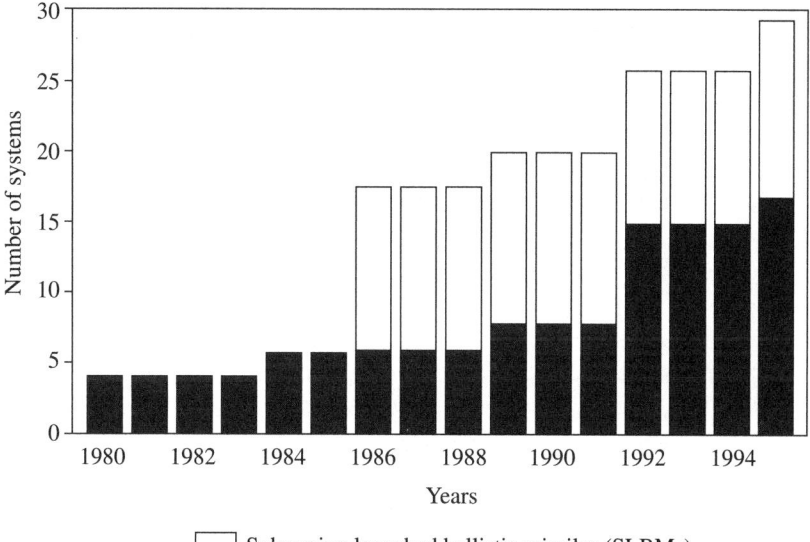

Submarine-launched ballistic missiles (SLBMs)
Intercontinental ballistic missiles (ICBMs)

B. Intermediate and medium-range

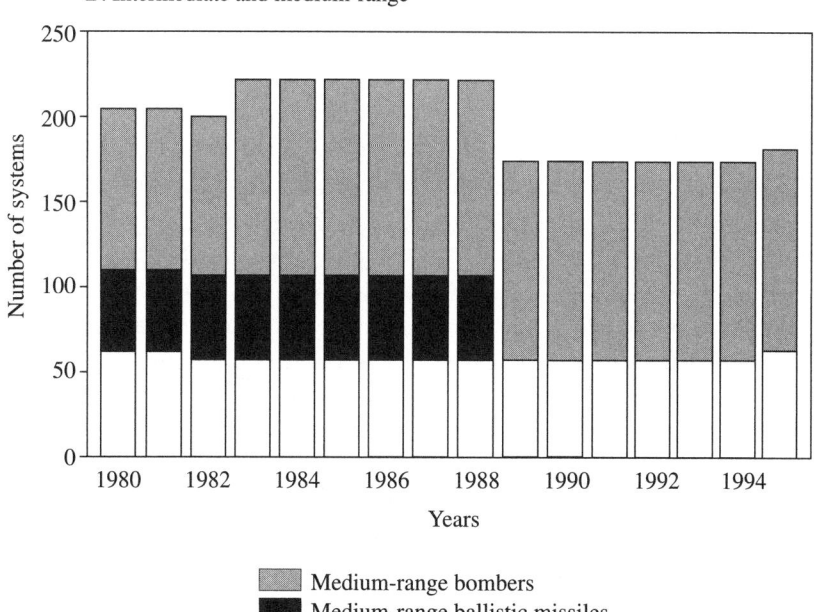

Medium-range bombers
Medium-range ballistic missiles
Intemediate-range ballistic missiles

Source: International Institute of Strategic Studies (IISS), *The Military Balance* (London, various years).

not only are smaller than many assume but according to IISS data may even have shrunk since 1988, dropping from nearly 250 to less than 200 deployed medium, intermediate- and long-range delivery systems in 1994 before rising slightly in 1995 (see figure 13–8).[40] Only long-range missile systems that can reach the European part of the former Soviet Union, and presumably any part of India, appear to have increased. This increase, however, was both surprisingly small and slow, a buildup of from four to seventeen ICBMs, and from zero to twelve submarine-launched ballistic missiles (SLBMs)—which may not be strategically operational[41]—over the course of fifteen years.[42]

The modest downsizing of China's nuclear force has gone hand-in-hand with modernization and thus probably reflects resource constraints rather than a specific new policy of numerical or geographical restraint. It would be regrettable, however, if a nuclear arms buildup in India and Pakistan caused China to revert to a buildup and redeployment of longer range nuclear weapons to target India. Given the major theater and strategic reductions in nuclear arms agreed to since 1987 by the former Soviet Union and subsequently between the United States and Russia, it seems only logical that China would perceive less incentive today to enlarge its nuclear arsenal.

Arms Suppliers

The pattern of major suppliers to India and Pakistan and their respective market shares have been broadly the same for the last two decades. The Soviet Union had acquired a dominant share in the supply of arms to India by the early 1970s, whereas the United States, which had been a significant supplier in prior years, was all but excluded. The shares for the period 1986–90 are roughly indicative of the previous fifteen years (figure 13–9). Soviet arms accounted for about 73 percent of the $17 billion Indian market in those years, with France and the United Kingdom the most important alternate suppliers, with 10 percent and 7 percent of the market respectively. The Federal Republic of Germany, the Netherlands and Sweden each had smaller but still significant shares of about 3 percent.

The principal supplier to Pakistan after the Korean War was the United States until the mid-1960s, when China also became a major supplier. The Soviet Union, with a brief exception in the 1960s, did not enter this market. Alternate suppliers over the years were Western European states, principally France and the United Kingdom. The shares of the respective suppliers to

Figure 13-9. *Major Weapons Suppliers—Value of Transfers to and Share of India's and Pakistan's Markets, 1986–90 (Millions of U.S. $)*

A. India

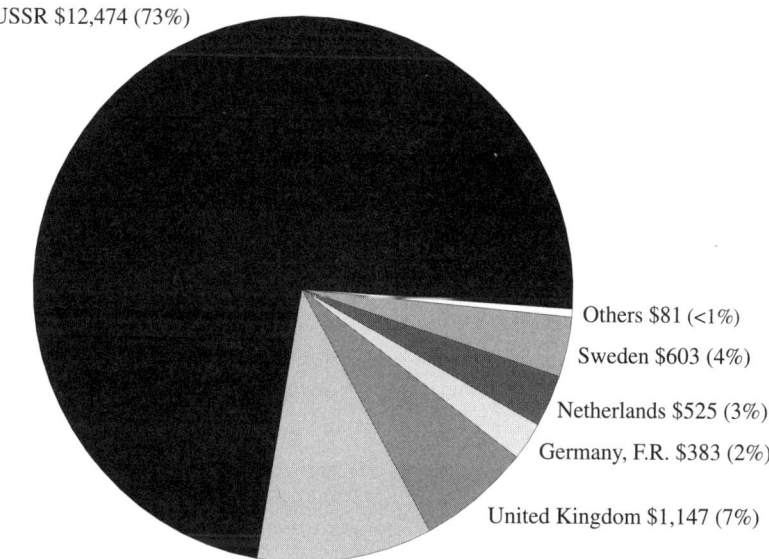

USSR $12,474 (73%)

Others $81 (<1%)

Sweden $603 (4%)

Netherlands $525 (3%)

Germany, F.R. $383 (2%)

United Kingdom $1,147 (7%)

France $1,776 (10%)

B. Pakistan

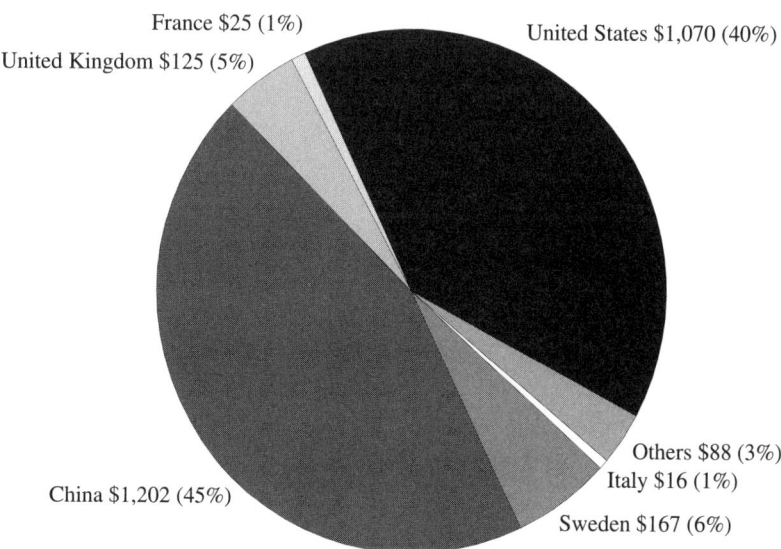

France $25 (1%)

United Kingdom $125 (5%)

United States $1,070 (40%)

Others $88 (3%)

Italy $16 (1%)

Sweden $167 (6%)

China $1,202 (45%)

Source: India—Stockholm International Peace Research Institute (SIPRI), *SIPRI Yearbook 1991: World Armaments and Disarmament* (Oxford, 1991) table 7.4, 209 and Pakistan—Stockholm International Peace Research Institute (SIPRI), *SIPRI Yearbook 1991: World Armaments and Disarmament* (Oxford, 1991) 210.

Pakistan's $2.7 billion market from 1986 to 1990 shown in figure 13–9 are roughly representative of the last decade, when China, with 45 percent of the market, overtook the United States, with only 40 percent, as Pakistan's dominant supplier.

The dominance of Chinese sales is more quantitative than qualitative: Chinese conventional weapons typically are much cheaper than weapons systems from the West but also inferior in performance. Chinese weapons also demand heavy maintenance. However, the level of technical skills needed to maintain such systems either is available in Pakistan, or can quickly be provided through cooperation and training with the supplier.[43]

It is uncertain what the pattern of suppliers to South Asia will be in the future. The collapse of the former Soviet Union disrupted its supplier arrangements with India, even for ammunition and spare parts. Russia, the principal successor state in the arms supplier role, has, as noted, been seeking payment in hard currencies rather than rupees (or bartered commodities) for future supplies.[44] This demand has raised the price of Russian military items closer to typical international levels and will squeeze India's limited hard currency resources. In the long run, India might therefore turn more often to Western sources. In the short run, its heavy dependence on Soviet (now Russian) equipment—in both its "order of battle" and defense production programs—has required that it try to restore reliable Russian supplies. Russia's promotion of military sales in Southeast Asia could enable India to bid on regional maintenance contracts for Russian equipment (for example, in Malaysia) to increase its own hard currency earnings.

The cutoff of U.S. military assistance to Pakistan as of late 1990, triggered by presidential action on the certification question required under the Pressler Amendment, virtually removes one of Pakistan's two primary suppliers from the long-term picture.[45] Unless Pakistan can find other suppliers—and Russia could enter this market—it will depend primarily on China and Western Europe for conventional arms in the foreseeable future.

Divergent Pressures

India and Pakistan are being pulled simultaneously in two different directions. One is toward increased militarization and a nuclear arms race. The other is toward the opportunity offered by the end of the Cold War and expansion of international trade to pursue economic growth and social progress, ameliorate the sources of bilateral conflicts, and cap or reduce military spending. In this context, it is difficult to judge the direction of likely trends on the demand side for conventional arms.

The pressure toward militarization comes from the inability of India and Pakistan to control by traditional political means the direction and outcome of the conflict over Kashmir and from their respective problems of domestic political turmoil. The problem is aggravated acutely by the introduction of ballistic missiles and nuclear weapon capabilities. These factors erode confidence in national security and set the stage for a continuing nuclear arms rivalry.

In mid-1990 India and Pakistan approached the brink of war over Kashmir, and the potential for a major confrontation over Kashmir remains. Warfare over this disputed territory could escalate into a nuclear exchange. Military planners would view the danger as sufficient to justify building more effective conventional defense forces. Some would go further and call for preparations for nuclear defense by developing, training and organizing nuclear-capable strike forces and drawing up contingency plans.

New military developments in neighboring regions could also encourage Indian and Pakistani militarization. Russia's willingness to supply Iran with advanced conventional arms packages—SU-24 fighter bombers, Kilo-class diesel submarines and modern tanks—could precipitate greater instability in the Gulf and exert pressure in turn on Pakistan and India to prepare for new military contingencies.[46] These pressures would become severe if, as the media once speculated, Russia were to sell Iran Backfire (Tu-26/Tu-22M) "nuclear" bombers.[47] Iran's maritime power projection based on the Shah of Iran's naval modernization in the 1970s was a significant factor in Indian naval planning and eventually was reflected in the scope of India's naval acquisitions in the 1980s.[48] In short, if military crises or compelling political pressures cause India and Pakistan to devote new resources to the import of conventional arms, it will be difficult to use supply-side measures alone to restrain a resumption of the conventional arms buildup each pursued in the 1980s.

On the other hand, India and Pakistan are also being pulled in the more positive direction of structural economic reform, with a view to unleashing foreign and domestic investment and stimulating rapid economic growth. The objective of sustained economic growth could be a strong incentive in both countries to hold defense spending down, limit military imports and investment, and avoid the drain of a nuclear arms race. Above all, they strengthen the case for the leadership in both countries to do everything possible to avoid incurring the costs of fighting another major conventional war.

The basic aims of policy for South Asia, in addition to security and arms control measures, should be to support these positive economic objectives, find means to strengthen the will of New Delhi and Islamabad to avoid armed conflict, and encourage the pursuit of political solutions that remove or reduce the sources of conflict. Arms control measures can help to reinforce policies supporting these basic aims.

In the short term, the most fertile areas for the development of arms control in South Asia for both conventional and nuclear arms proliferation are more likely to be found on the demand side than on the supply side. The weak purchasing power in the region, especially in hard currency terms, naturally restrains the demand for arms imports. Increasingly important, however, is the movement on both sides in South Asia to dampen the political causes of conflict, improve military stability with the help of CBMs and accept that the underlying causes of conflict must be addressed in serious negotiations.

This process is still in its first stages. It has a long way to go. To produce long-lasting results, it must be continued. There are signs that decision-makers in South Asia are beginning to respond realistically to the opportunities of the post-Cold War environment, not only on the economic front, but also on territorial and security matters as well. "Old quarrels" might yield to the "new realities."[49]

As regional measures between the two local powers begin to take shape, new supply-side measures that complement those on the demand side will need to be found. The Permanent Five (P-5) should, in addition to consulting on multilateral guidelines for arms transfers, address supply-side measures that link the regional security concerns of India and Pakistan locally with those of the United States, Russia and China outside.

Confidence-Building Measures

India and Pakistan initiated an important nuclear CBM in January 1991—a mutual commitment not to attack the other's nuclear facilities. This CBM provides a degree of transparency by requiring an annual exchange of data identifying each side's nuclear facilities and locations. The first exchange of data was implemented in early 1992. Pakistan's President Zia ul-Haq had proposed this measure to Indian Prime Minister Rajiv Gandhi in Delhi in 1985, and Gandhi announced the mutual pledge to the press before Zia's visit ended. Pakistan's interest in such an agreement probably stemmed from hints in the early 1980s that India had considered, and even discussed with Israel, military action to destroy Pakistan's sensitive Kahuta facility, the site of a gas-centrifuge uranium enrichment program.[50] The agreement helps allay authentic fears about the other side's intentions and can reduce the risk of military overreaction to non-threatening activities, such as routine military movements on the other side. More important for the long term, this original and apparently successful nuclear CBM is an important precedent for additional arms control measures.

Conventional military confidence-building initiatives gained impetus in South Asia in mid-1990.[51] Receptivity in New Delhi improved after the threat over Kashmir was defused, following the visit of U.S. Deputy National Security Advisor Robert Gates.[52] Before this time, India and Pakistan had experience with extremely narrow military CBMs.[53]

Proposals that India and Pakistan explore the use of CBMs to reduce military tension led to a bilateral dialogue on specific concepts. The two reached agreement on some concepts in April 1991, leading to ratification of two CBMs in August 1992.[54] One CBM agreement dealt with the prevention of airspace violations; the other provided for advance notice of military exercises, maneuvers and troop movements.

Although each of these conventional CBMs can help reduce the chances of a military miscalculation in peacetime and are encouraging steps, neither is ambitious or exceptional. Because CBMs must be ratified by the top political leaders, they must be engaged in their development. Such involvement can in turn become an inducement for wider public interest in arms control constructs. A public debate on the utility of CBMs is beginning in India.[55]

India has showed little enthusiasm for the nuclear CBMs suggested by other parties,[56] but it has moved forward with Pakistan on CBMs designed to preclude the introduction of other weapons of mass destruction into the region, such as chemical weapons (CW). At the August 1992 meeting of foreign secretaries, for example, the two signed a joint declaration banning CW, an action that anticipates their participation in the forthcoming, U.N.-sponsored Chemical Weapons Convention.

The declared CW ban is an important CBM for India and Pakistan. It could preempt military interest in CW in the subcontinent stimulated by the resort to chemical warfare between Iraq and Iran in the 1980s. It offset the erosion in India's image that resulted when it was revealed that Indian firms had sold CW precursors to other parties, notably Syria, which is believed to have a CW stockpile.[57]

India's diplomatic interest in conventional CBMs is spreading to other relationships. For example, Indian Defense Minister Sharad Pawar is reported to have proposed bilateral conventional CBM concepts to Beijing in his visit to China.[58]

Pakistan currently seems less enthusiastic over narrow, conventional CBMs than India does, because of a concern that India may use the concepts to control non-nuclear arms politically, mainly to improve its international image, while stalling serious negotiations on Kashmir, on substantive security issues and on nuclear weapon non-proliferation measures.[59]

Presumably, Pakistan has reason to be encouraged that India now implicitly acknowledges a relationship between the utility of CBMs and possible benefits of negotiations on Kashmir. CBM constructs on mutual troop withdrawal and monitoring could, for example, permit an end to the Indo-Pakistani Siachen Glacier conflict in Kashmir, where troops from each side stationed on the steep ice flows regularly die, sometimes from bullets but principally from exposure to the elements. They do so in defense of their nation's concept of where the undemarcated line of control should actually run.

There are two stabilizing conventional CBM constructs that India and Pakistan could pursue to complement their economic reform initiatives and reduce military tensions. One is an agreement in principle that each would benefit from placing a real limit, adjusted for inflation perhaps, on its military expenditures. Following this agreement, the two countries could begin to discuss where each should set its limit and how to monitor that limit. Public discussion of this issue could facilitate the beginnings of transparency in the defense budget. The second would be to agree in principle that each side should cap its opposing conventional forces and initiate a dialogue on acceptable lower limits and how to implement and monitor them.

Moving Forward

Breakthroughs in security and arms control arrangements often are preceded by long negotiations and incremental, interim steps. India and Pakistan appear to be on that track. The challenge is to keep the bilateral discussions moving forward constructively on several levels and to bring developments at each level together when circumstances permit. Perhaps for the first time in this region, the initiatives of the regional states offer significant opportunities to build on.

Multilateral Security Measures

P-5 deliberations on the question of what transfers of conventional arms to India and Pakistan are acceptable should go hand in hand with consultations on how additional security can be provided to the parties involved, particularly through multilateral cooperation. Regional leaders should have the opportunity to express their security concerns and visions for achieving them in both quiet diplomatic channels and international institutions, including appropriate organs of the United Nations. Under P-5 auspices, consideration could be given to developing a technical program to identify monitoring and peacekeeping assets that could be deployed to watch over implementation of CBMs, increase military transparency, reduce the chance of crises and enhance military stability. The success in Europe of the Conference on Security and Cooperation in Europe, and in the Gulf War of the liberation of Kuwait and subsequent peacekeeping operations to protect the Kurdish people in Iraq from internal war, gave new impetus to regional concepts for collective security. The implications for building a collective security framework for South Asia should be explored. Bilateral security instruments developed for South Asia in the Cold War could be reexamined side by side. The same is true of the charter of the South Asian Association for Regional Cooperation, with the focus on whether external security commitments could be refashioned and woven together in a collective security instrument for the region.

Supplier Policy Guidelines

Supply-side policies establishing restraint in transfers of conventional arms to the region may not need to emphasize quantitative restrictions, since the financial capacity of the region to purchase arms probably will be self-limiting in the near term. The near-term focus in conventional arms guidelines should be qualitative—restrictions on destabilizing weapons and delivery systems and on dual-use technologies that lend themselves to the development and deployment of nuclear weapons and other weapons of mass destruction.

P-5 statements indicate they will evaluate the stabilizing and destabilizing attributes of a given type of arms when considering requests for arms transfers. In the case of most conventional weapon systems, these attributes cannot be determined in the abstract. Whether receipt of a major system will be stabilizing or destabilizing depends, among other factors, on the recipient's security needs, the threat it faces and the mission intended for the weapon system. Even so, a certain degree of international consensus could be worked out when specific regional circumstances are adequately analyzed and discussed. For example, a moderate view on conventional arms criteria for South Asia might hold that: interceptor aircraft and air-to-air missiles, air defense artillery and missiles, and attack warning sensors (whether ground-based or airborne) favor territorial defense and therefore tend to be conducive to stability; long-range strike aircraft can be destabilizing, particularly if forward-deployed in large numbers; and large, ground-launched, surface-to-surface ballistic missiles are inherently offensive weapons and relatively destabilizing.

Qualitative criteria for the region—or at least for India and Pakistan—should embody a coherent set of normative standards and, other things being equal, should be applied consistently. Where other things are not equal, as when country A itself already produces a staple weapon system that country B also needs but can only obtain by foreign import, outright denial of the transfer to country B would be inequitable and could be destabilizing.

Arms Control Record and Eligibility

Policy decisions to transfer advanced conventional arms, weapons production technology and dual-use technology should take a recipient's arms control and non-proliferation record into account. A recipient's compliance with its own legal and political commitments to arms control and non-proliferation—for example, good faith compliance with implementation of the nuclear safeguards of the International Atomic Energy Agency (IAEA) in its facilities—is the starting point. In addition, however, a government's support for or obstruction of the norms of the international non-proliferation regime over time should be weighed, as well as the quality of its cooperation in developing new measures. Additional factors would be a recipient's openness to new CBM and arms control measures that could locally restrain the proliferation of nuclear and other weapons of mass destruction and thereby reduce the risks of nuclear war.

It follows that suppliers should be prepared to supply types and levels of conventional arms, commensurate with legitimate self-defense requirements, to those recipients that have a good record on arms control and non-proliferation. For recipients with uneven or poor records, the presumption should be that major future arms transfers—and, where relevant, arms production technology transfers—be conditioned on acceptance of new non-proliferation commitments. Sponsorship and participation in nuclear and conventional CBMs could qualify for consideration as part of the arms control record. Multilateral enforcement of non-proliferation norms by arms transfer conditionality could produce positive results in South Asia.

Demand Restraint Through Confidence-Building Measures

To complement supply-side initiatives, collateral efforts to encourage self-restraint in demand are needed. Conventional CBMs could help foster conditions favoring restraint in demand. In bilateral military talks on conventional defense sufficiency, for example, India and Pakistan could exchange basic data on their military forces, identify ceilings for force reductions at lower, stabilizing levels and define categories of heavy combat equipment for which numerical limits could be negotiated.

A further goal for such talks could be a commitment to exchange information on planned imports and production of arms and to consider mutual limits on the procurement of certain types of weapons.

Literal parity (that is, equal levels) of conventional forces between India and Pakistan alone is not a plausible objective. India also needs conventional defenses against China and has a much longer maritime perimeter to patrol. At the same time, agreed-upon ratios of deployed forces and permitted equipment levels by geographic sector would be conducive to military stability and is suitable for conventional CBM negotiation.

Two other conventional types of CBMs should be given near-term priority because of their potential value in limiting the danger of war from incidents that could easily be misjudged in light of the ongoing internal conflict in Kashmir. One is verifiable limits on the deployment of ground forces in defined border zones, with exchanges of observers or inspections to monitor compliance. Another is the development of a local "open skies" regime that allows each side to fly surveillance aircraft over its territory at regular intervals or on short notice. Overflights would help confirm by visual or photographic means a country's assurances that it is not preparing for attack and moving forces for hostile purposes.

Nuclear CBMs could be steps toward an effective solution of the threat of nuclear proliferation in South Asia. Perhaps the most politically useful nuclear CBMs at the turn of the decade would have been for India and Pakistan to declare a freeze on the production of nuclear material, allow monitoring of their facilities to verify the halt in production and publish the amounts of existing materials and the civilian purposes for which they will be used. Similarly useful would have been bilateral commitments to refrain from testing nuclear explosives, to place all new nuclear facilities under IAEA safeguards and to conduct reciprocal visits of scientists and technicians at nuclear facilities in each country. Arguably, however, both of these CBMs have been overtaken since 1993 by international negotiations for a zero-yield comprehensive test ban treaty and for a ban on the production of nuclear weapons material. Both of these regimes would entail international verification and thus would be more effective than regional CBMs.[60]

Conclusions

Five sets of policy objectives, two on the supply side and three on the demand side, should be evaluated and integrated with each other. On the supply side, new provisions for restraint in the transfer of conventional arms can be developed, building on the P-5 process initiated in 1991. Controls over transfers of conventional arms should be designed to complement the supplier regimes set up to control the dissemination of nuclear, dual-use and missile technologies. Equally important on the supply side, diplomatic efforts to improve regional security are needed. These should draw not only on traditional security commitments and calibrated assistance with military equipment, but also on collective security and international peacekeeping schemes, some of which have already demonstrated new promise in the post-Cold War environment.

On the demand side, initiatives in three areas are needed. The two that are most likely to be productive as interim steps are regional efforts to build military stability through CBMs, including adoption of non-proliferation constraints and policies of economic growth through the expansion of trade and international cooperation. The third initiative involves finding permanent, political solutions to the underlying causes of regional conflicts, so as to reduce the incentives for arms proliferation, propensity for crisis and inclination to resort to war. While it is natural that political solutions originate with local actors and be adapted to local interests, the external powers and international institutions can make significant contributions.

Notes

1. The author wishes to express his appreciation to colleagues of the Washington Council on Non-Proliferation and to Michael Krepon of the Henry L. Stimson Center and Selig Harrison of the Carnegie Endowment for International

Peace for their inspiration and helpful criticism, although they remain free of any responsibility for the final result.

2. See R. Jeffrey Smith, "Possible Nuclear Arms Test by India Concerns U.S.," *The Washington Post* (16 December 1995).

3. National reporting of defense expenditures is considered sensitive and therefore varies in openness and detail; it is not uncommon to understate or conceal defense costs. In the case of South Asia, the following points are worth noting: India's defense budget is published and reviewed by the Parliament but may exclude, for example, costs incurred for military retirement, paramilitary forces, defense-related space and nuclear research and development (R&D), and defense-related transportation and infrastructure. Pakistan makes less information on the defense budget available to its National Assembly and public, and the accounting of defense foreign assistance, foreign subsidies, and barter relationships is poorly understood.

4. The figures in rupees have been translated into U.S. dollar values. As a result, the effect of the relatively high inflation (local currency) does not appear in the trend lines. India's rupee expenditures for defense have increased at a much more rapid rate than their dollar equivalents, as devaluations have reduced the rupee's international exchange value. Pakistani rupees have exhibited a similar pattern.

5. The International Institute of Strategic Studies (IISS) indicates that Indian defense expenditures in billions of rupees and billions of dollar equivalents were as follows between 1990 and 1995:

	Rupees	Dollars	
1990	176.84	10.10	
1991	183.55	8.07	(estimated)
1992	174.10	6.70	
1993	217.80	7.14	
1994	233.00	7.33	(estimated)
1995	255.00	8.12	(defense budget)

See International Institute of Strategic Studies (IISS), *The Military Balance 1992–93* (London, 1993), 131, and the country pages for India in the subsequent annual volumes of *The Military Balance* through 1995/96.

6. See IISS, *Strategic Survey, 1993–1994* (London, 1994), 189.

7. Note the surge in major weapons deliveries to India in the period 1986 to 1989, depicted in figure 13–2.

8. For a thoughtful account of Indian military planning and expansion of forces through the mid-1980s, including the expansion of naval doctrines to deal with the problem of energy security and threats emanating from the Persian Gulf, see Raju G. C. Thomas, *Indian Security Policy* (Princeton, 1986), especially 131–194, 275–300.

9. The IISS notes, however, that according to a report of the Indian Ministry of Finance, imported defense equipment accounted for about 9 percent of India's $91.8 billion external debt in 1993 (about $8.3 billion), of which about 90 percent ($7.4 billion) was owed to Russia in unrequited rupee credits, along with a sizable

dollar-denominated component of about $1 billion. IISS, *The Military Balance 1995/96,* 154.

10. For insights into Russian and Ukrainian efforts to reestablish an arms sales relationship with India, see Molly Moore and John Ward Anderson reporting on India's economic and trade predicaments, part of a larger assessment of the impact of the September 1992 European currency crisis, in *The Washington Post* (20 September 1992).

11. The unusually wide fluctuations in Pakistan's line in the left side of the figure appear to reflect the disruption of the 1971 Indo-Pakistan war. An outcome of that war was that Pakistan's defense pay-out as a proportion of GDP rose sharply from about 4 percent to 7–8 percent. East Pakistan's secession as Bangladesh meant that, for the first time, the economy of West Pakistan had to absorb the 1972 costs of the Pakistani armed forces, which were recruited mainly in the West.

As a proportion of gross national product (GNP), the burden of Pakistan's defense expenditures has been on the high side of world averages, although far from the extremes found in the Middle East and Far East. It has been comparable with, but somewhat lower than, for example, Egypt's since 1985 (7.6–9.2 percent of GNP), and much lower in those years than Iraq's (21–25 percent), Iran's (possibly as high as 32 percent), Israel's (10–21 percent), Syria's (13–16.4 percent) and Jordan's (12.8–14.6). It was also much lower than that of the Democratic People's Republic of Korea (North Korea) (23–26.7 percent). See IISS, *The Military Balance, 1992–1993,* 218ff, table on "Comparisons of Defence Expenditure and Military Manpower, 1985–1991," and 108, which contains figures and notes on Iran.

12. See the section on China in Saadet Deger, "World Military Expenditure," in Stockholm International Peace Research Institute (SIPRI), *SIPRI Yearbook 1991: World Armaments and Disarmament* (Oxford, 1991), 156–159.

13. Recently, the IISS, SIPRI and ACDA began revising their estimates of Chinese defense expenditures sharply upward. For example, the IISS noted in 1992 that, whereas China's published defense budget for 1990 was the equivalent of about $6 billion, the real level of defense spending was somewhere in a range between $11 billion and $23 billion. See IISS, *The Military Balance, 1992–1993,* 145. Three years later, the IISS had approximately doubled its estimates of China's defense expenditures, proposing a figure of $28.5 billion for 1994. Extensive reasoning for the revised estimates is provided in a special chapter, "Chinese Military Expenditure," in IISS, *The Military Balance 1995/96,* 270–275.

14. U.S. Arms Control and Disarmament Agency (ACDA), *World Military Expenditures and Arms Transfers, 1990* (Washington, D.C., 1991). This issue provides figures only through 1989—there is usually a two-year lag in the assembly of data in the preparation of each volume. ACDA's later, revised estimate of $56 billion, based in part on purchasing power parity reassessments of Chinese defense expenditures, is evaluated further in IISS, *The Military Balance 1995/96,* and see the preceding note.

15. Pending U.S. military sales to China following the imposition of sanctions in response to Beijing's 1989 Tiananmen crackdown would be a relatively small part of the bigger picture. President George Bush announced on December 22, 1992, however, that four suspended foreign military sales would go ahead. They included: upgraded avionics packages for Chinese F-8 (J-8) fighter aircraft (an indigenous version of the Soviet MiG-23); equipment for a munitions production

line; four anti-submarine torpedoes; and two artillery-locating radars. See the report by Reuter published in *The Washington Post* (23 December 1992).

16. See IISS, *The Military Balance, 1992–1993,* 220, 145.

17. See note 13 above.

18. Bangladesh's defense budget doubled from $139 million to $278 million between 1985 and 1989, rose to $327 million in 1990, and reached $461 million in 1993. Driven by the internal Tamil-Sinhalese conflict, Sri Lanka's defense expenditures are likely to continue increasing: they went from $228 million in 1985 to $439 million in 1990 and reached nearly $500 million in 1993. Myanmar's defense expenditures increased from $228 million in 1985 to $292 million in 1990 and $403 million in 1993. Nepal's defense expenditures in 1993 were $41 million. See IISS, *The Military Balance* series for 1991–92, 1992–93, and 1995–96.

19. These calculations for 1990 are based on corresponding tabular data, standardized by the IISS on the basis of U.S. dollars and 1995 prices and exchange rates, contained in IISS, *The Military Balance 1992–93,* in the Asia and Australia section of the table, "Comparisons of Defence Expenditure and Military Manpower 1985–91," 218–221.

20. These regional profile calculations for 1994 are based on corresponding tabular data, standardized by the IISS on the basis of U.S. dollars and 1993 constant prices, contained in IISS, *The Military Balance 1995/96,* in the Central Asia portion of the table, "International Comparisons of Defence Expenditure and Military Manpower in 1985, 1993, and 1994," 264–269.

21. The massive Soviet arms transfers to Afghanistan from 1980 to 1990 are usually included in compilations of "South Asian" arms transfers. The resulting impression that South Asia became, after the Middle East, the second leading importer of major weapons in the Third World in the 1980s is misleading. The Congressional Research Service (CRS), for instance, estimated that transfers of Soviet arms to Afghanistan cumulatively were worth $14.7 billion between 1984 and 1991. Since this inflow occurred under Soviet occupation, with much of it brought in directly by Soviet forces for counter-insurgency purposes, it is not typical of the patterns of regional arms transfers or acquisitions. This issue, because it was not central to the main policy concerns of this project, is not addressed further in this chapter. For sources that for geographical reasons include Afghanistan in their statistics on arms transfers in South Asia, see Ian Anthony, et al., "The Trade in Major Conventional Weapons," *SIPRI Yearbook 1991,* appendix 7A; Richard F. Grimmett, *Conventional Arms Transfers to the Third World, 1984–1991,* Congressional Research Service, U.S. Library of Congress (Washington, D.C., 1992), table 1J, 59.

22. Figure 13–4 is based on SIPRI estimates of the value of major arms transfer deliveries to a recipient in a given year. SIPRI's method of evaluation includes only *major* weapons, not all military transactions. It emphasizes comparability, estimating the value of major weapons of a given type regardless of production subsidies, financing terms or actual payments. See Ian Anthony, et al., "The Trade in Major Conventional Weapons," *SIPRI Yearbook 1991,* table 7.2, 199; Aaron Karp, "The Trade in Major Conventional Weapons," Stockholm International Peace Research Institute (SIPRI), *SIPRI Yearbook 1988: World Armaments and Disarmament* (Oxford, 1988), table 7.2, 178.

23. Anthony, table 7.1, 199.

24. Grimmett, *Conventional Arms Transfers,* table 2K, 72. Grimmett's measurements are based on the value of new arms transfer *agreements* (or of deliveries under the terms of those agreements), and it thus differs from the SIPRI approach, which estimates the value of *major weapons delivered* based on normalized international prices for comparable items.

25. Indian deals with Russia in 1994–95 reportedly included the possible sale to India of the Gorshkov aircraft carrier and Indian procurement of Russian naval frigates, fresh consignments of the MiG-27 tactical strike aircraft (a possible carrier of a nuclear bomb) and MiG-29 fighter/interceptors, and refurbishment of India's former mainstay but now venerable fleet of MiG-21 interceptors. See IISS, *The Military Balance 1995/96,* 152ff.

26. During the 1990s, Pakistan's main procurements were from China (type-85 battle tanks), United Kingdom (naval frigates and Lynx helicopters) and France (submarines). Pakistan is reportedly cooperating with China to develop and produce a light combat aircraft (similar to the Lavi) and may turn to France for new Mirage aircraft as alternatives to the U.S supply of F-16Cs cancelled under the Pressler Amendment.

27. ACDA estimates the value of items actually delivered in a given year, as does SIPRI. However, ACDA includes transfers of small arms, a broader measure than SIPRI's, which uses only "major weapons." ACDA estimates for all Chinese arms imports show wide annual variations: 1983, $122 million; 1984, $504 million; 1985, $740 million; 1986, $610 million; 1987, $672 million; 1988, $312 million; and 1989, $110 million. See ACDA, *World Military Expenditure,* table II, 100. See also p. 31 for the assumptions underlying ACDA's valuation of arms transfers.

28. The Su-27 (Flanker) is an agile, long-range interceptor with "look down, shoot down" capability, decades ahead of the other aircraft in the Chinese inventory.

29. In an unsuccessful effort to survive as Russian Prime Minister against opposition in the Congress of Peoples Deputies, reform proponent Yegor Gaidar announced and also endorsed Russian arms agreements with China (worth $1 billion), India ($650 million), and Iran ($600 million). See Margaret Shapiro, "Russian Reformer Pledges Arms Sales," *The Washington Post* (3 December 1992). The deals with China could easily double China's average annual defense imports.

30. IISS, "China's Military Expenditures," *The Military Balance 1995/96,* 270–275, and see particularly table 4, 275.

31. For a recent American survey, see "The Defense Industry of India," in U.S. Congress, Office of Technology Assessment (OTA), *Global Arms Trade: Commerce in Advanced Military Technology and Weapons* (Washington, D.C., 1991), 153–162.

32. An earlier treatment with other details may be found in Rodney W. Jones, "Old Quarrels and New Realities: Security in Southern Asia after the Cold War," *The Washington Quarterly,* XV (1972), 115–120.

33. The export versions of these aircraft were not equipped by their respective suppliers specifically for nuclear armaments. Hence, they would require some modifications, but none that would be difficult for technicians in either country.

34. The most recent Western MBTs tend to be relatively heavy, and while they have advantages of speed, firing range, and upgraded armor, their weight can be a

disadvantage in the riverine, irrigated terrain of the Punjab on both sides of the India-Pakistan border. The MBT types India and Pakistan have acquired are quite suitable for this terrain.

35. This figure omits numerous coastal and smaller naval craft, the so-called "brown-water fleets."

36. For a political analysis of India's cabinet decision in 1967 to reject the NPT, and of Prime Minister Indira Gandhi's decision in 1969 to initiate the technical preparations for India's 1974 nuclear explosive test, see the author's chapter, "India," in Jozef Goldblat (ed.), *Non-Proliferation: The Why and Wherefore* (London, 1985), 101–123.

37. Pakistan's Foreign Secretary Shaharyar Khan reportedly confirmed Pakistan's nuclear weapons capability in a February 6, 1992, interview with *Washington Post* editors by stating: "The capability is there." Pakistan possesses ". . . elements which, if put together, would become a device." See R. Jeffrey Smith, "Pakistan Official Affirms Capacity for Nuclear Device," *The Washington Post* (7 February 1992). See also Steve Coll, "U.S. Nuclear Diplomacy in South Asia Faces Obstacles," *The Washington Post* (8 February 1992); Paul Lewis, "Pakistan Tells of Its A-Bomb Capacity," *The New York Times* (8 February 1992).

38. See Kanti Bajpai, et al., *Brasstacks and Beyond: Perception and Management of Crises in South Asia* (Urbana, Ill., 1995).

39. Evidently, the intensity of these concerns led the United States to send then-Deputy National Security Advisor Robert Gates to New Delhi in mid-1990 to discourage resort to war as a reaction to the violence in Kashmir. See C. Raja Mohan, "India Ahead of Pakistan at Last," *The Hindu* (Madras) (22 August 1992).

40. There are two caveats to factor into the data in figure 13–8. Section A on intercontinental systems, according to the IISS, may now include a handful of multiple-warhead ICBMs, so that the number of systems (launchers) in the figure may understate the number of deliverable warheads slightly. By the same token, if China successfully deploys multiple warheads on all new ICBMs, the size of its strategic forces would increase more rapidly than this depiction suggests. Second, the IISS does not provide a numerical estimate of newer, nuclear-capable, mobile medium-range ballistic missiles (MRBMs) (such as the M-9 missile, with a 600 kilometer range) that it reports China is deploying, and gives only a rough estimate of ten operationally deployed DF-21 IRBMs (with an 1,800 kilometer range) that it reports are coming into service. See IISS, *The Military Balance 1995/96,* 169–171, 176. Thus, Section B on intermediate- and medium-range systems also may understate the current size of China's substrategic nuclear capabilities. Moreover, the IISS reports the existence of Chinese strategic modernization programs involving a new generation of mobile ground-launched missiles (designated DF-31 and DF-41), a new submarine-launched missile (JL-2), and a new class of strategic missile submarine (Type 094 SSBN) that could eventually lead China to a more rapid numerical expansion of its deliverable strategic weapons. IISS, *The Military Balance 1995/96,* 169, 271. That said, it remains striking how slow, comparatively speaking, Chinese strategic modernization has actually been.

41. The IISS noted as recently as 1992 that "China has only ever completed one Xia [strategic missile submarine], which has proved unsatisfactory and does not patrol far from its home port, where it spends most of the time." IISS, *The Military Balance, 1992–1993,* 139.

42. For a historical treatment of Chinese ballistic missile development that implicitly reflects—despite what appears to be a quite different intended spin by the authors—the severe limitations of China's "strategic" capabilities, see John Wilson Lewis and Hua Di, "China's Ballistic Missile Programs: Technologies Strategies, Goals," *International Security,* XVII (1992), 5–40.

43. The Bush Administration reportedly wished to approve the sale of between 300 and 700 Allied-Signal jet engines to China in a deal valued at about $500 million. These engines would be installed in Chinese air force training aircraft and "possibly in more capable military aircraft that China intends to export to Pakistan." Part of the controversy about this decision was concern that these engines might also be used by China, with minor modifications, for cruise missiles that it could produce for export. See R. Jeffrey Smith and Dan Southerland, "U.S. Plans to Let China Purchase Jet Technology," *The Washington Post* (12 December 1992).

44. See the report by Umit Enginsoy in *Defense News* (15 September 1992).

45. The Pressler Amendment sanctions mandate a cutoff of "assistance." Nevertheless, the Bush Administration concluded it was permissible to sell Pakistan spare parts for U.S. arms in direct cash transactions, arguing the sales involved no subsidy or assistance. Senator Larry Pressler contended the intent of the legislation was to cover such cash sales as well as subsidized arms transfers. See note 26, above, regarding the Brown Amendment.

46. Western complaints that these Russian transfers to Iran were highly destabilizing may have caused Russia briefly to suspend the transfer of three diesel submarines. See Fred Hiatt, "Russia Shelves Sale of Subs to Iran," *The Washington Post* (26 September 1992), A15. However, subsequent reports indicate that President Yeltsin firmly backed these sales, and the delivery of the first submarine took place by November 1992.

47. The Backfire bomber has been sensitive in U.S.-Soviet strategic arms control negotiations because under certain conditions it is considered to be capable of striking over an intercontinental range. It was excluded from the Strategic Arms Reduction Talks (START) upon Soviet assurances regarding its capability and agreement to numerical ceilings on deployments involving air force and naval Backfire aircraft. IISS data indicate the Backfire's weapon payload is about 12,000 kilograms (comparable with the Soviet Bear strategic bomber) and that its standard (unrefueled) radius of action is about 2,400 nautical miles—four to six times the radius of most modern tactical aircraft. See IISS, *The Military Balance, 1992–1993,* 235.

48. See Thomas, *India's Security Policy,* 30–32, 36–40.

49. For an early assessment of the possible impact on South Asia of the end of the Cold War and the breakup of the Soviet Union, see Jones, "Old Quarrels and New Realities," 115–120.

50. A sabotage operation against Kahuta is the trigger for the next Indo-Pakistani war in the fictional but realistic scenario developed by Ravi Rikhye in *The Fourth Round: Indo-Pak War 1984* (New Delhi, 1982).

51. By then, CBMs had become a major arms control tool for mitigating the East-West confrontation in Europe, first at Helsinki in 1975, next at Stockholm in 1986 and most recently at Vienna in 1990. In the Middle East, the little publicized "Olive Harvest" overflight monitoring helped sustain confidence in the Israel-Syria disengagement agreement after 1974. This accomplishment demonstrated the efficacy of a high-technology, conventional CBM in one of the most volatile areas of

the world. See George D. Moffett, III, "Confidence-Building for Peace," *The Christian Science Monitor* (24 September 1992).

52. Mohan, "India Ahead of Pakistan at Last."

53. A telephone "hot-line" between Army commands was instituted under the Simla Accord in 1972, and it became practice to notify the other side through military channels when major military exercises were to be conducted. Although worthwhile, these measures are extremely narrow, non-binding and devoid of public transparency.

54. "Joint Statement on Talks with Pakistan," All-India Radio Network, Delhi, 19–20 August 1992, in U.S. Department of State, Foreign Broadcast Information Service (FBIS), Near East and South Asia (20 August 1992), 31–32.

55. See C. Raja Mohan, "CBMs in the Strategy for Peace," *The Hindu* (Madras) (27 August 1992).

56. Such as Pakistan's varied proposals, including those for a bilateral nuclear test ban, mutual inspections of nuclear facilities, and a regional nuclear weapons-free zone.

57. See Michael R. Gordon, "U.S. Accuses India on Chemical Arms," *The New York Times* (21 September 1992).

58. Gordon, "U.S. Accuses India on Chemical Arms."

59. At least, India has this perception. See Mohan, "India Ahead of Pakistan."

60. Unfortunately, India, which for decades had pressed the nuclear weapon states to join a comprehensive nuclear test ban and cease production of nuclear weapons, is now "backpedaling" from its own commitments—just as multilateral negotiations on agreements to codify these goals gained serious momentum. As India's commitment weakens, so does Pakistan's. See Michael Krepon, "Before South Asia's Next 'War Scare,'" *The Washington Post* (30 November 1995), A23.

Part Five

ARMS RESTRAINT:
WHAT SHOULD BE DONE?

FOURTEEN

"Conditionality": Linking Development Assistance to Military Expenditures

Nicole Ball

Bringing Defense Considerations Into the Development Dialogue

Development and security are closely intertwined. High levels of military spending, large military-industrial sectors, arms proliferation, the inability to resolve disputes peacefully and the use of security forces to prevent the emergence of representative political systems can reduce opportunities for development. In contrast, successful development, the fruits of which are shared in a reasonably equitable manner, can be a source of domestic and regional stability, further strengthening the capacity for economic and political development.

Throughout the Cold War era, foreign policy considerations routinely dominated development needs. As a result, the impact of military spending, politically active armed forces, weapons acquisition and conflict over development was largely ignored. In the wake of both the Cold War and the Persian Gulf War, the serious political and economic disadvantages for all countries of governmental priorities skewed in favor of the military sector became increasingly evident.

By the time the Cold War ended, some 40 million people had died as a result of conflicts in the developing world, and the military had absorbed over 5 percent of the developing countries' gross national product (GNP). In nearly 20 percent of the developing countries, military budgets in the late 1980s exceeded their combined expenditures on health and education—sometimes by wide margins. The countries that spent the most on their armed forces during the 1980s generally ranked the lowest in the United

Nations (U.N.) Development Program Human Development Index at the beginning of the 1990s.[1]

These factors alone justify exploring how the international community, and in particular the bilateral and multilateral lending institutions, can encourage aid recipients to transfer resources from the military sector to development. Rapidly growing demands for capital make it even more important that all available capital resources be used as efficiently as possible. The implication is, among other things, that military budgets—the details of which to this day are often not available to civilian government officials, legislators, the public or the international lending community—can no longer escape detailed scrutiny.[2]

The commitment of the aid recipient is essential to the success of reforms. Only rarely can governments be compelled to modify policies—on any issue—if they are seriously opposed to reform. There is a wide spectrum of opinion on the need for military reform among borrower countries. Some governments, such as the one that ruled Haiti between September 1991 and September 1994, are essentially uninterested in reform. Others—mostly in Africa, Latin America and the former Soviet Union—are seeking assistance for reforms they have already decided to implement. Countries such as Angola, El Salvador and Uganda are resolving disputes and have sought external financing to support troop demobilization, democratic institution-building and economic reconstruction. Others such as Argentina, Czechoslovakia and Russia have requested external financial and technical support to reduce the economic and political burden of their military sectors, for example, by assisting privatization and conversion of the defense industry.

This chapter begins by briefly reviewing the different kinds of economic leverage available to the international community to influence governments in the developing world. Drawing on the experience of the international lending community since 1989, a number of external strategies for promoting military reform are then outlined. The ways in which these strategies can be used to promote conventional arms control are the subject of the third section. The chapter concludes by proposing a series of steps that the international community should take to help developing countries achieve enhanced security at lower levels of expenditure.[3]

Types of Leverage

The international community has several economic tools to press for military reform in the developing world. The most commonly discussed are development assistance and other official financial flows, trade, investment and technology transfer.

Aid and Other Official Financial Flows

In the first half of the 1990s, the debate on using economic tools to promote change in the military sector of developing countries occurred primarily within the international lending community. Accordingly, it focused almost exclusively on aid and other types of official financial flows. Initially, the objective was to enhance development by reducing military spending, rather than to promote conflict resolution, arms control, non-proliferation, non-aggression or the creation of collective security mechanisms. As the lenders faced growing demands to finance post-conflict peace-building activities, attention increasingly focused on demobilization, police reform, political reconciliation and even conflict prevention.

The dependence of individual countries on aid varies considerably. The countries that rely most heavily on concessional aid—official development assistance (ODA)—tend to be the poorest ones, with the least diversified economies. As a group, the Sub-Saharan African countries are the most aid-dependent.[4] While there are good reasons to believe the military is absorbing too many resources in a number of the most aid-dependent countries, the same problem exists in other countries that receive very little or no ODA. Lenders may find themselves pressing for military reform in the economically weakest countries, while some of the major military powers in the developing world escape control. In India, Israel and Pakistan, ODA accounts for less than 4 percent of GNP, and Chile, the Republic of Korea (South Korea) and Saudi Arabia receive no aid at all. Yet these non-aid-dependent countries—many of which import substantial amounts of weaponry, have domestic arms industries and are capable of producing weapons of mass destruction—pose potentially greater threats to regional peace and stability than do the impoverished aid-dependent states such as Mozambique and Uganda.

Concessional aid is, however, not the only form that official financial flows from the industrial countries to the developing world can take. Some countries, only a small portion of whose national product derives from ODA, nonetheless require lender assistance to carry out specific projects of importance. These countries may also need credits from the International Monetary Fund (IMF) to stabilize their economies and IMF approval of their economic policies to open the door for non-concessional public and private lending. Thus, there may be some leverage even in negotiations with countries such as India, Pakistan and Chile.[5]

Trade, Investment and Technology Transfer

It is clear that official financial flows cannot bear the entire burden of promoting change in the military sector. The international community needs to consider using other economic tools, such as trade, investment and technology transfer, to influence government policies in the security sector.

SUPPLIER GROUPS

A number of defense technology supplier groups in the security sphere are intended to limit access to the know-how and material required to produce weapons of mass destruction and ballistic missiles. They are the London Club (nuclear), the Australia Group (chemical and biological warfare agents) and the Missile Technology Control Regime (MTCR) (longer range ballistic missiles).[6]

That these groups have been less than totally effective is evident from the capabilities that countries such as Iraq and Pakistan have acquired through purchases of restricted technology from their members. Industrial country governments are now interested in strengthening these supplier groups precisely because of their failure to police their own corporations adequately. The experience of the Gulf War and the subsequent revelations of Iraq's capabilities in the nuclear and chemical spheres caused, for example, the United States to push the Australia Group to increase to fifty the number of precursor chemicals subject to control and to control biological warfare organisms, certain toxins and dual-use equipment that can be used to produce biological weapons.[7]

Revelations of Iraq's weapons of mass destruction programs spurred governments to enforce existing legislation and guidelines more stringently, a crucial step in combatting clandestine efforts to acquire these weapons. In October 1991, for example, Japan Aviation Electronics Industry was banned from exporting for eighteen months because it sold Iran missile guidance and aircraft navigation components in violation of Japan's Foreign Exchange and Foreign Trade Control Law.[8] Eliminating programs to develop weapons of mass destruction is vital to creating regional security environments that are sufficiently stable to allow governments to engage in conventional arms control, force reductions and mutual spending cuts.

SANCTIONS

Trade and investment sanctions are one of the most extreme forms of pressure. Frequently these are unsuccessful in forcing governments to change policy.[9] When governments are not at all dependent on external financing and routinely seek to circumvent restrictions imposed by supplier groups, however, the only way to influence their policy may be to restrict other forms of economic activity. Sanctions are most effective when all of a country's trade and investment partners support them and all parties strictly adhere to them.

The larger the number of countries that must join in sanctions to make them work and the longer they must remain in force to effect change, the harder it is to obtain strict adherence. The sanctions the United Nations imposed on Iraq in August 1990 have never been airtight. Although they are still in force, they do not appear to have significantly influenced Saddam Hussein's policies. The various sanctions the United Nations imposed on South Africa for transferring weapons and weapon production technology were also only partially effective. Although the cost to South Africa of procuring weapons surely increased as a result of the 1963 and 1977 U.N. embargoes, South Africa was nonetheless able to procure a substantial amount of weaponry from abroad as well as from domestic industry.[10]

The strengthening of multilateral bodies such as the United Nations and the Organization of American States (OAS) might lead to increased use of economic sanctions. However, it is unlikely the world community will impose comprehensive economic sanctions with any frequency. For one thing, the widespread moral outrage at Iraq's invasion of Kuwait and South Africa's repression of the majority of its population will probably be duplicated only rarely. For another, it is difficult to justify cutting off all or most economic contact with very poor countries. The OAS-imposed embargo on Haiti following the military overthrow of that country's elected president was widely criticized in view of the hardships it imposed on Haiti's poorest citizens.[11]

Efforts specifically targeted to prevent countries from acquiring certain military capabilities might be more broadly acceptable than comprehensive economic sanctions against entire populations. Suppliers of defense technology could penalize governments and firms acquiring material directly involved in the manufacture of prohibited weapons by cutting off access to highly desirable but non-essential goods, such as sophisticated computers, or by limiting but not halting trade, such as revoking most-favored-nation status.[12]

External Strategies for Reform

During the Cold War, the international lending community generally turned a blind eye to the potential competition between development financing and military budgets. Several of the major bilateral donors—notably the United States and the Soviet Union—provided economic assistance specifically to enable strategically important client states to maintain military forces larger than domestic resources alone could support. It has always been evident that official financing for non-military purposes provided by any source—the World Bank and the IMF included—enables governments to spend more on the military if they are so inclined.[13] It was not until the Cold War ended, however, that lenders began to tackle this issue head on and explore how they could prevent their funding from leaking to the military sector.

There are some important differences in the strategies that governments, as bilateral lenders, can adopt and those open to the multilateral lenders such as the World Bank and the IMF. The Bank and the Fund now argue that military expenditure is an economic as well as a political-strategic issue. However, their mandate is to strengthen the economies of member governments, and the objective of enhancing security will have to be secondary, at best. National governments, on the other hand, are able to address overtly political issues. They can use economic tools to seek to alter government policies in areas such as weapons procurement and conflict resolution as well as the level of military budgets. They may also have the capability to evaluate the security needs of other governments, although aid agencies do not normally have this ability.[14] Thus, Japan can insist that the Democratic People's Republic of Korea (North Korea) dismantle the Yongbyon plutonium facility as one criterion for normalizing political and economic relations.[15] The World Bank, however, is likely to rely more on persuasion, support for reform-minded governments and pressure in the economic and social sectors designed to squeeze the military budget.

The following is a survey of lender strategies that either have been applied or currently are under consideration. While they are described here as discrete, these strategies are in practice complementary. Circumstances in the borrower country will play an important role in determining which strategies are used and the combination and sequence in which they are applied.

Persuasion

Persuasion has played a central role in recent efforts to bring military budgets and policies into the development dialogue. The objective is to convince developing country governments that military reform is in their own best interests. Persuasion can take the form of: public statements; seminars, meetings and training sessions for the employees of both lending institutions and borrower governments; and private policy dialogue.

Some of the earliest statements about the conflict between military spending and development financing came from senior management at the World Bank and the IMF. In September 1989, for example, then-World Bank President Barber Conable noted, "It is important to place military spending decisions on the same footing as other fiscal decisions, to examine possible trade-offs more systematically, and to explore ways to bring military spending into better balance with development priorities."[16] Public statements such as those serve notice to the recipients of international financing that the rules of the game are changing. Henceforth, military spending will be viewed as an economic as well as a strategic/political issue.

The willingness of the Bank and the Fund to address military spending encouraged bilateral donor governments to begin formulating their own policies. In December 1993, for example, the Organization for Economic Development and Co-operation's (OECD) Development Assistance Committee (DAC), the coordinating committee for the major bilateral donors, adopted a statement on participatory development and good governance that contained a section on reducing excessive military expenditures. DAC members agreed to examine how they could strengthen the capacity of civilians to oversee military affairs, support countries in reducing the size of their armed forces and their military budgets, and encourage governments to reduce excessive military expenditures.[17] In turn, the growing acceptance by bilateral donors that official financing can be used to reduce competition between military spending and development has provided important support for Bank and Fund officials who find it difficult consistently or repeatedly to take positions that vary significantly from those of their major shareholders.

Putting governments on notice that lenders are concerned about their patterns of resource allocation should be a precondition for pressure. Some governments may be well aware that their military spending is excessive and want to change their priorities. Their leadership may not, however, control military budgeting in their countries and therefore need assistance in raising the issue with their armed forces. External support, beginning with policy dialogue, can be very fruitful in such cases. Furthermore, it may be desirable to reward countries that make steady progress in reducing military budgets, even if expenditure remains relatively high for a period of years. Therefore, a policy that relies primarily on pressure—for example, making assistance conditional on meeting specific, rigid criteria—rather than on support for reform is unlikely to achieve its desired ends.

Policy dialogue is important for another reason. The degree to which borrowers feel ownership of reform programs crucially affects their success or failure. It is desirable that lenders avoid imposing a particular course of action on borrowers. Even governments that are contemplating inaugurating their own reform and would welcome external support are sensitive in this regard.

It is important to identify multilateral fora in which military expenditure and other related issues can be raised in a non-confrontational fashion. The Global Coalition for Africa—whose members include African governments, bilateral donors and multilateral organizations—has addressed the possibility of reducing military expenditure as one element of improved governance.[18] In Central America, consultations between regional governments and industrial countries under the San José accords (which involve the European Community and its members) or under the Partnership for Democracy and Development (which involves all OECD governments) could be broadened to include opportunities for lowering military spending.

Support

External support—which can take various forms, depending on the nature of the reforms pursued and the borrower's own resources—can facilitate efforts by borrower governments to implement change in the military sector.

Financial support can help governments absorb the extra costs associated with reform, such as compensating soldiers released from the armed forces or workers laid off as a result of the rationalization of the defense industry sector. In 1991, Argentina and the World Bank began negotiating a loan that would support the privatization of firms in the defense industry that produce goods primarily for the civilian market.[19] Lenders increasingly are working with governments in Central America, Africa and Asia to underwrite programs designed to integrate demobilized soldiers into the civilian economy.

Technical support can provide skilled manpower or equipment to carry out tasks, such as the destruction of weapons, the integration of former soldiers into the civilian economy or the privatization/conversion of military industries. For example, the OAS provided demobilized Nicaraguan resistance soldiers with construction materials and farm implements to assist their transition to civilian life, while bilateral donors and non-governmental organizations have provided training to enhance their skills.[20]

Pressure Without Conditions

The World Bank and the IMF have stated very clearly they will not apply specific military-related conditions to their lending.[21] If policy dialogue is to be a credible tool for these two institutions, however, they must have some means of motivating borrowers to reexamine their funding priorities.

In a number of borrower countries with high levels of military spending and poor development outcomes, the World Bank has begun to use structural adjustment lending to squeeze military budgets so that adequate funding will be available to meet the government's economic and social goals. The Bank sets expenditure targets in the economic and social sectors that cannot be met without a significant reallocation of domestic resources.[22] The intention is to force governments to shift funds from sectors the Bank views as economically unproductive, including the military, to ones that are productive.

A similar strategy is available to the IMF when it sets targets for cutting fiscal deficits. Some governments may need to identify such substantial budgetary savings to meet the IMF's deficit reduction targets that they will find it difficult to avoid cutting spending on the military sector. While it remains to be seen how successful strategies of this nature can be in actually reducing military budgets, the experience with Uganda in the early 1990s suggests that a coalition of bilateral and multilateral lenders can pressure even very reluctant governments into making some concessions.[23]

Specific Conditionality

Many borrower countries have made it clear they oppose the extension of conditionality into the military sphere. They argue that demanding reductions in military spending without reference to security needs is dangerous. In their view, only the country concerned has the capacity to define these needs and decide how resources should be allocated to meet them.

Control over security policy, the armed forces and military budget has traditionally been among the most closely guarded prerogatives of states. Nonetheless, during the Cold War, the major powers routinely influenced the security sector in many developing countries. While some developing country governments were critical of this external involvement, others welcomed it as a means of strengthening their position vis-à-vis their opponents, both domestic and foreign. Virtually every country in the developing world has received some portion of the billions of dollars of military hardware, training, technology and aid that the United States, the former Soviet Union, their allies in Eastern and Western Europe and a few oil-rich developing countries provided on concessional terms during the Cold War.[24]

The collapse of Soviet power in Eastern Europe, the subsequent disintegration of the Soviet Union and the revelations about Iraq's weapons of mass destruction capabilities following the Gulf War altered to some degree both the nature and content of external involvement in the security sectors of developing countries. The major powers are now calling for conflict resolution, non-proliferation and lower military expenditure. Although the arms trade continues apace, subsidies are no longer as widely available as in the past. As a result, governments that were content to use external support to build up their military establishments and pursue local and regional disputes now object to efforts to limit the size and armament of their armed forces. They argue that this constitutes an infringement on their sovereignty.

The lack of consultation associated with conditionality is an important component of developing country concerns. It is widely agreed by lenders and borrowers that far greater efforts must be made jointly to assess evidence that reform is in the self-interest of particular governments. At the same time, it is important that lenders not allow their desire to persuade rather than force borrowers to change policy to serve as a rationale for maintaining the *status quo* in the military sector. The losses in terms of human, financial and material resources that conflicts, excessive arms procurement and high military budgets have incurred in the developing world over the last four decades render the sovereignty argument increasingly invalid. At times it may be necessary to place conditions on economic cooperation to force a government to reconsider a policy that a majority of the international community views as unacceptable.

Conditionality can take several forms and be applied with varying degrees of stringency. Most lenders are reluctant to identify specific military expenditure targets that borrowers *must* meet to obtain financing. Suspension of aid and other forms of sanctions have, however, been applied, particularly by the United States, in response to specific activities such as military coups d'état, efforts by non-nuclear powers to obtain nuclear weapons and failure to negotiate in good faith. Thus, Washington sought to encourage the Israeli government to negotiate seriously with the Palestinians and several of its Arab neighbors by threatening to place conditions on the $10 billion in housing loan guarantees Israel requested in 1991.

Bilateral donors may increasingly link specific conditions to particular *actions* by individual countries to promote non-proliferation, conflict resolution and regional arms control in the developing world. In contrast, bilateral lenders will probably use specific conditions related to the level of a country's military *expenditure* only rarely. Whereas a consensus may be evolving that lower military budgets are preferable, there still is no agreement on what constitutes an appropriate level of expenditure in this sector. The size and composition of the military force (and hence the military budget) necessary to guarantee security vary from country to country. Lenders are therefore disinclined to insist that borrowers reduce military spending by a specific amount. Instead, they tend to speak of measurable, sustained progress toward lower levels of expenditure.

"Carrots" Instead of "Sticks"

Lenders may choose not to impose specific conditions on their aid but to give preference to governments whose performance is favorable (allocative conditionality). A reward-based strategy is grounded upon the premise that positive reinforcement is more likely to change behavior than punishment. While this assessment may be valid for governments with a fairly firm commitment to reform, it is unclear how much leverage this strategy provides with less committed borrowers. Furthermore, desirable as it may be to offer the carrot rather than the stick, allocative conditionality virtually always requires official lenders to *shift* resources from one country to another. Japan, whose aid budget has been expanding more rapidly than most other DAC members, may find it easier to reallocate resources than most other bilateral donors. The United States in particular faces substantial limits on its capacity to reallocate assistance. A substantial portion of aid resources are earmarked by Congress for specific countries, and the aid budget is shrinking drastically.

One reason why allocative conditionality is particularly attractive to bilateral lenders is that it enables them to register their displeasure with a borrower's policies but does not immediately jeopardize existing projects. It may take several years for a shift to be felt as old projects are completed. This delay gives borrowers the opportunity to adjust their policies to obtain new funding. Another method of allocating resources to indicate dissatisfaction with the borrower government without actually cutting assistance is to channel funding through non-governmental organizations. Apparently some donors, notably Canada, the Netherlands and the Nordic countries, increasingly favor this tactic.

Linking Economic Assistance and Conventional Arms Control

The new willingness within the international lending community to encourage borrower countries to reduce military expenditure as one way to increase financing for development and improve development outcomes has led some policy-makers and analysts to examine the opportunities for using external financing to promote a broad range of security objectives. Security specialists have been particularly interested in identifying ways to incorporate the financial weight of the World Bank and the IMF into these efforts. With regard to conventional arms control, the Bretton Woods institutions can play an important supporting role even though they are restricted by their respective Articles of Agreement from taking political factors into account in lending decisions.

First, they can engage both supplier and recipient governments in policy dialogue. On numerous occasions in the early 1990s IMF Managing Director Michael Camdessus spoke out about the opportunities the post-Cold War era holds for reducing arms transfers, particularly subsidized sales: "We must not ignore the international trade in armaments. It is most desirable to avoid a recurrence of a situation in which substantial holdings of offensive weapons—far beyond the justified needs for defense—can be readily accumulated and indeed financed on easy terms. One very practical step would be to tighten the rules for granting export credits for arms sales."[25] Senior Bank and Fund officials can deliver a similar message to selected governments privately.

Second, by insisting on greater transparency in military budgeting, including tracking military-related debt, the Bank and Fund can help borrower governments identify the economic trade-offs involved in arms purchases.

Third, by providing appropriate technical and financial support for the conversion of military industries in borrowing countries, the World Bank can reduce the imperative to export weapons to maintain economic viability.

Last, as part of its efforts to increase the efficient use of resources, the World Bank can assist member governments in evaluating the economic costs associated with domestic arms production.

Bilateral donors can employ all of these strategies and more. Unlike the Bank and the Fund, they can take a country's arms trade policies into account when allocating financial assistance. In April 1991, Japan announced its intention to apply several military-related criteria when determining aid levels: (1) trends in military spending; (2) production or possession of weapons of mass destruction and ballistic missiles; and (3) participation in the arms trade. In June 1992, it incorporated these criteria into Japan's Official Development Assistance Charter. Germany considers, among other military-related variables, (1) arms imports as a share of total imports, (2) arms production as a share of industrial output, (3) export dependence of the domestic defense industry, and (4) the amount of technologically sophisticated weapons a country procures through either domestic production or trade.[26] Governments that rank poorly risk seeing their aid reduced.

Bilateral lenders can also make their development assistance conditional on meeting certain conventional arms-related criteria in the same way the United States cut economic and military assistance to Pakistan pending changes in its nuclear program. The question is: what form should such criteria take? The international community might agree that failure to adhere to international arms control agreements, such as the Nuclear Non-Proliferation Treaty (NPT), or efforts to develop prohibited weapons clandestinely, as India, Iraq, Israel, Pakistan and others have done or attempted to do with nuclear arms, invite the imposition of conditionality. There are, however, no equivalent treaties or prohibited weapons in the conventional field.

In the early 1990s the five Permanent Members (P-5) of the U.N. Security Council discussed limiting their sales to the Middle East and enunciated a set of general guidelines for the export of conventional weapons.[27] No enforcement mechanism exists or is currently contemplated, for the simple reason that each of the five is seeking to increase its share of the global arms market to reduce the impact of declining domestic procurement following the conclusion of the Conventional Forces in Europe (CFE) Treaty and the demise of the East-West conflict.

If suppliers were to reach an agreement to limit arms sales—to certain regions, of specific weapons or in terms of overall volume—the bilateral lenders could insist that all countries receiving their aid abide by that agreement and attempt neither to import nor export restricted equipment. Unless the sellers were willing, however, to abide by similar restrictions in their own procurement and deployment policies, the perennial question of evenhandedness would immediately arise. Control regimes that focus solely on limiting trade, while leaving the major producers free to procure whatever they want, lose much of their force, irrespective of their merit. One possibility would be to focus efforts to limit the trade of conventional weapons on those systems whose deployment has been restricted under the CFE Treaty.

Perhaps more important, the reasons why governments procure weapons need to be addressed if enduring limits on conventional arms transfers are to be achieved. The countries purchasing the most weapons are the ones engaged either in domestic or external conflicts or in longstanding arms races rooted in latent conflicts. Rather than linking official financing to some as yet unnegotiated agreement limiting conventional arms transfers, the international lending community might more fruitfully use its assistance both to encourage settlement of these disputes and ensure that settlement accords are implemented. Such accords, negotiated with the assistance of the international political community, should include mechanisms for reaching agreement on reductions in the level of armaments in individual countries and/or regions. The bilateral lenders could agreed to withhold all or some significant portion of their assistance to parties to a conflict until the parties agree to enter into serious negotiations. External financing could also be linked to carrying out verifiable measures such as agreeing to a cease-fire, signing a peace accord, officially ending hostilities, opening certain facilities for mutual or international inspection, reducing troop levels, implementing peace agreements and negotiating regional limits on certain categories of weapons.

Next Steps for The International Community

The international lending community has made substantial progress since the days when it virtually never raised military issues in the context of development policy or financing. It is no longer possible to ignore the impact that armed forces have on a country's prospects for sustained development. According to the DAC, military expenditures "will be . . . of continuing concern and attention in the years ahead, and can be expected to feature prominently in aid allocation decisions by donors, both bilaterally and collectively in such fora as Consultative Groups, etc."[28]

The formulation and implementation of policies to encourage change are nonetheless still at an early stage, and there is considerable uncertainty among lenders about how far and how rapidly they can and should proceed, particularly in the areas of arms control, conflict resolution and collective security building. The following proposals suggest the next steps the international community institutions might take to enhance both development and security in borrower countries, with a special focus on conventional arms control.

(1) *All lenders should actively seek to create an enabling environment for military reform in the developing world.* Now that the potential competition between military spending and development has been acknowledged, all lenders should adopt policies that will actively promote reform in the military sector, rather than simply punishing bad behavior, passively rewarding good behavior, or limiting their involvement to reform-minded governments.

At times it may prove necessary to attach specific conditions to economic interactions with borrower countries. The core of lenders' policies should, however, be policy dialogue and financial and technical support designed to assist governments in changing their behavior and policies in the military sector, in placing the military sector on the same footing as other portions of the government, and in altering the balance between expenditures on the armed forces and those on development. The objectives of these policies should include: establishing greater transparency and accountability in the military sector; encouraging civilian control over the armed forces, paramilitary groups and police; supporting reductions in the size and armament of military forces; and promoting the creation of new security arrangements that will enable governments to provide enhanced security at lower levels of expenditure.

One of the most important contributions lenders can make in this regard is to integrate the military sector into normal development practices, that is, strengthening the capacity of governments to manage the public sector, improving the efficiency of public sector expenditures and supporting market-oriented reforms. Each have military components that all too often are ignored. With regard to conventional arms control specifically, they would include such measures as:

—Greater transparency on military-related debt and the cost of weapons procured through domestic production and international trade.

—Incorporation of assessments of the economic costs and benefits of defense industries into economic plans.

—Financial and technical assistance for conversion of the defense industry or economic diversification following the closure of defense plants.

(2) *The international community should recognize the need for close cooperation, based on genuine consensus, among the bilateral donors,*

multilateral lending institutions, developing-country governments, and relevant political institutions to create the conditions under which enhanced security at lower levels of expenditure can be achieved. Developing countries can be divided into two broad groups for the purpose of devising strategies to facilitate the transfer of resources from the military sector to development-oriented projects and programs. The first group spends demonstrably more on the military than is warranted by internal and external security threats. Examples are Argentina, Egypt and Uganda. These are the countries with which lenders feel most comfortable discussing military reform, particularly that subset whose governments have themselves decided to reduce the size of their military sector. Even with the most reform-minded governments, however, lenders must take care to engage borrowers in a genuine dialogue about the desirability of restructuring their military forces as well as the assistance that lenders can offer governments to carry out military reforms.

In addition, although the security environment may generally be conducive to substantial reductions in military budgets, including reductions in weapons procured, the international political community may need to assist these countries in reaching arms control or limitation agreements or establishing regional security mechanisms that will reduce the likelihood that military spending and arms purchases will increase substantially in the future. To promote conventional arms control, assistance from the lenders could be linked to such activities.

The second group of countries consists of states that will have to alter their behavior toward other countries or their own citizens before the economic burden of the military sector can be significantly reduced. Examples are countries in South Asia and the Koreas. From the very beginning the pursuit of military reform in these countries requires the collaboration of all members of the international community.

While political institutions will of necessity take the lead in searching for solutions to the political issues dividing these countries, lenders should use their financial and technical resources to encourage the changes in policies and patterns of behavior that are a prerequisite for significant reductions in military budgets and arms purchases. The changes would include:

—Negotiating with domestic and regional opponents to resolve outstanding security issues.

—Embarking on a process of confidence-building to raise mutual levels of trust.

—Restructuring armed forces so that they match existing security needs.

—Engaging in regional negotiations with the participation of arms suppliers designed to reach agreement on acceptable regional levels of armaments and reduce pressures by suppliers to import weaponry.

—Creating domestic political structures and regional security mechanisms that will prevent future disputes from threatening internal and interstate security.

—Implementing peace agreements.

Although the bilateral donors and multilateral organizations such as the United Nations may be best suited to link official financing with arms control and conflict resolution, the Bretton Woods institutions can clearly participate in certain activities, such as helping devise and implement job creation and retraining programs for demobilized soldiers. In addition, it would be worth exploring whether the multilateral development banks (the World Bank and the African, Asian and inter-American regional development banks) could, under certain circumstances, make development assistance contingent on the cessation of hostilities and the inauguration of serious negotiations between disputing parties. The World Bank, for example, participated in the November 1991 decision of the Consultative Group for Kenya to withhold new loans as a result of Nairobi's lack of progress in moving toward multi-party democracy. If in particular cases bilateral lenders agree that a high level of military spending or an ongoing conflict constitutes a serious obstacle to development, and that only negotiations to resolve underlying security imbalances can free up resources urgently needed by the economic and social sectors, action by a multilateral development bank might be justifiable on economic grounds.

(3) *Bilateral donors and U.N. agencies should agree that countries receiving their financial support will conform to certain internationally agreed-upon standards in the security sector.* These standards should reflect, insofar as possible, multilateral treaties and conventions and internationally accepted norms of behavior. As a first step, they might include:

—Adherence to international arms control treaties, such as the NPT, the Biological Weapons Convention, the Chemical Weapons Treaty and the Environmental Modification Convention, or to internationally recognized alternative arrangements, such as the full-scope nuclear safeguards agreement concluded between Argentina and Brazil with the International Atomic Energy Agency.

—Agreement not to contravene international arms control treaties and regimes by covertly developing the capability to produce weapons based on prohibited technologies or by assisting other countries to develop such weapons.

—Respect for the territorial integrity of other countries by supporting the principles of non-aggression and non-intervention.

—Participation in dialogues designed to increase regional security and stability.

—Willingness to negotiate in good faith as necessary to end longstanding conflicts, both internal and external.

—Participation in the U.N. standardized military expenditure reporting exercise and the U.N. arms transfer register.

—Willingness to subject expenditure on the military sector to the same scrutiny and discipline as other public-sector expenditure.

(4) *The bilateral donors should affirm that the same principles will govern security policies in OECD member states as in the developing world.* The use of economic tools to affect policy always runs the risk of producing charges of unwarranted interference in the affairs of those states that are the object of reform efforts. One way in which the bilateral lenders can gain the confidence of borrowers is to accept that the same norms apply to *their activities* in the security sphere as they wish to see developing countries apply.[29] Reciprocity is desirable not only on grounds of fairness, but also because an evenhanded approach to all countries makes it more difficult for borrowers to avoid policy reform. In contrast, allowing one category of countries—primarily the rich industrial states—special privileges will undermine policy reform.

The arms reduction currently underway in the industrial countries has given greater authority to lender calls for military reform in the developing world. In 1991 the European Community noted that, "in a period in which donor countries are engaged in a process leading to levels of armament not exceeding sufficiency levels, development co-operation with governments which maintain much larger military structures than needed will become difficult to justify."[30]

Nonetheless, there are further steps the industrial countries can and should take. Military cutbacks should continue in many countries, as military spending in most of the developed world remains substantially higher than is justified, given the benign security environment most developed countries now face. To the extent possible, future international arms control agreements and supplier groups should be universal in membership and non-discriminatory in scope. When military intervention in the developing world is necessary, it should be multilateral, not unilateral, in nature and given legitimacy in some broadly accepted way, for example, by occurring under the umbrella of the United Nations or a regional organization.

One of the areas of greatest disparity between industrial country objectives in the developing countries and their own behavior is that of conventional arms. All DAC members export weapons.[31] Several of them either export only a very small amount of arms or sell only to other industrial countries. However, some of the largest DAC members, notably France, the United Kingdom and the United States, export a substantial amount of arms to the developing world and are currently seeking to expand their share of the market in the face of declining domestic orders. Most developing countries do not produce a significant amount of weaponry, and even fewer are major players in the arms export market.

A portion of the arms transfers from the DAC countries are subsidized by their governments, notably those of the United States. While all governments have the right to equip armed forces to defend their sovereignty, and many have no means of acquiring weapons other than through imports, countries that aggressively promote arms sales seriously undermine their ability to promote military reform. IMF Managing Director Camdessus has urged arms-exporting countries to eliminate the subsidies associated with arms transfers. Such a move—which could have an impact on the long-term viability of some arms producers in the industrial world—would offer concrete proof that the industrial countries do not expect borrowing countries to make all the concessions.

Support for this line of thinking has come from Japan. At a meeting of the DAC in early April 1992, the Japanese government noted that the "provision of military export credit to developing countries with accumulated debt could aggravate their debt problems, and thus act as a constraint on development. In such a case, donors will be encouraged to stop or reduce such provision to those developing countries."[32]

The industrial countries should also agree to include the *production* of weapons in the U.N. arms transfer register as soon as possible. They should explore the possibility of concluding international agreements limiting access to certain categories of conventional weapons for *all* countries, not just the developing world. Finally, they should encourage and participate in regionally-based negotiations among recipients that would result in agreements to reduce arms procurement.

Conclusions

The end of the Cold War has produced significant changes in relations between and within states. Democratization, conflict resolution, non-aggression, collective security, multilateralism and compromise are beginning to replace authoritarianism, war, military alliances, unilateral action and confrontation as internationally accepted modes of behavior.

Yet many of the problems of the Cold War era are still very much with us. The proliferation of chemical, biological and nuclear weapons and ballistic missiles remains a danger. Strong industrial pressures persist to maintain and, if possible, expand exports of conventional weaponry. Low rates of economic growth and social progress, along with numerous unresolved conflicts, including serious ethnic, religious and national disputes, threaten long-term stability in many parts of the developing world, Eastern Europe and the former Soviet Union.

From the perspective of the industrial world, the defining relationship of the 1945–90 period was the hostility between the United States and the Soviet Union. The Cold War admittedly left its mark on many inter- and intrastate relations in Africa, Asia, Latin America and the Middle East, but its influence in these parts of the world was frequently exaggerated. Local and regional conflicts derive from local and regional disagreements, many of which predated the Cold War and many of which have survived it.[33]

It is unrealistic to assume that conflict can be eradicated. It can, however, be channeled into non-military avenues, and the norms defining acceptable behavior on the part of governments, both toward their neighbors and toward their own citizens, can be changed. The challenge of the 1990s and succeeding decades is to ensure that emergent trends toward international cooperation and domestic participation are maintained and strengthened.

The international official lending community clearly has a role to play by using its resources to enhance both development and security. The lenders can help to change the norms relating to secrecy in the military sector, the domestic role of the military, the spread of weapons and weapons-production technology, and the appropriate means of settling disputes within and among states. Where the objective is related primarily to development, for example, transparency and accountability in the budgetary process, lenders can act at their own initiative. Where the objective is primarily to alter the security environment, for example, conventional arms control, official lenders will need to work in tandem with international, regional and national political institutions.

In the final analysis, military reform in developing countries will occur only if governments in those countries accept the need for change. They must take a hard look at their priorities and devise policies that will promote reduced tensions and increase development. They will have to strike compromises both domestically and among themselves. The industrial countries can underline their own commitment to the emerging norms by making them the cornerstone of their own policies. They can also help instill new modes of behavior by conditioning relations with other countries on adherence to these norms. Ultimately, however, it is the people and governments of developing countries that must seize the opportunities presented by recent changes in the international system to create a more equitable and prosperous future for themselves.

Notes

1. The data suggest there is a particularly large discrepancy between military spending and outlays on the social sectors in Angola, Bolivia, Brunei, Chad, China, Egypt, Ethiopia, Iran, Iraq, Israel, Jordan, Oman, Pakistan, Peru, Sudan, Syria, Uganda, and the United Arab Emirates. See Robert McNamara, "The Post-Cold War World: Implications for Military Expenditures in the Developing Countries," in Lawrence H. Summers and Shekhar Shah (eds.), *Proceedings of the World Bank Annual Conference on Development Economics, 1991,* Supplement to the World Bank *Economic Review* and the World Bank *Research Observer* (Washington, D.C., 1992), 122–124; World Bank, *The Challenge of Development: World Development Report 1991* (New York, 1991), 142. See also United Nations Development Program (UNDP), *Human Development Report 1994* (New York, 1994), especially 47–60.

2. Throughout this paper, the terms "international official lending community," "official lenders," and "lenders" are used interchangeably for both multilateral lending institutions such as the World Bank and bilateral aid agencies such as the U.S. Agency for International Development (U.S. AID). The terms "bilateral donors," "bilateral lenders," or simply "donors" denote—according to context—either the governments that extend aid or their bilateral aid agencies.

3. The issues raised here are relevant to the debate on official lending to the new republics of the former Soviet Union. Nonetheless, this paper does not specifically apply them to the successor states.

4. Development Assistance Committee (DAC), *Development Co-Operation, 1994 Report* (Paris, 1995), table 39, H8-H9. ODA is defined as grants or loans undertaken by the official sector, with promotion of economic development or welfare as the main objectives, and at concessional financial terms (if a loan, at least 25 percent grant element). In addition to financial flows, technical cooperation is included in aid. "It consists almost entirely of grants to nationals of developing countries receiving education or training at home or abroad, and payments to defray the costs of teachers, administrators, advisors, and similar personnel serving in developing countries," 114.

5. In addition, it is important to recall that the leverage donors can exert over recipients can at times far outweigh the amount of ODA provided, since at the margin even a modest amount of aid can be critical.

6. For a description of these supplier groups, see Zachary S. Davis, *Non-Proliferation Regimes: A Comparative Analysis of Policies to Control the Spread of Nuclear, Chemical and Biological Weapons and Missiles* (Washington, D.C., 1991), 91–334.

7. "Non-proliferation Efforts Bolstered," *U.S. Department of State Dispatch* (July 20, 1992), 569–571; U.S. General Accounting Office (GAO), "Export Controls: Multilateral Efforts to Improve Enforcement," GAO/NSIAD-92-167 (Washington, D.C., 1992).

8. Robert Thomson, "Japanese Company under Export Ban in Iranian Missiles Row," *Financial Times* (26–27 October 1991). Japan Aviation, a subsidiary of NEC, apparently began exporting military equipment to Iran in 1983.

9. Gary C. Hufbauer and Jeffrey J. Schott, *Economic Sanctions Reconsidered: History and Current Policy* (Washington, D.C., 1985), provide a useful survey of the successes and failures of efforts to impose sanctions since World War I. See also Margaret Doxey, *International Sanctions in Contemporary Perspective* (New York, 1987).

10. See, for example, Douglas McDaniel, *Economic Sanctions: Issues Raised by the Sanctions Against Iraq,* 92–370 F (Washington, D.C., 1992); Signe Landgren, *Embargo Disimplemented: South Africa's Military Industry* (Oxford, 1989).

11. See, for example, Kenneth Freed, "Haiti's Wealthy Elite Turn Blind Eye to Plight of Poor," *Los Angeles Times* (Washington edition) (29 September 1992).

12. Current U.S. legislation provides for sanctions against both foreign individuals and companies under the MTCR (Title XVII of the National Defense Authorization Act of 1990) and against foreign individuals, companies, and countries under the Chemical and Biological Weapons Control and Warfare Elimination Act of 1991 (Title V of Public Law 102-138, 28 October 1991).

13. A well-documented example of fungibility occurred when Iraq used export credits and loan guarantees approved by the Reagan and Bush administrations in the late 1980s to free up resources for arms procurement. See Murray Waas and Douglas Frantz, "Bush Had Long Supported Aid for Iraq," *Los Angeles Times* (24 February 1992); Douglas Frantz and Murray Waas, "Secret Effort by Bush in '89 Helped Hussein Build Iraq's War Machine," *Los Angeles Times* (24 February 1992); Paul Houston, "House Banking Committee to Probe Reports on Aid to Iraq," *Los Angeles Times* (25 February 1992); Douglas Frantz and Murray Waas, "U.S. Loans Indirectly Financed Iraq Military," *Los Angeles Times* (25 February 1992).

14. See, for example, the comments by Pierre Landell-Mills, senior policy advisor in the Africa Technical Department at the World Bank, in "Seminar Panelists Discuss Ways and Means to Reform Military Spending," *IMF Survey* (Washington, D.C., 1992), 375.

15. Growing concern over North Korea's nuclear program in the wake of the Gulf War prompted Tokyo to announce on 30 October 1991, that before it would recognize or trade with Pyongyang, North Korea would have to dismantle its plutonium processing plant at Yongbyon. Lawyers Alliance for World Security with the Washington Council on Non-Proliferation, *North Korea: Do They or Don't They Have the Bomb?* (Washington, D.C., 1992), 16.

The Japanese position may have been undercut by the inauguration of large-scale plutonium imports from reprocessing plants in Europe to fuel its breeder reactor program. Because Japan will be accumulating plutonium more rapidly than it can consume, Japan's East Asian neighbors expressed concerns that Tokyo may one day reverse its decision to produce nuclear weapons. See, for example, William Walker and Frans Berkhout, "Japan's Plutonium Problem—And Europe's," *Arms Control Today* (September 1992), 3–10.

16. Barber Conable, Joint Bank/Fund Annual Meetings, World Bank (Washington, D.C., 25 September 1989).

17. Organization for Economic Co-operation and Development (OECD), *DAC Orientations on Participatory Development and Good Governance,* OECD/GD(93) 191 (Paris, 1993), 19–21.

18. Global Coalition for Africa, *Reducing Military Expenditure in Africa,* Document GCA/AC.2/04/4/92, prepared for the Second Advisory Committee Meeting, Kampala, Uganda, May 8–9, 1992. For a report of the meeting at which this paper was presented, see Ernest Harsch, "African Reforms under the Spotlight," *Africa Recovery* (August 1992), 10–11.

19. "Layoffs Planned for Military Industry Employees," *Buenos Aires Herald* (23 September 1991), 7, reproduced in U.S. Department of State, Foreign Broadcast Information Service (FBIS), FBIS-LAT-91–186 (September 25, 1991), 26–27.

20. Santiago Murray, "Building Towards Reconciliation," *Americas,* XLIV (1992), 52–53.

21. Jonathan E. Sanford, *Multilateral Development Banks: Issues for Congress,* IB87218 (Washington, D.C., 1992), 13.

22. The Bank's third review of structural adjustment lending programs concludes that the Bank needs to give greater attention to the composition of expenditure in order to influence development outcomes, rather than merely specifying expenditure targets. World Bank, Country Economics Department, *Adjustment Lending and Mobilization of Private and Public Resources for Growth,* Policy and Research Series 22 (Washington, D.C., 1992).

23. Uganda, where military expenditure absorbed 30 percent or more of the government budget in the late 1980s and early 1990s, even as the external and internal threats diminished substantially, was subject to considerable pressure from lenders in 1991 and 1992 to reduce the military budget. In May 1992, the Army Council announced plans to demobilize a large portion of the National Resistance Army. The implications of this decision for the military budget are at present uncertain, since salaries in the military sector are unreasonably low—as in other portions of the public sector—and the military would like to upgrade its equipment. Keith Richburg, "Rule Brings Stability to Uganda; Aid Donors Look for Looser Rein," *The Washington Post* (3 March 1992); "Army Council Issues Criteria for Reducing Army," Radio Uganda, 5 June 1992, reproduced by the FBIS, FBIS-AFRJ-92–100 (8 June 1992), 10–11.

24. For a review of security assistance during most of the Cold War period, see Nicole Ball, *Security and Economy* (Princeton, 1988), 237–294.

25. Michael Camdessus, Address to the Board of Governors of the International Monetary Fund, 1991 Joint Bank/Fund Annual Meetings, Press Release 4 (Washington, D.C., 1991) 12.

26. Japan's Official Development Assistance Charter, approved June 30, 1992, calls for Japanese ODA to be provided

. . . in accordance with the principles of the United Nations Charter (especially sovereign equality and non-intervention in domestic matters), as well as the following four principles. (1) Environmental conservation and development should be pursued in tandem. (2) Any use of ODA for military purposes or for aggravation of international conflicts should be avoided. (3) Full attention should be paid to trends in recipient countries' military expenditures, their development and production of mass destruction weapons and missiles, their export and import of arms, etc., so as to maintain and strengthen international peace and stability, and from the viewpoint that developing countries should place appropriate priorities in the allocation of their resources on their own economic and social development. (4) Full attention should be paid to efforts for promoting democratization and introduction of a market-oriented economy, and the situation regarding the security of basic human rights and freedoms in the recipient country.

On 5 November 1992, the Director-General of Japan's Economic Cooperation Bureau, Takao Kawakami, indicated that Japanese policy would emphasize transparency in the military sector, trends in military spending, policy dialogue to convince governments of the need for reform, and multilateral cooperation. "Japan's ODA Policies for a Peace Initiative," address by Takao Kawakami, Director-General, Economic Cooperation Bureau, Ministry of Foreign Affairs, Tokyo Conference on Arms Reduction and Economic Development in the Post-Cold War Era (5 November 1992) (mimeo). See also Japan International Cooperation Agency, *Annual Report 1991,* (Tokyo, 1991) 19.

On Germany's policy, see Veronika Buttner and Joachim Krause, "The Arms Policies of Third World Countries: A New Topic in the Development Debate?" *Internationale Spectator* (The Hague), XLVI (1992), 670–676.

It is unclear to what extent these guidelines have actually influenced aid flows from these two countries.

27. U.S. Department of State, "Meeting of the Five on Arms Transfers and Non-Proliferation," (London, 17–18 October 1991) (mimeo).

28. Development Assistance Committee (DAC), *Development Co-operation 1991 Report* (Paris, 1991), 19.

29. William Hartung makes a similar argument in "Curbing the Arms Trade," *World Policy Journal,* XI (1992), 242.

30. Council of the European Communities, General Secretariat, Press Release, 1538th Session of the Council on Cooperation and Development, Brussels, 9555/91 (Press 217), Preliminary Version (28 November 1991), 16.

31. The 21 DAC countries in descending order as arms exporters are, based on data in billions of 1993 U.S. dollars for 1983–93: United States, $149.0; United Kingdom, $48.8; France, $43.9; Germany, $22.5; Italy, $9.6; Canada, $7.6; Spain, $5.7; Sweden, $4.4; Switzerland, $3.8; the Netherlands, $3.7; Belgium, $2.8; Japan, $2.1; Austria, $1.7; Portugal, $1.1; Finland, $1.0; Australia, $0.9; Norway, $0.5; Denmark, $0.3; Ireland, $0.01; New Zealand, $0.006; and Luxembourg, $0.005. U.S. Arms Control and Disarmament Agency (ACDA), *World Military Expenditures and Arms Transfers 1993–1994* (Washington, D.C., 1995), table II.

Japanese legislation sharply limits the ability of Japanese arms producers to export. Japan sells only unarmed land vehicles, naval vessels and aircraft to foreign armed forces. These platforms are, however, frequently equipped with weapon mounts. Such transfers are recorded by ACDA and SIPRI as arms exports.

32. Statement by Nobuyuki Sugimoto, Director of Multilateral Cooperation Division, Economic Ccooperation Bureau, Ministry of Foreign Affairs, report of the Informal Meeting on Participatory Democracy and Good Governance, Paris (9 April 1992), 4.

33. For example, McNamara, "The Post-Cold War World," 97–98; Mark Fineman, "Lines in the Sand Drawing Gulf Nations into Disputes," *Los Angeles Times* (Washington, D.C. edition) (13 October 1992).

FIFTEEN

Toward An International Regime for Conventional Arms Sales

Andrew J. Pierre

FROM THE comprehensive geographic and functional analyses of the previous chapters, two fundamental conclusions can be drawn. First, the proliferation of conventional weapons will be a critical dimension of national order, regional stability and international security in the decades ahead. Second, managing and restraining that dispersion will be an extremely daunting task, given the manifold and often clashing strategic, political, economic and bureaucratic interests involved. The challenge is to create an international regime, multifaceted and varied in its approaches, that will bring some order to the unfettered global arms trade.

Systemic Conditions

The post-Cold War world environment encompasses certain systemic conditions that are conducive to restraint, and others that weigh against it. Among the factors that might encourage restraint are the following:

—The end of superpower rivalry in the Third World has significantly reduced the value of arms transfers as instruments of foreign policy. For decades the provision of arms was the currency of diplomacy. Arms went to friends and allies, with the supplier often paying for or subsidizing them. Arms substituted for soldiers, as with the Nixon Doctrine, which sought to replace American forces overseas with weapons provided by the U.S. foreign military assistance program. For the Soviet Union, arms sales were an even more important tool. Moscow had very little else with which to appeal to Third World countries, not having funds for economic assistance or the appeal of an attractive ideology. With the end of the Cold War, the major arms manufacturing states lost most of their political incentives for supplying weapons.

—Arms proliferation in all its dimensions—nuclear, chemical, biological and conventional—is now widely perceived as *the* principal military threat to international security, today and in the decades ahead. The distinction between some of these components of proliferation—especially between weapons of mass destruction and conventional arms—is becoming increasingly blurred with the growth of highly sophisticated technologies and the spread of ballistic missiles. Conventional arms are the most widely available weapons, are used in all conflicts, are becoming more deadly and can threaten regional stability. In a world no longer dominated by bipolar nuclear deterrence, and in which the growth in deep-seated ethnic and national animosities often leads to armed conflict, the dispersion of lethal non-nuclear weaponry is of particular concern.

—The principal conflicts of the nineties have demonstrated the grave risk of having too high a level of conventional arms in a local or regional conflict. The decade began with the Persian Gulf War, which illustrated to the arms-supplying nations the folly of transferring a quantity of weapons totally out of scale with a nation's true security needs, in such a manner as to destabilize an entire region. More than any other event, the Persian Gulf War led to a world-wide consciousness-raising of the dangers of arms proliferation. Baghdad's accumulation of $83 billion worth of arms from abroad, its stockpiling of Scud missiles, its clandestine nuclear program, its attempt to build a super-gun, and the difficulties the United Nations Special Commission (UNSCOM) has had in eliminating its weapons of mass destruction—all these factors underline the dangers of proliferation. Significant quantities of arms found their way into Bosnia, including heavy artillery, in spite of an arms embargo. The availability of large amounts of firearms was a factor in the civil war in Somalia. Rwanda has tragically shown the awesome lethality of small arms, even primitive ones, in a genocidal intra-state conflict.

—The former Soviet Union has collapsed as a major arms supplier, in spite of a few headline-catching sales Russia has made to countries such as Iran and China. Over periods in the 1980s the former Soviet Union was the world's largest arms supplier, and for several decades it and the United States controlled more than 70 percent of the market. The total value of the

Soviet Union's (now Russia's) agreements with the Third World has, however, fallen dramatically with its political and economic disintegration, from $27.3 billion in 1987 to a low of $1.3 billion in 1993. Although the value rose to $6.0 billion in 1995, it is uncertain whether this increase will continue, since much of it was based on what could be a one-time large sale of SU-27 aircraft to China.[1] Russia has terminated grant military assistance to its former client states. Relatively few arms-purchasing countries are interested in buying arms from Russia or Ukraine, except existing stockpiles at fire sale prices. The unreliability of supplies and spare parts is a consideration, and Russia's military technology has not kept up with that of the West in this decade. Moscow has, in general, been relatively cooperative in the arms restraint initiatives of the P-5 talks and the Wassenaar Arrangement, the U.N. Register of Conventional Arms and the Middle East multilateral arms control talks. On the other hand, it must be noted that some of the defense enterprises in Russia and the Ukraine have at times pursued quite independent approaches, making export deals on their own as best they could, unimpeded by foreign policy implications or concerns for restraint.

Balanced against these positive conditions are a number of systemic factors that suggest that achieving arms restraint will grow increasingly difficult over time.

—Military technologies are increasingly being spread across national borders in a variety of ways that lead to a transnationalization of arms industries. The high expense of research and development (R&D), the relatively small number of expensive and sophisticated weapons that countries can afford to purchase, the financial difficulties in maintaining autonomous weapons-producing companies—all are leading to cross-border mergers of arms manufacturers. Linked to this situation is the strong trend toward co-production and joint assembly arrangements. Technology transfer is multiplying not only through bilateral agreements between governments but also on a company-by-company basis. The net result will be to make "full-scope" national controls elusive and often beyond the ability or even intent of governments. In short, the growing globalization of defense industries is substantially complicating the establishment of effective international controls.[2]

—Dual-use items—civilian technologies that when exported can have a significant utility in the development of military capabilities—are another challenge. Advanced computers, machine tools, sophisticated measurement systems and certain chemical processes are the most worrisome of these many items. Many of these technologies are important to the development of modern industries, and withholding them from developing countries is perceived as discriminatory and paternalistic.[3] Nevertheless, the very considerable success Iraq has had using dual-use civilian technologies to enhance its military capabilities demonstrates the potential dangers of such exports. Most developed countries have established some controls over dual-use exports, but the trend is toward deregulation of export rules rather than toward greater control. This is due to the growing international economic competitiveness, the distaste of private industry for excessive government regulations that interfere with trade and profits and the growing complexity of maintaining controls. As long as there was a Cold War and the enemy could be clearly identified, export controls were easier to apply and justify, as was done through the Co-ordinating Committee on Multilateral Export Controls (COCOM), which was terminated in March 1994. But such controls are far more difficult under the more ambiguous, changed circumstances of the world in the nineties. In the future it should still be possible to achieve very selective controls on certain dual-use items of clear danger, but only if they are agreed upon multilaterally. This is one of the aims of the new Wassenaar Arrangement, established in 1996 (and discussed later).

—More countries in the developing world are significantly expanding and upgrading their defense industries technologically so as to make their arms products competitive in the world arms trade. Countries such as Argentina, Brazil, China, the Democratic People's Republic of Korea (North Korea), Egypt, India, Israel, the Republic of Korea (South Korea), South Africa and Taiwan have not only created arms production capacities but in many cases are now seeking to reduce their technological dependence on other nations. As a result, the major arms exporters of the past are losing some of their control over the arms market and technology of production. This reality complicates the creation of multilateral restraints. Increasingly these countries will have to be taken into account in regulating all but the most advanced weapons systems, which are still the province of only a few suppliers.

—Defense industries in the advanced, developed world have entered a period of severe contraction because of the decline in defense budgets that became possible with the end of the Cold War. The domestic markets for their arms products are shrinking, and due to industries' delayed response to curtailments in national weapons acquisition, the full extent of the downsizing that will be required is only now being fully recognized. The situation varies from country to country. In the United States downsizing was well underway by 1995 through mergers of major defense companies and their movement into non-military production. The process is less advanced in Western Europe and the relevant states of the former Soviet Union. A major consequence of the economic pressures on the defense industries has been far more aggressive attempts to sell arms overseas to offset reductions in national domestic procurement. Most of the defense corporations in the United States as well as the other developed countries are aggressively seeking to sell more arms overseas. Overseas sales will not, however, be the panacea the American defense industry is looking for. Traditionally it has not exported more than 15 percent of its production, and it is highly unlikely this amount could increase to more than 20 percent. Nevertheless, given the heightened international economic competitiveness of the 1990s and without sufficient progress toward defense conversion, it is extremely difficult to deny companies lucrative sales. President George Bush's decision, in the heat of the 1992 election battle, to sell 72 F-15s worth $9 billion to Saudi Arabia and 150 F-16s worth $5.8 billion to Taiwan is a perfect illustration of this point. The Clinton Administration has also strongly supported the overseas marketing efforts of the U.S. defense industry. European firms have traditionally exported a far larger proportion of their military production, and they stand to be even more hurt by the defense cutbacks. Thus the pressure to use arms exports to maintain jobs and prevent financial losses is even greater there than in the United States.

Clearly, creating multilateral restraint on the proliferation of conventional arms will not be easy. At best any restraint is likely to be imperfect and less than comprehensive. Nevertheless, it is a goal the international community should vigorously pursue. Many of the major challenges to international security in the twenty-first century will involve arms proliferation and regional security. Because conventional arms are linked in a variety of ways with weapons of mass destruction—and are the arms actually widely used in conflict—it is essential they not be neglected.[4]

Early Attempts at Controls: Emphasis on the *Recipients*

Before looking at the developments of the 1990s and making suggestions for regime-building in the future, it is important to remember that controlling arms transfers is not a wholly new idea. As far back as the Middle Ages, nations were reaching informal agreements on the selling of arms, such as the understanding among the Christian nations of Europe not to transfer weapons to the "infidel" Turks. Provisions in the General Act for the Repression of the African Slave Trade of 1890 prohibited the introduction of all arms and ammunition other than flintlock guns and gunpowder into a vast zone of Africa.[5] The League of Nations published a statistical yearbook on the trade in arms from 1925 to 1938 that was the first attempt to create "transparency" through public disclosure of arms imports and exports. That yearbook is the antecedent of the U.N. Register of Conventional Arms that came into operation in 1993.

The traditional and deeply held view among the arms-supplying states right up to the Gulf War was that *the recipient states* within a region were principally responsible for reaching agreement on conventional arms limitations. Throughout the 1970s and 1980s American and British representatives at the Geneva Conference on Disarmament ritualistically talked of the need to reduce conventional arms by limiting purchases. "This would be the most satisfactory way," British Minister of State Lord Goronwy-Roberts told the conference. "If the demand is not there, the suppliers will not be able to export." He reasoned that once recipients took the initiative and established an arms control regime, the outside powers would respect it.[6] Secretary of State Henry Kissinger adopted a similar approach in a 1976 memorandum sent to some members of Congress who had expressed an interest in curbing American arms sales at the time of the large transfers to Iran. Agreement among the suppliers to regulate weapons sales, he argued, would amount to a "cartelization" of the world arms trade. But historically cartels are only effective when the members share common interests. In this case the major suppliers of arms lacked common interests. The most promising restraint proposals, Kissinger concluded, will therefore "be those that derive from the initiatives taken by the leaders in the regions concerned."[7] The only regional restraint established by the purchasers of conventional arms to date, however, is the Ayacucho Declaration of 1974 that the eight Andean states of South America entered into, and it was not successful. Concerned about deteriorating economic conditions—attributable to some degree to the high expenditures for armaments—the nations committed themselves to "create conditions which permit effective limitations of armaments and put an end to their acquisition for effective warlike purposes in order to dedicate all possible resources to economic and social development."[8] Discussions subsequently took place among the military in the region. But this initiative did not prevent sizable and competitive arms purchases by Chile, Ecuador and Peru, while Brazil never really became an active participant.

The principle of recipient-side responsibility for instituting multilateral restraints was also dominant in the scholarly community until recently. A 1988 study by the Stockholm International Peace Research Institute (SIPRI), *Arms Transfer Limitations and Third World Security,* concluded that "initiative must come from the recipient side."[9] The lonely exception, posed in 1982, was the author's own analysis that the "best approach would be a supplier-initiated, regionally-oriented framework for managing the process of arms sales to the Third World."[10]

Prior to 1991, all the world's major suppliers of arms had *never* been brought together to discuss weapons transfers. Only twice before had there been remotely similar attempts, leaving aside specific arms embargoes, such as those on Bosnia, India/Pakistan or South Africa, which were temporary and not intended to be a process or regime. Following the Tripartite Declaration of 1950, France, the United Kingdom and the United States set up the Near East Arms Coordination Committee to regulate the flow of weapons. For five years it was quite successful, but the effort came to an end in 1955 when the Soviet Union, and its then-vassal Czechoslovakia, sold large quantities of arms to President Gamal Abd-Al Nasser after his takeover of Egypt. Note that one of the world's major suppliers was not a participant. When the Carter Administration first took up the idea of what later became the bilateral Conventional Arms Transfer Talks, but before suggesting it to Moscow, Vice President Walter Mondale raised the possibility in several European capitals of initiating multilateral talks among the major suppliers. The response was skeptical, and the Europeans never participated. When the talks between the United States and the Soviet Union ran into difficulties, the political ballast the Europeans might have provided was missing.[11] Clearly the lesson of these two experiences is that *all* the major players must be present if multilateral arms restraint is to succeed.

The Persian Gulf War and The Recognition of *Supplier* Responsibility

In the aftermath of the Persian Gulf War, a number of proposals were made regarding ways to develop multilateral restraints. In the United States some senior members of Congress such as Representative Lee Hamilton (D-Ind), chairman of the House Foreign Affairs Subcommittee on Europe and the Middle East, and Senator Joseph Biden (D-Del), chairman of the Senate Foreign Relations Subcommittee on European Affairs, took the lead in pressing for attention to the issue. A number of resolutions calling for a moratorium on arms sales to the Middle East were tabled.[12] Former Secretary of State James Baker said in testimony to the Senate that "the time has come to try to change the destructive pattern of military competition and proliferation in the Middle East and to reduce the arms flow into an area that is already overmilitarized."[13]

After considerable dithering, on May 29, 1991 President Bush announced a Middle East Arms Control Initiative that called for the five major suppliers of arms to develop guidelines for restraints on destabilizing transfers to that region. It sought to establish a code of "responsible" conventional arms transfers and a freeze on the acquisition of missiles by and their ultimate elimination from the region. Other provisions dealt with the control of chemical, biological and nuclear technology but did not add anything substantially new to American policy. Following the presentation of this initiative, however, then-Secretary of Defense Richard Cheney emphasized the importance of continuing arms sales to support America's friends and allies, and the Bush Administration announced a number of such sales. Indeed, in the first year after the Gulf War the United States announced $21.4 billion in arms transfers to the Middle East, a figure that rose to $64.7 billion by 1996.[14] In the 1992–95 period, the United States substantially increased its market share of arms sales to the Middle East to 56.4 percent.[15]

Not to be outdone, only a few days after Bush's announcement, French President François Mitterrand unveiled his proposal, which Paris had been developing for some time. It called for a global, rather than Middle East-focused, approach. It sought military balances on a region-by-region basis, with the amount of arms in each region gradually phased down to the lowest possible level. This action was to be supplemented by confidence-building measures (CBMs) similar to some of those already agreed upon under the Conventional Forces in Europe (CFE) Treaty. The Mitterrand initiative also sought to extend the Missile Technology Control Regime (MTCR) to more nations and to ban all chemical and biological weapons.

Britain's efforts focused upon Prime Minister John Major's espousal of a United Nations arms register, although London indicated it would also participate in multilateral discussions on arms transfer restraint. As for the Soviet Union, the winding down of the Cold War led to a substantial change in attitude. As early as the fall of 1988, in his address to the United Nations, President Mikhail Gorbachev spoke about the need to restrain the flows of weapons. After the Gulf War, the Soviet Ambassador to the United Nations suggested consultations among the Big Five on this issue. The Group of

Seven countries, meeting in London in July 1991, issued a statement supporting both the start of multilateral talks and the U.N. Register of Conventional Arms.

What these proposals had in common, given the historical experience noted earlier, was quite revolutionary: acceptance of the principle that primary responsibility for creating some system of international controls over arms transfers rested with those nations that produce and supply the weapons. The prior view of the supplier states—that it was up to the recipient states—was nothing less than a recipe for inaction. States within a region where there is a high level of tension or conflict, such as the Middle East, are the last ones that can be expected to begin developing arms restraint initiatives. Even though they might see some advantages in regional arms control, the initial impulse must come from the suppliers. It is, after all, the producers of weapons that can turn the tap of arms flows on or off. Therefore, they also have the main responsibility.

The logic of assigning primary responsibility to the major suppliers derives from the oligarchic market situation: in 1995, for example, six of the largest suppliers of arms to the developing world (United States, Russia, France, Great Britain, Germany and China) delivered nearly 90 percent of the arms and a probably higher proportion of the "big ticket" items such as supersonic combat aircraft, heavy armored tanks and missiles or their components.[16]

The P-5 Suppliers Meetings

These proposals led to the P-5 talks, three rounds of meetings of the five major arms suppliers.[17] At the first meeting, held in Paris in July 1991, the P-5 "recognized that indiscriminate transfers of military weapons and technology contribute to regional instability" and acknowledged that they are "fully conscious of the special responsibilities that are incumbent upon them to ensure that such risks be avoided, and of the special role that they have to play in promoting greater responsibility, confidence and transparency in this field."[18] Accordingly, they agreed to work on modalities of consultation and exchange of information on arms transfers, giving priority to the Middle East, and also to review their national systems of controls so as to encourage restraint.

At the second round, held in London in September 1991, the P-5 agreed to inform each other about the transfer of seven categories of arms to the Middle East—battle tanks, armored combat vehicles, large caliber artillery, combat aircraft, attack helicopters, warships and certain missile systems. When deciding whether to transfer weapons, each country pledged to consider carefully whether the proposed transfer would meet legitimate needs for self-defense in the recipient country, serve as an appropriate and proportionate response to the military threats confronting it or enhance the capability of that nation to participate in regional or U.N. collective security arrangements. The P-5 further agreed to guidelines according to which they would avoid transfers likely to:

    a) prolong or aggravate an existing armed conflict;

    b) increase tension in a region or contribute to regional instability;

    c) introduce destabilizing military capabilities in a region;

    d) contravene embargoes or other relevant internationally agreed restraints to which they are parties;

    e) be used other than for the legitimate defense and security needs of the recipient state;

    f) support or encourage international terrorism;

    g) be used to interfere with the internal affairs of sovereign states;

    h) seriously undermine the recipient state's economy.[19]

Such guidelines are, of course, so broad as to be totally dependent on interpretation in the case of each transfer. While they create a standard and project an intent of restraint, they are completely subject to whatever determination each state makes.

By the time the third round took place in Washington on May 28–29, 1992, substantial areas of disagreement had surfaced. Little progress was made in this round except for an agreement on "interim guidelines" for the export of technologies related to weapons of mass destruction. In reality, the document simply restated principles on which there was already broad, conceptual agreement. At a meeting of military experts prior to the third round it had proven difficult to reach agreement on the details of which specific weapons systems were to be included in the general categories outlined above. The Chinese, who from the start were reluctant participants in the talks, insisted that limitations on all ballistic and surface-to-surface missiles should be matched by equally broad limitations on advanced fighter aircraft, as they could also carry weapons of mass destruction. Also contentious was the definition of the Middle East. This issue was important because, although the guidelines agreed to were global in nature, the procedure for notification focused on that region. The United States was anxious to include Libya, as well as Algeria, Morocco and Tunisia. The Chinese argued that, if these countries were to be included, Turkey and Cyprus also should be.

The main unresolved issue, however, was establishing the point at which a country would be obliged to notify the others of a sale or transfer. In London the P-5 countries had agreed to "exchange information for the purpose of meaningful consultation." Clearly this provision did not include the right to veto a sale or transfer. But meaningful consultation had to include *prior* notification of a sale or transfer so as to give any other of the five nations a right to question its wisdom or the opportunity to challenge it directly.

Prior notification generated a number of objections. Initially, both the British and French saw it as incompatible with the normal processes of discussion with parliamentarians and believed it could even raise constitutional issues. Some officials in both countries acknowledged, however, that these problems could be readily resolved through legislative changes or modifications of arms export regulations. China opposed any significant prior notification, preferring that notice be given at the time of transfer or, at the earliest, thirty days beforehand. Russia, which replaced the Soviet Union during the negotiations, had yet to elaborate its arms export procedures, but their validity would in any case be questionable given the uncertainty of Moscow's control over the nation's defense industries, and even over the existing stockpiles of weapons. The United States was the only country for which prior notification was not a significant issue, since Congress by law had to be informed of intended sales.

Following President Bush's decision in September 1992 to sell 150 F-16s to Taiwan, in apparent violation of the U.S.-China Joint Communiqué of 1982 that banned U.S. arms sales to Taipei if they exceeded previous weapons levels qualitatively or quantitatively and also promising a gradual reduction in arms sales, Beijing decided to boycott the P-5 talks. France followed the American precedent soon after by also selling aircraft to Taiwan, something it had been reluctant to do until then. The fourth round, scheduled for Moscow, was postponed indefinitely and in the end never took place. Although Washington was denounced for its arms sales and Radio Beijing told it to "eat the rotten fruits of its deals," interestingly the Chinese never formally withdrew from the process.

If the Clinton Administration had actively sought to revive the P-5 talks during its first year in office, as some outside observers urged, it may well have succeeded. The Russian foreign ministry, which quietly discussed this possibility with China, suggested the talks be given a new name, have a slightly altered agenda, and be opened to one or more additional suppliers, to make their resumption acceptable to Beijing. Germany would have been a logical candidate, given its growing arms sales. Indeed, the talks could have resumed without Beijing's participation, with its chair kept in place but empty, since at that time China accounted for only 4 percent of the world arms trade. London and Paris waited to see what Washington would do. The Clinton Administration's priorities were elsewhere, however. Although it engaged in a very active non-proliferation policy, the Administration's focus was on nuclear matters. It put arms transfer policy on the back burner, relegating it to what became a very lengthy interagency dialogue. At the same time, parts of the Administration actively sought to promote American arms sales abroad. By the time the Clinton Administration's arms transfer policy was established and announced in February 1995, the P-5 process was all but forgotten.

In retrospect, the member countries never allowed the P-5 talks to pursue their potential. Several of the participating countries began some exchanges of information on weapons transfers. Reportedly some specific objections to intended Russian sales led to their cancellation. The talks never went as far as discussing regional limitations, however, nor was the consultative process ever fully agreed upon. The P-5 talks were a promising but interrupted process. Viewed in a historical perspective, they were an important innovation and breakthrough, for they shifted responsibility for restraints to the suppliers. Fortunately there was to be a second chance with the creation in 1996 of the Wassenaar Arrangement.

Components of an International Regime

Developing internationally agreed upon standards, not to mention rules, for dealing with the proliferation of conventional arms is a long-term endeavor. In many ways this task is far more difficult and ambitious than dealing with the proliferation of nuclear weapons. Unlike nuclear weapons, which are still limited in numbers and possessors, conventional weapons have proliferated widely and are available readily. The questions for the future are *which* weapons are to be transferred (or restrained), in *what* quantities and to *whom*? Also unlike nuclear weapons, the manufacture and sale of conventional arms can bring large economic benefits and provide significant levels of employment. But most importantly, unlike with nuclear weapons, there is no agreed upon norm for avoiding the proliferation of conventional arms. Almost all of the governments of the world believe that the dispersion of nuclear arms is undesirable. No similar judgment can be made with respect to non-nuclear weapons. Conventional arms can be used aggressively or defensively. Their availability can be such as to encourage a conflict, or they can deter a conflict. It all depends upon the particular case at hand and the expectations that decisionmakers hold about its future.

For these reasons there will never be a comprehensive regime for dealing with the proliferation of conventional arms similar to the Nuclear Non-Proliferation Treaty or the Chemical Weapons Convention. What the international community can seek is to build a composite of approaches that are complementary and supportive. Some could be *multilateral,* as in agreements among a group of arms suppliers or weapons recipients, and some could be *unilateral,* as in the national export policies of individual countries. Some could be *global* in orientation, while others could be *regional* or even *nation-specific.* Although treaties that *prohibit* certain types of conventional arms proliferation are not inconceivable, in light of the ubiquitous and dynamic nature of global arms flows an international regime is more likely to consist of mechanisms for *managing* and *restraining* the transfer of armaments.

Such an international regime, it is now generally recognized, should have three elements:

—*Transparency.* The public release of information about arms transfers and greater sharing of information between governments can ensure, or at least improve the likelihood, that proposed transfers are in fact desirable. Internally, within nations, greater transparency can put the spotlight of legislatures, the media and public opinion on arms acquisitions so as to see if they make sense. In some countries arms have been acquired for reasons of prestige or glamor more than for reasonable need. Externally, the exchange of information between governments can serve as a confidence-building measure. The U.N. Register of Conventional Arms is essentially a transparency measure based upon the concept—by no means always valid—that the requirement of public disclosure of a country's arms imports and exports can serve to prevent or reduce unwise transfers.

—*National regulation and oversight.* The degree of government regulation over arms sales, as well as over arms manufacturing companies, varies considerably around the world. Even more varied is the role of parliaments in oversight over the decision-making process concerning arms. Most industrialized nations have a set of regulations that cover the activities of firms and individuals, but they are enforced with differing degrees of vigilance. In the case of the sale of dual-use technology to Iraq in the late 1980s by the British firm Matrix Churchill, officials at the highest level appear intentionally to have turned their eyes away from the existing regulations. Laxity crept into German export procedures as a result of the devolution of regulatory responsibility to state authorities in the various regions. Controls were inadequately applied or were neglected until this situation led to an outcry and subsequent reforms in the early 1990s.[20] With the breakup of the Soviet Union there is evidence that governmental control disintegrated and that many weapons-producing firms, desperate to maintain their employment and revenue, took matters into their own hands and sought to export arms without the approval of authorities in Moscow or Kiev.[21] American statutory regulations are comparatively comprehensive in their coverage: the State Department maintains a Munitions Control List, while the Defense Department maintains a Military Critical Technologies List. Critics nonetheless point out the need for a tightening of existing regulations and the creation of greater transparency in U.S. military sales. The role of the U.S. Congress in the transfer of American arms is far greater than that of the parliaments in the other major supplier nations. Under the International Security Assistance and Arms Export Control Act of 1976, all planned military sales over set ceilings are to be reported to Congress. In addition, an annual submission of prospective transfers based upon the indicated interests of potential purchasers must be made annually. Nevertheless, the Congress does *not* approve a sale; it can only block one by concurrent resolution passed by both Houses of Congress. Historically Congress has been reluctant to take that step. For many countries, enhanced parliamentary oversight of national arms export policies and practices, or the creation of oversight where it does not exist, would be highly desirable. Such unilateral measures should be seen as critical underpinnings of an international regime.

    —Multilateral controls. The establishment of actual controls or restraints over arms sales gets to the heart of the issue and is by far the most important of these three elements. Controls may need to vary widely, depending on the nature of the military technology. Light arms, which can be critical in localized ethnic conflicts, present a different control challenge than do major weapons systems such as supersonic aircraft, which may affect regional balances. Controls may be placed on certain types of armaments, such as missiles or landmines, or on specific technologies, such as Stealth technology. They may be quantitatively based or qualitatively oriented. In most instances, controls over the transfer of conventional arms need to be implemented on a regional rather than a global basis. They will be in response to political and military tension or conflict in an area such as the Middle East or East Asia, rather than to an abstract or universalized need.

Most controls will only be effective if they are multilateral in their composition. Most types of armaments today are made by more than one country. The risk is real that if only one country exercises restraint, another supplier will be tempted to take its place. The aim of multilateral restraints should therefore be a joint effort on the part of all the major suppliers. However, this will only be fully achieved when there is a common political and strategic appreciation of a situation that provides the basis for a policy of restraint.

Transparency: The U.N. Register of Conventional Arms

The principal form of transparency for conventional arms proliferation to date is the U.N. Register of Conventional Arms. On December 9, 1991 the U.N. General Assembly passed a resolution on "Transparency in Armaments," which was sponsored by Japan and the European Community. It called upon the U.N. Secretary General to establish a register of conventional arms, to include certain categories of arms imported and exported on a country-by-country basis. States were also invited to provide as "background information" data on their overall military holdings and on their procurement through national production, as well as information on their relevant national policies for the export and import of arms. Arms transfer data are to be provided for seven categories of weapons: battle tanks; armored combat vehicles; large caliber artillery systems; combat aircraft; attack helicopters; warships; and missiles and missile launchers. The rationale for choosing these weapons was that they are potentially the most destabilizing because they are particularly suitable for surprise attack and large-scale offensive action; they are also easily identifiable. The first five categories correspond to the weapons covered in the Conventional Forces in Europe (CFE) Treaty. Information for each year is to be provided by April 30 of the following year; thus the first Register covered 1992 and used information submitted by April 30, 1993.[22]

The concept of an international register for arms transfers is not new. As noted, the League of Nations published an annual *Statistical Yearbook,* which gave the values of arms imports and exports based on official national statistics. The first volume covered twenty-three countries, the last volume sixty countries plus sixty-four colonies, protectorates and mandated territories. The yearbook had many imperfections with regard to the listing of categories of arms, as well as serious problems of comparability across the data the various nations provided. The yearbooks were of only limited utility, and there is no evidence they had any impact on the actions of governments. Their publication came to an end with the gathering war clouds.

In the first three years of the U.N. Register, which covered arms transactions in calendar years 1992 through 1994, a total of ninety-one, eighty-eight and eighty-four, respectively, of the organization's 186 (in 1994) members made submissions to the United Nations. By November 1995 a total of 111 states had made submissions at least once. Of the twenty-five leading arms exporters in 1990–94 as listed by SIPRI, all but North Korea made a submission in 1994. The response of arms-importing countries in the developing world was more mixed, with eighteen of the top importers during 1990–94 providing reports. In the critical region of the Middle East/Persian Gulf, Israel and Iran sent reports, but Egypt (except in 1992), Kuwait, Syria and the United Arab Emirates did not, as was also true of Saudi Arabia, the world's largest purchaser of arms. In Asia, India, Malaysia, Pakistan, South Korea and Thailand (except in 1992) sent in their reports, but North Korea did not. Taiwan, which is a large arms importer but not a member of the United Nations, abstained from the Register, although it had been invited to join in. The region with the worst level of participation was sub-Saharan Africa, with only five of forty-eight possible countries reporting in 1994. Nevertheless, by one estimate, in the Register's second year it accounted for 95 percent of the known trade in major conventional weapons.[23]

Even so, the information provided is far too limited. Although, for example, the number of combat aircraft or tanks is given, their types or characteristics are not provided, so that a *qualitative* evaluation is impossible. Their value and cost are not specified. There are vast discrepancies in how the nations reported their imports and exports. The lack of uniformity in how each nation prepared its reports is evident when attempting to cross-reference reported exports and imports. For example, Greece reported imports of 28 combat aircraft but not of armored combat vehicles or heavy artillery from the United States in 1994, whereas the United States reported exports of 120 armored combat vehicles and two artillery pieces and no combat aircraft to Greece. One reason for the data's inadequacy is that, to encourage as much participation as possible, the United Nations gave its members only very limited guidelines for completing the submissions. It did not initially even provide a precise definition of what constitutes an arms transfer or when a transfer takes place.

In the debates at the United Nations on establishing the Register, as well as in more recent discussions on its future, a number of proposals have been made to expand it to include other aspects of arms and military technology. Options include broadening the scope of the Register to encompass additional categories such as small arms, including cluster bombs, landmines and high-technology systems with military application. Egypt has even proposed including weapons of mass destruction. Another suggestion is to disaggregate the seven categories so as to make them more specific and useful. It has also been proposed that there be an annual submission of information on military holdings and procurement through national production, rather than the present informal and optional presentation of any "background information." Many of these suggestions have been contentious and difficult to deal with.

The debates at the United Nations over the Register can be viewed as a struggle between transparency and opacity. There is a risk that if the Register is expanded too aggressively, some nations that now comply will discontinue their submissions. Ambitious proposals were therefore sent for further discussion and evaluation to the U.N. Committee for Disarmament in Geneva. An international Group of Experts selected by governments was also charged with reviewing the implementation of the Register thus far and making recommendations for its future development. Reporting in October 1994 after examining at length various ways to strengthen and expand the Register, it was unable to come to any consensus on recommendations. This outcome was severely disappointing to proponents of the Register. If the Register is to become a lasting and reasonably effective instrument, capable of making a significant contribution to international security, its modest start must definitely be built upon now. A new Experts Group is due to be convened in 1997.

The underlying assumption of the U.N. Register is that promotion of transparency will encourage prudent restraint by arms exporting and importing nations. Disclosure of information is thought to lead to greater public scrutiny. Domestically it should promote governmental accountability and monitoring. Internationally, it should reduce the risk of misunderstanding, misperception or tensions resulting from the lack of information on arms transfers. Transparency can be achieved when there is a certain amount of "openness." Nations engage in openness when they have a general policy of making information on arms transfers and military matters public. Thus, transparency is dependent upon openness and is carried out by systematically providing information on arms transfers.

A corollary of this reasoning is that greater transparency will serve as a confidence-building measure and can become part of arms limitation agreements. Hence, the CFE Treaty of 1990 calls for the exchange of a great deal of information on armaments so as to reduce mistrust. Transparency will be critical to anything roughly similar that some day might be instituted in the Middle East.

It must be noted that there is already a great deal of transparency in arms transfers. SIPRI publishes data on arms sales on an annual basis in its *Yearbook*. The U.S. Arms Control and Disarmament Agency (ACDA) publishes *World Military Expenditures and Arms Transfers,* an annual compilation of the value of completed arms transfers. The U.S. Library of Congress puts out an annual publication that covers arms sales by supplier countries, *Conventional Arms Transfers to the Developing World.* The International Institute for Strategic Studies (IISS) and Jane's Information Group publish additional data. A few governments provide similar information on their own activities, although on a limited basis.

The U.N. Register therefore adds only modestly to the already published data, although it has the advantage of being based on official reports of governments. If the Register were successfully expanded, it could provide useful information in a comprehensive, comparable and authoritative manner. Whether it would add to the store of knowledge within the intelligence communities of governments, especially those governments that accumulate data of special interest to them on the military capabilities of states that concern them, is uncertain. Of course, not many governments have at their disposal an agency remotely comparable to the U.S. Central Intelligence Agency (CIA).

The degree to which the U.N. Register can contribute to international confidence will depend upon its universality and comprehensiveness. Ideally, all suppliers and recipients would provide information on arms transfers. In addition to the categories of weapons already agreed upon, the data should include weapon components, technical support and training arrangements. Qualitative information should be included. The Register should also identify the date of sale, as well as of the actual transfer (which is often several years later), and the financial terms. Since these data can only be provided on a voluntary basis by states, it is unlikely the Register would be conducive to a system of verification or inspection. Some degree of verification could be achieved, however, through official cross-checking and analysis by the U.N. Secretariat after the importing and the exporting countries submit the data.

Experience has shown, however, that some states have been reluctant to provide information that allows their military capabilities to be evaluated.

Other states resist providing data that allows the levels of their expenditures on military hardware to be assessed. This appears to be the case for the governments of some developing countries that fear that some "conditionality" to the multilateral economic assistance they receive could be applied on the basis of what others judge as "excessive" expenditures on defense.

The provision of information to the U.N. Register is made after the fact, that is, after a weapons transfer is made. There is no prospect for registering a pending sale prior to its completion. This is a critical point. Publication of information of a pending sale, were it possible, might arouse governments and public opinion in such a manner as to work to prevent completion of sales adjudged to be undesirable. But this will not occur in a body of more than 180 member nations with their diversity of interests. It could, on the other hand, be achieved in confidence among the principal suppliers through the new Wassenaar Arrangement.

In the final analysis, the U.N. Register of Conventional Arms raises critical questions. Will it prevent transfers that would destabilize a region or a country? Will the regular airing of information reduce excessive expenditures on arms by recipient states?

There can be no definitive answer to these questions. The impact of the U.N. Register will depend on its effect upon public opinion and, through such opinion, upon the decisions of governments. It is therefore desirable that the Register be given the widest public exposure. The value of increased transparency is, to a considerable extent, dependent upon the extent to which publicizing data would embarrass or shame governments, if such is warranted. In many countries there is still little public accountability on military matters.

At the same time, it is important to remember that arms sales are often perfectly legitimate and desirable instruments of foreign policy. Whether greater publicity would deter what some might see as destabilizing or otherwise undesirable sales or purchases will depend upon how a particular arms transfer is viewed in the specific context of its time. Similarly, whether a purchase of weaponry is deemed excessive or unnecessary, and is thereby discouraged, will depend on the openness of a society and the relationship of the government and its informed public.

Despite these normative uncertainties, in a number of regions the U.N. Register of Conventional Arms could become a confidence-building measure. In particular, it could provide early warning of an excessive accumulation of arms that could become destabilizing, as was the case of

Iraq. Even if the importing country acquiring weapons refuses to cooperate, or cheats in its submission to the Register, the exporting country transferring the arms may have made an accurate report. Cross-checking of information could thereby reveal some transfers, unless both the recipient and supplier states agreed on secrecy, something that is quite possible in very sensitive cases. Ultimately, the U.N. Register will also serve as a public reminder of the benefits of demilitarization through the curtailment of the acquisition of conventional weapons and a reduction in the size of the armed forces.[24]

National Regulation and Oversight: The Major Suppliers

In the quest for international controls of arms sales, it is often forgotten that control starts at home with national systems and policies of regulation and oversight. Unilateral measures of restraint should form an important part of a conventional weapons non-proliferation strategy.

United States

In the United States the Congress has taken on an important role in the arms transfer process through the International Security Assistance and Arms Export Control Act of 1976. All exports of "significant military equipment" of more than $14 million, or of other weapons and military services of over $50 million, are subject to a thirty-day period of Congressional review (fifteen days for sales to countries in the North Atlantic Treaty Organization [NATO], as well as to Australia, Japan and New Zealand). Once the executive branch decides on a sale, it sends a Letter of Offer, which defines the equipment, estimated cost and conditions of the sale, to Congress. To disapprove a sale, both the House and Senate must pass a resolution. If the president vetoes the resolution, a two-thirds majority in both Houses must vote to override. The president can still bypass such a concurrent resolution with a presidential waiver, issued on national security grounds. Congress is also given advanced notification of expected arms sales at the beginning of each fiscal year through the Arms Sales Proposal, known as the "Javits List." This necessary notification, which is contained in a 1978 amendment to the above-cited law, requires listing all prospective sales of over $7 million in arms exports or defense articles and services of over $25 million for the coming year.[25]

This legislation gives the American Congress a greater role in arms sales than that of the parliaments of other major suppliers. At times it has served to deter presidents from submitting an arms sale when informal soundings by administration officials led to the conclusion that it had little or no chance of getting through. On occasions when a president has proceeded with notification of a controversial sale, based on a reasonable possibility of getting it through, major battles have occurred. Two of the most noteworthy involved sales of F-15 fighters to Saudi Arabia in l978 and again in 1993.[26] In practice, there has not been a single occasion when a sale has reached the point of being vetoed by a concurrent resolution of the Congress. For this reason, it has been proposed that the law be changed to require Congress to approve all sales above a specified dollar threshold, essentially reversing the present procedure.[27] Such a change, however, would in recent years have required a floor vote on around 125 sales a year. To ease the legislative burden the change would create, it should be possible to group the sales together.

In spite of the end of the Cold War, a large portion of American arms sales are still negotiated, as they have been for decades, through the Defense Security Assistance Agency (DSAA) of the Department of Defense, rather than as commercial sales between American companies and foreign governments. This practice is a historic residue of the time when U.S. arms exports were largely political instruments for supporting friends and allies engaged in the East-West competition. Military security assistance was a major element of the foreign aid program. The DSAA became a substantial bureaucracy, with 80 percent of its operating budget financed through a 3 percent administrative, or "recoupment," fee it charged over and above the costs of the weapons it arranged to procure and then transfer to foreign governments. This fee, which is still charged, ostensibly is intended to recoup for the United States part of the R&D costs of the weapons. By the early 1990s the fee yielded $330 million per year. In 1994 the DSAA used the proceeds to support an overseas field staff of 2,085 U.S. government civilian employees and 5,766 U.S. contract employees in as many as 105 countries. Their mission is to help shape a specific request by matching the military assistance needs of the host country with American equipment. Inevitably this process leads to an incentive to promote sales and transfers of American weapons. As the U.S. Office of Technology Assessment (OTA) has observed, "because the operating budget of the agency is tied to the volume of weapons transferred, there is a powerful incentive for DSAA personnel to make as many sales as possible."[28]

In the changed world environment of the 1990s and beyond, one in which non-proliferation must be a key objective, the impact of all arms sales should be carefully scrutinized, especially their potential longer term impact within the regions to which the weapons are transferred. The Department of State is charged with this task, but it is not in the best position to undertake the job, since its natural inclination as an institution is to seek good bilateral relations with countries. Turning down a weapons request from the leaders of a friendly nation is difficult. In addition, the Department of State has been under pressure not to deny arms sales that are economically lucrative. In January 1990 its Office of Munitions Control was renamed the Center for Defense Trade and given a clear mandate to support American business abroad. The result is a reluctance to apply a longer term, strategic perspective when doing so involves a shorter term, tangible cost.

Accordingly, ACDA might be given an enlarged role within the U.S. government in the making of arms sales decisions. After seeking advice from the Departments of State, Defense, Commerce and other agencies in developing its recommendations, ACDA would have final authority, under the president, for decisions. The agency might also be mandated to present a "weapons transfer proliferation impact statement" for all U.S. arms sales of a certain size or over a specific dollar amount.

Arms transfers are an important component of foreign policy, and thus it is not surprising that groups seeking to influence American diplomacy use them to further their aims. Supporters of both Israel and the Arab nations have pressured Congress for arm sales on behalf of their friends. Advocacy groups concerned about the abuse of human rights have also sought to link arms transfer policy with their aims, arguing that no weapons should be transferred to regimes that repress their own citizens. The Human Rights Watch has established an Arms Transfer Project that exposes such situations through fact-finding missions and reports on conflicts occurring in such places as Angola, Lebanon and Rwanda.

In 1993 legislation was introduced in Congress to create a "Code of Conduct" to guide American arms transfers. Sponsored by Republican Senator Mark Hatfield (Ore) and Democratic Representative Cynthia McKinney (Ga), the proposed code required that recipient nations meet certain criteria in their domestic and international practices before being eligible to receive U.S. military assistance and arms transfers. Nations that lacked democratic institutions and the rule of law, failed to respect human rights, were engaged in certain acts of armed aggression and were not full participants in the U.N. Register of Conventional Arms would be ineligible. Eighty-three members of Congress co-sponsored the bill, which was also supported by forty-five arms control, human rights, religious and veterans groups that comprise an Arms Transfer Working Group. The "No Arms to Dictators" coalition is not likely to be fully satisfied: in May 1995 it lost the vote in the House of Representatives (157 to 262) and in July 1996 in the Senate (35 to 65). Nevertheless, such activities are indicative of the saliency of the human rights issue and the question of Congressional oversight on arms sales within the American political debate.

Indeed, such causes help spawn, and create transnational linkages with, similar groups in other countries. The British American Security Information Council, Saferworld and the World Development Movement proposed in May 1995 a European Code of Conduct for the Arms Trade. The primary intent was to have the Council of the European Union adopt common arms export criteria that took into account human rights considerations. Further impetus for an international code of conduct came in February 1996 from a group of luminaries organized by Oscar Arías Sanchez, a former president of Costa Rica and winner of the Nobel Peace Prize in 1987, who had enlisted the support of fellow Nobel laureates. A broadly based international coalition, with participants in some twenty countries, now supports and lobbies for an International Code of Conduct on the Arms Trade.

Russia

The situation in Russia is totally different than that in the United States. Its Parliament has virtually no oversight capacity on arms sales. Ever since the demise of the Soviet Union, various parts of the Russian bureaucracy have been in competition for dominance in decision-making on arms sales. The resulting struggle is related to the fact that the extremely sharp drop in weapons orders by Russia's armed forces—approximately 80 percent—has left the large and diverse defense industry in dire condition. Overseas arms sales appear to many in the military-industrial complex as the most promising, and probably only, way out of the economic morass.

Under the Soviet regime, the Defense Council of the USSR, composed of senior Politburo members and key ministers, made all decisions to begin or terminate arms supply relationships with foreign countries. Since the demise of the USSR, a number of reorganizations[29] have been carried out that reflect the fierce internal competition for control of Russia's arms exports and access to the income from sales. The Ministry of Foreign Affairs, which significantly influenced decisions on arms sales when weapons transfers were an important instrument of foreign policy, has been unable to maintain its power. With the new dominance of economic factors in Russia's arms exports, other agencies have acquired strength. The finance and economic ministries of Russia have sought to maximize arms sales for their hard currency revenue. Apart from raw materials and energy in the form of oil and gas, Russia has not had much to sell but armaments. The Defense Ministry has promoted arms exports when it was assured the resulting revenue would be used for the care and maintenance of the armed forces. With the greatly reduced funding of the military, first priority has been given to salaries, housing, schools and social needs rather than new weapons for Russia's forces. Weapons exports, it has been argued, are essential for both urgent domestic social needs and for purchasing equipment for the military services.

Because plant directors and weapons designers found that the ministries in Moscow often kept the revenue derived from arms exports and did not pass it down to them, they have sought a free hand to sell on the world market to save their plants from closure. The "defense cities," which are fighting severe economic depression and social disarray, exercise a powerful influence through regional administrators, who are also in the Council of the Federation, the upper house of the Russian Parliament. They have been "free-lancing"—seeking themselves to sell their products abroad independently. Although they lobbied and have been able to gain President Boris Yeltsin's limited approval of this unorthodox practice, their success has been limited by their lack of expertise in negotiating in the international arms market.

To bring order to the chaotic diversity of agencies involved in Russia's arms sales, each often pursuing its own agenda in competition with the other(s), in January 1994 President Yeltsin created a super-agency with the aim of coordinating arms exports. *Rosvooruzhenie* brought together the three principals in the arms export structure—*Oboronexport* (General Defense Export Corporation), *Spetsvneshtekhnika* (General Technical Department) and *Gusk* (General Department for Cooperation of the Ministry of Foreign Economic Relations)—and placed them under the personal watch of one of Yeltsin's advisors, General Yevgeny Shaposhnikov, former commander of the strategic forces of the Commonwealth of Independent States.

The clear mandate and main thrust of the new agency, however, have been the aggressive promotion of arms exports to recoup Russia's past position in the world's arms market, rather than accountability or control. In its first years *Rosvooruzhenie* sought to achieve its goal by adopting Western promotional techniques such as slick color brochures depicting arms and an expensive English language magazine, *Military Parade,* and by creating offices in more than thirty countries. As a result, the level of arms exports rose by about 50 percent in 1995–96, and Russia appears to be gaining a larger slice of the market. Whether *Rosvooruzhenie* can overcome the downward trends in the world arms market remains to be seen.

Important elements of legal regulation and political oversight of Russian arms exports are still lacking. Export regulations are based upon presidential decree rather than legislation. In 1992 the Parliament began drawing up a basic arms export control law with the guidance of four committees—Defense and Security, Industry and Energy, International Affairs, and Budget and Taxation—but did not complete the work prior to its dissolution in September 1993. The new Duma was slow to address the issue, even though there is support among officials for instituting legislation and among parliamentarians for creating some form of oversight.[30]

Western Europe

In Western Europe, among the principal arms manufacturing and exporting countries—France, Great Britain and Germany—there is strong regulatory control at the executive level, with very little role for the parliaments. In *France,* the arms industry is still relatively large: it has direct employment of 250,000, accounts for 5 percent of total industrial employment, and provides for an additional 100,000 jobs at the subcontractor level. Nearly four-fifths of the defense industry is partially state-owned, although privatization is underway. Given this heavy state involvement, it is not surprising that France's elite bureaucracy keeps a tight hold on export decisions. These are made by the Prime Minister upon the advice of an interministerial committee (*Commission Interministérielle pour l'Etude des Exportations de Matériels de Guerre*) and implemented by the Ministry of Defense. The French parliament, the Assemblée Nationale, has no role in arms sales. In 1978 a socialist deputy, Charles Hernu, proposed legislation that would have required the government to inform and consult with the parliament before proceeding with arms sales. However, when he became the Minister of Defense, he no longer espoused this idea. After the Persian Gulf War demonstrated how French arms, once exported, could be used against French forces, an attempt was made to force the government to report its arms sales practices to the Assemblée Nationale. As a result the Minister of Defense made some voluntary disclosures, but the process has not been maintained.

Practice in the *United Kingdom* is not very different. The House of Commons has no formal role except for the right of its members to question a minister. But when a minister is asked about an arms sale by a Member of Parliament, he is customarily told that it is not the practice of Her Majesty's Government to provide information on arms exports! Parliamentary committees meet sporadically but have no oversight powers. Within the Whitehall establishment, the Ministry of Defence typically drives much of the process surrounding arms sales, although the Foreign Office does undertake political and strategic analysis of a proposed arms sale when warranted. The Defence Ministry's Defence Export Support Organization seeks to promote and assist sales while also ensuring that they fall within political guidelines. When there are differences between the ministries involved in a sale, which include the Department of Trade and Industry, the issue is brought to the Strategic Export Working Party for resolution and, if necessary, to the Cabinet.

But organizational charts and known procedures may be misleading. A rare look into the inner workings of the British system, and how it can in certain circumstances totally circumvent not only Parliament but the Prime Minister as well, came out of the publication in 1996 of the 2,000-page report of an investigation by Sir Richard Scott, a high-ranking judge, into the handling of arms sales to Iraq from the mid-1980s to the beginning of the Gulf War. When several business executives from the Matrix Churchill company, makers of machine tools, were put on trial in 1992 for violating export laws by selling artillery-related machinery to Baghdad, they insisted that they had done it with the approval of British authorities. Indeed, it turned out to be the case that British intelligence had set up the operation as a way of gathering information on Iraq. It also turned out that a series of statements in the House of Commons by several senior ministers, including William Waldegrave, then a Foreign Office minister, were misleading or false. Waldegrave had denied that the 1984 policy of even-handedness towards Iran and Iraq had been altered, whereas in conjunction with Lord Trefgarne, a defense minister, and Alan Clarke, a trade minister, it had been decided in 1988 to relax restrictions on exports to the Iraqis, who were prepared to pay cash, without informing the senior cabinet ministers who subsequently made clearly erroneous statements to the Parliament. The Scott Report exposed, the *Economist* observed, "an obsessively secretive government machine, riddled with incompetence, slippery with the truth and willing to mislead Parliament."[31] As for Sir Richard, he concluded that "the present legislative structure, under which Government has an unfettered power to impose whatever export controls it wishes and to use those controls for any purposes it thinks fit, should, in my opinion, be replaced as soon as possible."[32]

As for the *Federal Republic of Germany,* following the revelations that led to some international protest about German companies having sold dual-use and other militarily relevant technology to Iraq and Libya in the 1980s, national legislation was adopted to create stronger export controls. Although parliamentary committees were involved in this process, they did not reserve a regulatory role for the Bundestag in the future.

Clearly, the governments of the major arms supplier nations of Western Europe have not encouraged an expansion of parliamentary roles in the oversight of weapons exports. Still, they have had to weigh what should be the proper role of the *European Union.* Proposals have been made in recent years for a common European Union arms export control policy. Article 223 of the Treaty of Rome, which established the European Union, specifically preserved for member states all measures involving trade in arms, munitions or war materials, thereby excluding these from the organization's commercial and trade rules. Nevertheless, interest has developed within the European Parliament, the European Commission, and some public interest groups in creating oversight in the defense sector in a manner comparable to the rules covering the civilian economy. This is particularly needed, it is felt, because of the transnational collaborative nature of much defense production in Europe. The unilateral restraint of one nation can be under-done by the laxity of another.

Particular attention has been given to dual-use technologies that have a civilian and military application and therefore do not fall squarely within Article 223. Saferworld, a British advocacy group, has been lobbying for an international code of conduct to govern such technologies. The European Union's Council of Ministers in early 1995 adopted a policy on the export control of dual-use equipment and technology that, while not as restrictive as some might have preferred, does create uniformity of practice among the supplier members. In addition, the European Parliament adopted a resolution in 1992 calling for a European code of conduct on arms exports. The previous year the European Council established a European Political Cooperation Working Group on Conventional Arms Transfers, and the protocol to the Maastricht Treaty of 1993 identifies arms exports as a field where "joint action" procedures for common foreign and security policies could apply.

A European agenda for arms transfers could range from coordinated national controls to consultations on actual or planned sales to common licensing and export regulations. The extent to which such measures are adopted will essentially depend on the degree to which Europe moves toward further economic and political integration.[33]

China

China presents a somewhat different picture with respect to regulation and oversight. The arms industry consists of a complex network of nationalized corporations that are linked to the People's Liberation Army (PLA). The modernization of the Chinese military is dependent to some extent upon the revenues that can be achieved through arms exports. This is one key economic motivation for an aggressive export policy. At the same time, the dominant influence over these corporations has been retained in the hands of the leadership of the Communist Party. It is widely believed that the families of these leaders, especially the Red Princes and Princesses, as the more entrepreneurial sons and daughters of the aging leaders are called, use arms exports as a way to acquire vast personal fortunes. This income is said to be another compelling incentive for weapons exports.

Nevertheless, it would be an error to assume that Chinese arms sales are out of control. As the size of its exports increased in the late 1980s, Beijing established an export control agency, the State Commission for Arms Export Administration, with representatives from the appropriate ministries. The agency must approve arms sales and grant export licenses. Thus a formal system of control does exist. When China wishes to desist from lucrative sales, as when it decided to forego weapons sales to Iraq, the interests of the PLA and the princes were overruled. China's arms sales in recent years have often proven to be contentious when viewed from abroad, but they are likely to have reflected the distribution of power within the nation and the wishes of its leaders.[34]

Multilateral Controls

Ultimately, the management of arms transfers will depend on the development of some type of international controls. The simple, oft-cited reason is that if country A does not sell, country B will be only too happy to oblige. Far too often, however, this supposed axiom of international behavior has been used in an overly facile manner to justify the making of arms sales. It overlooks the basic fact that some, and at times all, the countries that can

Table 15-1. *Arms Deliveries to Developing Nations and the World by Suppliers, 1995 (as percent of total deliveries)*

	Developing Nations Deliveries *(percent)*	World Deliveries *(percent)*
United States	44.08	44.42
United Kingdom	20.80	17.35
Russia	11.09	10.97
France	7.39	7.79
Germany	3.70	4.25
China	2.77	2.12

Source: Richard F. Grimmett, *Conventional Arms Transfers to Developing Nations, 1988–1995,* Congressional Research Service, U.S. Library of Congress (Washington, D.C., 1996), 58 and 83.

supply a particular weapon or component may have a common interest in avoiding that transfer.

The creation of international controls should be feasible, at least in theory, because of the still limited number of countries that are major arms suppliers. Despite the growth of new arms industries around the world, the market for the more significant weapons systems is still oligopolistic. In 1995, six countries accounted for just about 90 percent of arms deliveries to developing nations and almost 87 percent of total global arms deliveries (see table 15-1).[35] This monopoly of the few has been quite constant, with the same six countries accounting for just under 84 percent of world arms deliveries over the 1988 to 1995 period. During that period, the United States and the Soviet Union/Russia alone were responsible for slightly over 54 percent of world arms deliveries (see table 15-2). Data from SIPRI are corroborating: they show the same six countries supplying 89 percent of arms exports in 1990–94.[36]

Because the principal suppliers do have the capacity to create controls, the willingness of President Bush to allow the P-5 talks of 1991–92 to collapse for the sake of the votes of workers at the General Dynamics aircraft plant in Fort Worth, Texas, and the Texas economy more generally, was a grievous mistake. Although the Chinese were ambivalent about agreeing to discuss their arms sales with the major suppliers, the U.S. decision to sell 150 F-16s to Taiwan tipped the scales in Beijing toward those who wished to end the talks. Also unfortunate was the incoming Clinton Administration's failure to seize the initiative by attempting to revive and reinvigorate the P-5 process, as discussed. To its credit, however, the Administration later sought to fill the vacuum with an ambitious and promising initiative.

Table 15-2. *Arms Deliveries to the World by Supplier, 1988–95 (in millions of constant 1995 U.S. dollars)*

	1988	1989	1990	1991	1992	1993	1994	1995	TOTAL 1988–1995
United States	10,561	8.717	10,256	10,259	11,524	11,159	10,038	12,549	85,063
Russia	27,017	22,330	17,216	6,795	2,689	3,238	1,530	3,100	83,915
France	2,456	2,836	5,968	2,411	1,936	1,149	1,428	2,200	20,384
United Kingdom	6,017	5,789	5,279	5,151	5,056	4,804	5,303	4,900	42,299
China	3,684	3,190	2,295	1,534	1,076	1,253	714	600	14,346
Germany	2,210	1,536	1,836	2,630	1,183	1,775	1,428	1,200	13,798
Italy	614	236	230	329	430	418	102	0	2,359
All Other European	8,351	4,726	3,328	1,973	3,227	1,567	1,326	1,000	25,498
All Others	5,649	4,017	2,295	2,082	1,829	2,089	2,448	2,700	23,109
TOTAL	66,559	53,377	48,703	33,164	28,950	27,452	24,317	28,249	310,771

Source: Richard F. Grimmett, *Conventional Arms Transfers to Developing Nations, 1988–1995,* Congressional Research Service, U.S. Library of Congress, (Washington, D.C., 1996), 82.

The Wassenaar Arrangement

The best hope for creating a supplier-based, multilateral regime now lies with the Wassenaar Arrangement on Export Controls for Conventional Arms and Dual-Use Goods and Technologies (for a time known as the New Forum).[37] Thirty-three countries formally approved this new regime on July 11–12, 1996, after over two years of very quiet, sensitive and often difficult negotiations.[38] The Arrangement's significance lies in its potential, if elaborated and wisely developed, of filling the gap of unrestrained conventional arms flows in the global non-proliferation effort. As U.S. Under Secretary of State for Arms Control and International Security Lynn E. Davis observed, "For the first time there is a global mechanism [the Wassenaar Arrangement] for controlling transfers of conventional armaments, and a venue in which governments can consider collectively the implications of various transfers on their international and regional security concerns."[39]

Named after the small town in the Netherlands where much of the Arrangement was negotiated, it resulted from two needs. One was to replace COCOM, which during the Cold War years served as a screen whereby the West could prevent the transfer of strategic goods and military technologies to the Warsaw Pact, with a new organization that would restrict the sale of sensitive military and dual-use technologies when such transfers could upset regional balances or damage international security.

The second was to fill the void in international efforts to manage conventional arms transfers resulting from the collapse of the P-5 negotiations.

The Wassenaar Arrangement commits the participating states to meet on a regular basis and to exchange information that will lead to a common understanding of any risks that might result from the transfer of arms and dual-use technologies. The intention is to prevent excessive acquisition of weapons or sensitive items that could threaten peace or stability in a region. Many framers of the regime clearly had the destabilizing impact of Saddam Hussein's Iraqi purchases during the 1980s in mind, as well as subsequent concerns regarding the so-called pariah states of Iran, Iraq, Libya and North Korea. On the basis of this information exchange, the Wassenaar states will seek to coordinate their national control policies for transfers to non-members of the Arrangement.

The Arrangement, spelled out in a document entitled *Initial Elements,* has two principal components, one relating to dual-use goods and technologies and the other to conventional arms.

The *dual-use* component divides certain goods and technologies into a Tier 1 basic list and a Tier 2 sensitive list. A sub-set of the Tier 2 list identifies very sensitive items for which extreme vigilance is to be exercised. Wassenaar members will give each other notice of transfers to non-Wassenaar states and of denials of licenses when this occurs for reasons relevant to the purposes of the Arrangement. For Tier 1 items, members will make the notification of denials on an aggregate basis twice a year. In the case of Tier 2 denials, notice will be given individually for each denial, preferably within thirty days but no later than sixty days from the date of denial. This notification does not impose any obligation on another participating state to refrain from a similar transfer.

Despite considerable discussion about requiring consultations prior to a state's decision to undercut another state, no agreement could be reached. However, a participating state "will notify, preferably within thirty days but no later than within sixty days, all other participating states of an approval of a license which has been denied by another participating state for an essentially identical transaction during the last three years."[40] This provision assures that Wassenaar members will subject the decisions of governments to scrutiny and may force them to justify their actions. It also provides an opportunity for governments to share additional information that may dissuade the exporting government from proceeding with the transfer.

The *conventional arms* component provides for an exchange of information every six months on completed arms transfers involving the seven categories of weapons used in the CFE Treaty and the U.N. Arms Register. However, the Arrangement calls for more information than the treaty and register do, such as identification of the models and types of weapons. A Wassenaar member may request additional information on specific transfers. Emerging trends in weapons programs and the accumulation of particular weapons systems, when a matter of concern, may be included in the data exchange. Notably lacking, however, is any provision for notification *prior* to a sale or transfer, which would give members time and opportunity to challenge the transaction. The United States pressed for such a provision without success.

Because of the modest nature of the provisions in the conventional arms component and the reluctance of some of the major suppliers to open their arms sales process to the view of the entire Wassenaar membership, an informal "small group" consisting of the United States, Russia, Germany, France, Great Britain and Italy met several times in 1995 to explore more robust procedures, including "meaningful consultations" on transfers before nations reach decisions. In addition, a variety of proposals were floated, including ways to develop common approaches to regions of instability such as the Middle East and South Asia, means of preventing the introduction of advanced weaponry in areas where it does not currently exist, and even some Russian ideas for market-sharing arrangements. Because of the differences in views about these initiatives, the inner group achieved little, and in the first half of 1996 attention shifted to completing the larger Wassenaar Arrangement. With this now completed, the "small group" should be further developed as an important component of the Wassenaar undertaking.

A small secretariat has been established in Vienna. Plenary sessions are to be held once or twice a year, with working groups meeting more frequently. Decisions are made by consensus, a potential weakness in the regime. Membership is open on a global and non-discriminatory basis to countries that meet the agreed criteria: they must be producers or exporters of conventional arms or dual-use equipment, have fully effective national export controls, and follow appropriate non-proliferation policies, including adherence, where applicable, to the guidelines of the Nuclear Suppliers Group (NSG), MTCR, Australia Group, and related treaties and conventions (NPT, CWC, Biological Weapons Convention [BWC] and Strategic Arms Reduction Treaty [START] 1, including the Lisbon Protocol).

Wassenaar must be viewed as an embryonic, evolving regime that needs to be further implemented and developed. For this reason its framers named it an "Arrangement" and based it on an accord of "Initial Elements." Participating countries need to set precedents and develop rules to flesh out the initial framework, all of which will take time. The MTCR serves as an example of how an initially rudimentary regime can be incrementally strengthened over time. However, it should also be noted that the patterns set in this initial phase can determine much of the final result.

Wassenaar has the makings of a reasonably good transparency regime. It falls short, however, with respect to creating controls. Looking toward the future, there is a need to establish some mandatory *prior* notification and consultations on weapons sales to regions of tension. The "small group" of nations, which account for some 90 percent of global arms shipments by value, should engage in in-depth discussions of regional military balances and internal conflicts before they decide on arms transfers.

Another gap in the Arrangement is that it does not cover small arms, which are widely used in ethnic conflicts. They also need to be addressed. With respect to dual-use items and sensitive technologies, a provision calling for "no undercutting" without intensive consultations, especially with countries that have foregone a lucrative deal, is needed. Prior agreement might be reached on certain advanced technologies, such as Stealth, that no state should export.

Beyond such measures, it is important that a greater understanding is reached regarding the geopolitical nature of the Wassenaar endeavor. It would be a grave mistake if the Wassenaar Arrangement came to be perceived as focusing only on pariah states. Some countries have objected to explicit "targeting" and view the United States as overly concerned with certain nations. Neither should the Arrangement come to be seen as biased against technology transfers to the developing world. Member states should maintain a dialogue with potential recipients of arms and technology, and those that share the aims and criteria of the Arrangement should be brought into it. China, in particular, should be engaged in some manner.

At this still early stage, the purview of the Wassenaar Arrangement is limited to transfers to non-Wassenaar states, the argument being that the regime brings together a group of "like-minded" nations on matters of non-proliferation. In time it will be essential to broaden its responsibilities to include transfers *between* Arrangement members, especially if the membership is expanded significantly. It should be clear that the Wassenaar Arrangement will only succeed if its promising beginning is carefully nurtured and it is encouraged to become a stronger and more fully effective regime.

The European Union and The Organization for Security and Cooperation in Europe

Two additional organizations that include major supplier nations—although neither covers all of them—and that have addressed the development of common criteria for arms exports by their members are the European Union (EU) and the Organization for Security and Cooperation in Europe (OSCE). Their interest is testimony to the significant numbers of citizens of their member countries who believe that restraints should be imposed on the global arms trade.

The EU, in addition to its guidelines for the regulation of dual-use exports, as discussed, has looked at conventional weapons proliferation more broadly. The European Council of Ministers has adopted eight criteria to govern conventional arms exports. They include: respect for human rights in the recipient country; the behavior of the buyer country with regard to terrorism and international law; the compatibility of the arms purchases with the buyer country's need for economic and social resources; and legitimate needs for defense while preserving regional peace, security and stability.[41] The Inter-Governmental Conference on the Maastricht Treaty will address this issue further in 1997, no doubt balancing the economic needs of the European defense industries with the arguments for adopting restraints.[42]

In 1993 the OSCE, formerly the Conference on Security and Co-operation in Europe (CSCE), adopted an even more detailed and extensive list of criteria to govern arms transfers than that of the EU.[43] The significance of its action lies in its membership—the fifty-five participating countries include all the major arms-producing countries except China, and a majority of the secondary suppliers. Included are the countries of Eastern Europe and the former Soviet Union. Many of these have limited experience with weapons export controls, yet are seeking to maintain arms industries developed during the Cold War years. To assist in the development and implementation of such controls, the OSCE could sponsor workshops at which the experience of some member countries could be shared.

Supplier-Based Limitations on Arms Sales to The Middle East and Persian Gulf

The Middle East and Persian Gulf region account for the largest purchases of weaponry. The Iran-Iraq war of the 1980s and the Persian Gulf crisis that resulted from Iraq's invasion of Kuwait in 1990 stimulated an increased desire for armaments in the region. The ease with which Iraqi forces overran Kuwait created a sense of vulnerability that led to a heightened demand for arms not only by Saudi Arabia, but also by most member states of the Gulf Cooperation Council. Potential threats from a hostile Iran have also stimulated a desire to upgrade defense capabilities. Much of this demand has focused on the advanced weaponry, mainly American, used in the Persian Gulf War, which, correctly or not, greatly impressed the populations and leaders of the region. During 1992–95, the Middle East and Persian Gulf region accounted for 53.5 percent of all arms sales to the developing world and 65.6 percent of actual deliveries. Half of the top ten purchasers of arms in the developing world in 1995 were in the region, with Saudi Arabia receiving $8.3 billion worth.[44] The U.S. Department of Defense expects the Middle East and Persian Gulf to continue to be the world's largest market for arms at least until the year 2000.[45]

Accordingly, this region must be seen as the one most urgently requiring arms restraint and where supplier countries might seek to achieve some multilateral restraints. The existence since the Madrid conference of 1991 of ongoing multilateral arms control talks on the Middle East (the Arms Control and Regional Security Working Group) is also a positive inducement, although thus far the talks have focused on confidence-building measures, rather than on arms limitations.

Three authoritative proposals merit attention. A study group of the Henry L. Stimson Center warned of the dangers of opening the floodgates wide with an unbridled introduction of arms that could alter the relatively stable balance of power now existing in the region. It urged that the flow of arms transfers be governed by a strategic logic geared toward maintaining military stability and peace, rather than by considerations of hard currency earnings, a desire to keep weapons production lines warm or job security. The Stimson group favored a reduction in the transfer of arms into the region and saw that task as being primarily the responsibility of the big five suppliers. Specifically, it proposed that:

—Weapons or sub-systems that are not currently in the region, or that are in the region in only limited amounts, should not be transferred to the Middle East.

—Weapons in the five categories of equipment covered under the CFE Treaty—armored combat vehicles, artillery, fighter-attack aircraft, helicopters and tanks—should not be transferred to the region in quantities that exceed current ceilings. These are the arms that are viewed as primary weapons for launching surprise attacks and initiating large-scale offensive actions. One-for-one replacements in these five categories would be permitted, but recipient nations would have to destroy the weapons that are being replaced.

—No surface-to-surface missiles should be transferred to the Middle East.[46]

The U.S. Congressional Budget Office (CBO) made a considerably more ambitious proposal. Rather than relying on broad principles, the CBO suggested imposing mandatory, quantitative limits on the transfer of arms to the Middle East. Recognizing that only a handful of countries have been, or are likely to be, the principal suppliers of weapons to the region, the CBO candidly called for binding limits, to be set through a suppliers cartel. These limits could be imposed in one of several ways:

—*Export limits.* Each major supplier could be restricted from selling weapons beyond a certain number or certain dollar value. Exports by each supplier might be limited to a level equaling one-half of past sales.

—*Import limits.* Suppliers could restrict the number or dollar value of weapons that any one Middle Eastern country could import from them. This restriction would include suppliers' contributions to the co-production of weapons in the recipient country. A ceiling of $700 million in imports by each country was suggested.

—*Export and import limits.* Suppliers could initially impose modest limits on their exports to the Middle East to minimize the affront that limits might cause to importing nations in the region. More restrictive checks on imports could gradually be imposed to ensure that no country could amass too large a stock of weapons.[47]

RAND conducted a study for the U.S. Department of Defense that focused on the Persian Gulf. It concluded that a regime to control the transfer of weapons to the Persian Gulf is indeed feasible, should encompass all the states of the region, and should be supplier-focused. Such a regime should seek to prohibit rather than just regulate when certain transfers would affect the regional military balance in a manner inconsistent with U.S. strategic interests.

RAND presented several criteria for determining which weapons systems the regime should include:

—The weapons should exert "high leverage" on battlefield outcomes, for example, their effects should be disproportionately high compared with their numbers.

—The weapons should have "low substitutability," for example, arms that have no substitutes of comparable capability that are freely available to others. The purchasing country should not be able to circumvent the control regime by acquiring a comparable system from a supplier not a party to the regime.

—The weapons should have a "low opportunity cost," for example, the cost of the foregone sales should not comprise a significant fraction of the total dollar value of the supplier's arms sales. The logic is that to the extent the costs of complying with the multilateral control regime can be kept low, the greater is the likelihood that the supplier state will find it in its interest to join and abide by the regime.

Recognizing that a control regime should have some method for dealing with sales that are in dispute among the suppliers, the RAND study proposed a "market-stabilization mechanism" that could either compensate suppliers for losses incurred as the result of foregone sales or penalize a supplier for violations of the regime.[48]

What is most interesting about these three studies is that they all begin by accepting the premise that cooperation among the principal five suppliers is not only desirable but feasible. This is still not widely accepted today. The indigenous armament industries in countries such as Egypt, Israel and Iraq, although not inconsequential, are not viewed as of sufficient importance to be able to negate the effectiveness of a suppliers' cartel.

Each study can be faulted. The Stimson Center report is excessively cautious in concentrating on avoiding a military build-up by maintaining arms ceilings at current levels, rather than calling for a progressive build-down. The CBO study takes too facile an approach toward imposing quite drastic cuts and limitations. And the RAND study's criteria for weapons systems that would be subject to curtailment are so restrictive as to cast some doubt on the workability or effectiveness of its basic proposal.

Nevertheless, all three studies are predicated on the belief that with the termination of the Cold War—and an end to most of the East-West competition in the Middle East and Persian Gulf that made cooperation among the outside powers so difficult—it should be possible to work multilaterally and cooperatively to dampen the arms race in the region.

A Look at The Demand Side

Ultimately, a viable process of restraint will also require the cooperation of prospective recipient countries, at least in many instances. Even if a suppliers' cartel would work because only a limited number of countries can sell a particular weapons system, a supplier state may be reluctant to impose limitations upon a country in which it has other important interests that could be harmed. Clearly, a would-be purchaser might view the denial of a sought-after arm as an unfriendly, if not hostile, act. In the developing world especially, denial can be viewed as discriminatory and evoke memories of colonial imperialism.

For such reasons, the management of arms flows should, to the extent possible, draw upon the self-interests of the recipient states within a region. National security within a context of regional stability is the goal of most states. The excessive accumulation of arms, or of weapons with a high level of lethal capabilities such as accurate missiles, can be viewed by *all* the parties within a region or sub-region as something to be avoided.

Unfortunately, regional arms control for conventional weapons has only been minimally developed thus far. In Latin America, as noted, eight Andean states agreed in the Declaration of Ayacucho of 1974 to limit their arms acquisitions, but implementation of the agreement ran into difficulties. The only successful regional arms limitation agreement to date is the CFE Treaty of 1990, even though, after the dissolution of the Soviet Union, it needed revision to account for the changed circumstances Russia is facing in the location of its armed forces. Ironically, this conventional arms limitation accord has also had the unintended and somewhat undesirable side effect of bringing about a large-scale transfer of arms from Central Europe to Greece and Turkey.

Nevertheless, there are grounds for believing that creative diplomacy could yet lead to some significant regional arms control agreements. The end of the Cold War has removed the policy straightjacket that governed most arms transfers of both the Soviet Union and the United States for several decades. During the past era the primary goal was political support for the East or West's respective friends and allies, within the framework of an ongoing superpower rivalry. Indeed, many arms transfers were not sales at all but rather direct grants or transfers subsidized by long-term loans that it was understood the recipients were unlikely to repay. Now more attention must be paid to intra-regional factors and the local context. The future world will be more chaotic, with threats of ethnic conflict and nuclear proliferation, yet with outside powers more reluctant to become involved. In these circumstances, nations within a region must rely more on themselves to create mechanisms for conflict prevention.

Such policy proposals are beginning to be heard for the Middle East, where considerable interest has developed in the possibility of regional arms control. Among the ideas being discussed are an array of confidence-

and security-building measures (CSBMs), restrictions on the acquisition of certain high-technology weapons, ceilings on specified levels of combat arms, limits upon weaponry along certain borders and "open skies" inspections to enhance confidence and verify agreements.[49] In his contribution to this book, Israeli political scientist Gerald Steinberg outlines an imaginative proposal for a "freeze" at current levels on major platforms of weapons, that is, tanks, helicopters and fighter aircraft, that the Middle Eastern nations could agree upon. This agreement is possible, he argues, because of the relative stability of the military balance.[50] It should be an important component of a broader Middle East peace treaty.

Writing from an Arab perspective, in his chapter Egyptian strategist Abdel Monem Said Aly comes to many of the same general conclusions as his Israeli counterpart. He sees a growing trend toward arms control and conflict prevention in the Arab-Israeli context. A moratorium on the acquisition of high-technology weapons should be part of the peace negotiations, as should restrictions on their deployment. Said Aly also emphasizes the importance of creating additional transparency to reduce mutual suspicion.[51]

Turning to South Asia, Rodney Jones notes in his chapter that the surge in arms imports since the 1980s is part of a defense modernization that includes the introduction of ballistic missiles and nuclear weapon capabilities. At the same time, he points out the domestic pressures for increasing economic development and social programs, for toning down the antagonism between India and Pakistan, and for capping or reducing military spending. He sees regional arms control as requiring long-term, patient and continuing efforts on the part of both suppliers and recipients. Supply-side efforts should focus on qualitative rather than quantitative restraints, avoiding, for example, the supply of long-range strike capabilities that could destabilize the region or be used for nuclear delivery. Demand-side initiatives should involve collateral efforts to encourage regional self-restraint. These might include confidence-building measures such as military talks between India and Pakistan on the requirements of conventional defense sufficiency, identification of ceilings to which various categories of arms could be reduced, exchange of information on planned national production of arms, and verifiable limits on the deployment of ground forces in contested border zones.[52]

Where there is no regional coherence or few incentives to limit arms purchases, the creation of demand-based restraints will be far more difficult. In his chapter on the Asia-Pacific region, Andrew Mack makes clear

that it is the lack of regional cohesion, as well as an undeveloped understanding of the benefits of arms restraint, that make the task more daunting
than in the other regions discussed in this book.[53]

Multilateral controls, to be effective and successful, will in many instances have to address both the demand and the supply sides of the equation.
There will have to be supply-side cooperation and demand-side discussions. These should lead to dialogue between producers and purchasers in
what should be a complementary, even symbiotic, relationship. Such a
dialogue must go well beyond the technicalities of arms transfers to include
fundamental security concerns, means of reassurance and broad foreign
policy considerations.

Still, it must be acknowledged and understood that suppliers have the
right to say "No sale."

Technology Control Regimes

In addition to focusing on the actions of suppliers and recipients, efforts
toward arms control for conventional arms should directly address some of
the technologies involved. Technology-based controls could be established
at either the global or the regional level. In this sense, they would not be
alternatives to what has been discussed thus far, but would be part of a
composite of multilateral controls.

Restraint through denial or restriction of technology is a well-accepted
concept in many of the non-proliferation regimes: nuclear (Non-Proliferation Treaty and the Nuclear Suppliers Group); chemical and biological
(Chemical Weapons Convention, Biological Weapons Convention and the
Australia Group); and missiles (Missile Technology Control Regime).

Conventional arms, thus far, have been the glaring omission, with no
creation of a similar control regime. The Wassenaar Arrangement will
fill that troublesome gap in part with its lists of dual-use goods and
technologies to be monitored. However, beyond the notification procedure, it offers little in the way of specific restraints on weapons-related
technologies.

Selective restraints might be sought on:

—Weapons that are not yet widely dispersed and whose spread could be
deemed as "destabilizing" within a region, for example, Stealth aircraft,
anti-satellite systems, advanced precision-guided munitions, surface-to-
surface missiles, modern anti-ship missiles, intelligence sensors and real-
time targeting systems.

—Weapons capable of causing great suffering to civilian populations, such as napalm, cluster bombs, fragmentation weapons and landmines.

—Weapons that terrorists seek, examples being easily transportable missiles and night vision capabilities.

—Weapons that are viewed as excessively cruel or "inhumane" even in warfare, for example, blinding lasers.

—Weapons of torture and police equipment that are likely to be used primarily for internal repression and violation of human rights.[54]

Some progress has been made in the past two years in a few of these areas, an indication that restraint measures can be adopted when there is sufficient public interest and support.

The banning of *anti-personnel landmines*—their production, stock-piling, sale, export and use—has been the object of a grassroots movement, very active in the United States as the result of the work of the Arms Project of the Human Rights Watch, and complemented by the International Campaign to Ban Landmines, a coalition of over 450 veterans, human rights, arms control, and developmental and medical groups in forty countries.[55] The results have been mixed. At the urging of Senator Patrick Leahy (D-Vt) and Representative Lane Evans (D-Ill), in 1993 Congress imposed a one-year moratorium on American exports of anti-personnel (as contrasted with anti-tank) landmines, and it has renewed it each year thereafter. By April 1996, forty-seven nations, including a number of industrialized ones and NATO members, announced similar moratoria or export restrictions.

In his address to the United Nations in September 1994, President Bill Clinton called for the "eventual elimination" of all landmines. But the Pentagon objected to limiting the production, stockpiling and use of anti-personnel landmines on grounds that they are needed for certain military missions. Leahy and Evans then sponsored legislation that would give the Department of Defense three years to find military alternatives to such landmines, after which U.S. forces would be prohibited from using them.

The issue came to a head in the spring of 1996 after U.S. Ambassador to the United Nations Madeleine Albright, upon her return from a visit to Angola where she saw hundreds of children whose limbs had been ripped off in landmine explosions, called for a comprehensive review of U.S. policy. The Defense Department split on the issue, with some senior officials favoring a total ban on anti-personnel landmines, while two of the military services argued for their retention for use along the demilitarized zone in Korea and in the desert in the Persian Gulf. They also argued that mines were a good way to "channel" enemy troops on the move and an inexpensive way to protect American soldiers. They were, however, willing to give them up in the year 2010, by which time they hoped more humane alternatives would be developed.

Sentiment against landmines grew with the deployment of American troops to Bosnia and the constant fear of accidental detonations there. A group of 15 prominent retired military officers, including Gulf War commander General Norman Schwartzkopf and former Joint Chiefs of Staff Chairman David Jones, endorsed an early ban on landmines as a "humane and militarily responsible step." President Clinton, unwilling to risk a breach with the Joint Chiefs of Staff, attempted to weave a compromise by retaining "smart" landmines, which self-destruct after a set period (although there is a 10–20 percent failure rate) and eliminating "dumb" ones. The exception would be South Korea, where they would stay activated indefinitely. This political compromise pleased few, and the issue is certain to be reopened.

Lack of a strong American initiative was also detrimental to the international effort to get rid of anti-personnel landmines. Under the auspices of the United Nations, experts began meeting in late 1994 to review and improve upon the 1980 U.N. Convention on Conventional Weapons, also known as the Inhumane Weapons Convention, which established guidelines, up to then largely unsuccessful, to regulate the use of landmines. Provisions being promoted in 1995–96 for the Convention included a requirement that landmines be built to self-destruct after a period of time and that they be limited to areas that are clearly marked and fenced in. By mid-1996 little had been accomplished, however. A significant number of countries favored a total ban on landmines, while others, such as China, India and Russia—all exporters of such—were unwilling to phase them out. Clearly there is room for a renewed international initiative on landmines.[56]

Blinding laser weapons are still in the developmental stage but close to initial production. They are designed to counter battlefield surveillance by disrupting optical devices ranging from binoculars to gunner's sights to infrared sensors. But they can also be used as weapons that can permanently blind humans by burning the retina and optic nerve. As with landmines, the view is emerging that they are excessively cruel and their use repugnant to the public conscience and that they should therefore be banned.[57] (Whether being blinded is actually worse than being maimed or even killed by a "legitimate" weapon of war is a debatable proposition.) The International Committee of the Red Cross has warned that blinding lasers could become inexpensive and be available to terrorists or even ordinary criminals.

The small number of countries doing research on blinding lasers have been reluctant to give up the option to manufacture them. They appear, however, to be more willing to agree that the lasers not be employed. Sweden has proposed that a protocol banning the use of blinding laser weapons be added to the 1980 U.N. Conventional Weapons Convention.

The proliferation of *small arms and light weapons* has recently received attention because of the growth in ethnic conflict and civil strife. The United Nations convened a special panel of governmental experts in 1996 to address the issue. From Cambodia to Bosnia, Kashmir to Somalia, Rwanda to Chechnya, weapons such as light mortars, machine guns, assault rifles and hand grenades have been used extensively. Hundreds of thousands, possibly millions, of innocent civilians have been killed, far more than were ever doomed by weapons of mass destruction or some of the classic conventional arms such as tanks and ships. The world is awash in an abundance of small arms with virtually no systematic attempts to create controls.

To what extent anything meaningful can be done about this phenomenon is uncertain. Much of the global trade in small and light weapons runs through commercial or irregular channels that governments cannot monitor effectively. Some of it is attributable to covert assistance by governments themselves. Certainly a tightening of customs procedures in many countries and a global watch of illicit arms trafficking could be useful and important steps, as Michael Klare suggests in his chapter in this volume.[58] But the usual methods of arms control will be quickly overwhelmed if they are not accompanied by political efforts to address such problems at their roots. In the more chaotic world of the post-war era, the containment and resolution of local conflict usually will depend primarily upon adequately addressing the economic, social, political, ideological and religious sources of the conflict.[59]

Controls on Co-Production, Licensing, Offsets and End-Use

The production and distribution of arms have changed a great deal in the past twenty years. Very rarely are whole weapon systems manufactured and sold "off-the-shelf" or out of inventories as, for example, automobiles are. Weapons are often manufactured through licensing arrangements or co-production agreements, whereby the purchasing state either has a share in the manufacturing itself or participates in the assembly of the product. As part of the deal, it may demand the technology to make parts of the weapon system locally. When in 1991 South Korea agreed to purchase 120 F-16s, it was able to require that it manufacture most of the airframe for 72 of the jets and assemble another 36 from imported kits, leaving only 12 to be made entirely in the United States. The OTA found, in *Global Arms Trade,* that these practices result in the dispersion of valuable technical and engineering knowledge that the purchasing state can subsequently use to develop its own indigenous production capability.[60]

Similarly, weapons deals today frequently have provisions for "offsets," or side deals, which are a form of countertrade. "Direct offsets" may require that the supplier transfer technology components or information on military technology to the purchaser as part of the overall package. "Indirect offsets" involve a commitment that the supplier accept partial payment in the form of goods, which may be totally unrelated. The U.S. General Accounting Office (GAO) estimated in 1996 that since the mid-1980s American companies have entered into offset agreements valued at over $84 billion.[61] To clinch a $1.8 billion sale of F-18 fighters to Spain, McDonnell Douglas offered $1.5 billion worth of offsets. The Missouri-based company agreed to market a wide range of Spanish products in the United States, including sunflower seed oil, sailboats, zinc, steel coils and marble. It also helped publish and distribute a picture book on Spain designed to promote tourism by Americans. And, yes, McDonnell Douglas helped to establish a Domino's Pizza franchise in Barcelona! These types of compensation have become a substantial and necessary part of the competition for arms sales, leading to bidding wars based on offsets, even though the defense companies see them as a royal nuisance.[62]

Altogether, these practices also result in the spread of military technologies to other countries and contribute to the proliferation of conventional arms in ways that are difficult to quantify because information is incomplete. William Keller found that "the distinction between technology and finished weapons has diminished significantly. Technology transfer is no longer an add-on to close a major sale, it has become an object of competition itself."[63] Certainly such transfers should be included in any comprehensive regime of arms restraint. Even a more limited effort toward restraints should take these increasingly prevalent practices into account. Today this is not being done, and it makes for a large lacuna.

The major arms-producing countries should seek to reach agreement among themselves on restrictions for co-production and licensing arrangements as part of their international arms sales. Although seemingly difficult to do, there is a powerful incentive: to the extent that the suppliers are currently agreeing to transfer arms-producing capacities overseas, they are helping create industries that subsequently often become direct competitors. Once the new producers fulfill their own military needs, they will seek export markets to keep their manufacturing lines busy. The unfettered use of co-production agreements accounts for a significant portion of the output of new developing world armament industries. It also contributes to the global overcapacity in defense production. Organized labor in the major suppliers' countries, it should be noted, will not object to curtailing the export of jobs overseas. Many defense corporations also do not like these agreements but feel they are necessary to compete. These corporations should, therefore, welcome such curtailments.

Restrictions on co-production licensing would be a highly appropriate undertaking of the new Wassenaar Arrangement. Members should also seek to regulate or manage the growing number of direct offsets that lead to the transfer of military technology. Defense contractors, as noted, see indirect offsets as an enormous bother. But how strongly can anyone object to distributing Swiss chocolates in exchange for Switzerland's agreement to purchase advanced fighter aircraft, to cite a real example? Direct offsets, on the other hand, have serious, if often unintended, proliferation consequences that currently are totally escaping international scrutiny.

Moreover, as the U.S. Department of Commerce concluded in 1996, offsets are of doubtful economic value:

> Because of the superiority of U.S. technology and weapons systems, U.S. defense companies usually have an advantage over foreign companies in terms of the types of direct and indirect offsets they can provide. However, this superiority presents a double-edged sword. As the world's preeminent supplier of weapons and high cost/high technology hardware, U.S. corporations are also highly vulnerable to offset demands. Their traditional consent to such impositions is a sign of competitive pressures. . . . Defense offsets may create or enhance foreign competitors, exacerbate already excessive defense production capacity, displace U.S. firms, and reduce U.S. employment.[64]

Also in need of international scrutiny is the lack of adequate end-use certification. The United States and several West European suppliers require, as part of their agreements to transfer arms or military technology, that the recipient country certify that it will not at a later date re-export the items to a third party. At times there are additional restrictions as to the circumstances under which, and against which country, the weapon can be employed. Enforcement of certification is, however, extremely lax. The labor and time involved could be enormous, and the investigation would run counter to the interests of the original recipient country, as well as the bilateral relations between supplier and purchaser.

An interesting case in point involves American transfers to Israel. South Africa (during the apartheid era) and China, among others, are alleged to have received unauthorized Israeli re-exports of U.S. defense technology. Pretoria is believed to have been sold ballistic missile technology and Beijing to have received Patriot missile know-how as well as other sensitive items. Nothing ever came of internal U.S. investigations, however.[65] Monitoring of re-transfers and enforcement of end-user certifications would also be appropriate subjects for the Wassenaar Arrangement.

The Clinton Administration's Approach

If there is to be progress toward achieving a regime of international restraints, American leadership is essential. As the world's top arms supplier, the United States made slightly over 44 percent of global arms *deliveries* in 1995 (down from just under 52 percent in 1994), with the next largest supplier, the United Kingdom, transferring only a little more than 17 percent. Russia, which as the Soviet Union had the lead in the late 1980s, dropped to just under 11 percent in 1995 (up from almost 7 percent in 1994). The American dominance becomes even more stark when looking at arms sold to the developing nations, which sales in recent years have comprised 72 percent of the global arms trade on average. The United States accounted for almost 60 percent of arms transfer *agreements* with the developing world in 1993 and has been responsible for approximately half during the 1990s.[66] Given America's position in global arms trade, U.S. policy is absolutely critical to the success of any serious initiative for controlling conventional weapons proliferation.

During the 1992 presidential campaign, then-Governor Clinton spoke of the need to curb the arms trade. He pledged to review U.S. policy and to take up the issue of arms restraint with the other major arms-selling nations. This statement was part of a proclaimed activist approach to non-proliferation in general.

Initially conventional weapons were included in statements dealing with the spread of nuclear, chemical and biological weapons. Before very long, however, the new Administration focused its efforts on weapons of mass destruction. As it began the internal discussions that would lead to a Clinton non-proliferation policy, the issues involving conventional arms were intentionally set aside. As a National Security Council official confided, these issues were harder to get a handle on than weapons of mass destruction. There were more angles to them, relating to economic, political and security policy, and involved more actors.[67] An additional factor was that the Administration also was keenly interested in improving the nation's industrial competitiveness and in supporting American exports, an important consideration that ran counter to restraint on weapons sales.

Not until after the Administration had been in office for over two years, on February 17, 1995, did the White House announce the Clinton Conventional Arms Transfer Policy. Presidential Decision Directive 34 received relatively little attention because it stated remarkably little that was new. Rather, it was a crystallization of the practices that had developed in the intervening years since the Carter Administration, which was the last to give priority to restraints on conventional arms. Arms transfers were to be seen as legitimate instruments of U.S. foreign policy; they were deserving of support when they "enable us to help friends and allies deter aggression, promote regional security, and increase interoperability of U.S. forces and allied forces." Decisions on arms transfers were to be made on a case-by-case basis, according to a list of criteria which was so broad as to be able to justify most any sale on the grounds of advancing the national interest.[68]

The directive spoke in approving if unoriginal terms of the need to promote restraint, transparency and control of arms transfers. It committed the United States to seek increased participation in the U.N. Register of Conventional Arms, to support regional arms control initiatives such as those being examined by the Organization of American States and the Association of South East Asian Nations, to encourage the adoption of confidence-building measures such as those under way in the Middle East and Europe, and to seek a successor regime to COCOM. These were sweet words, intended perhaps to please the *arms control* community, but except for a post-COCOM regime they had little in the way of new initiatives to back them up. In the internal debate leading to the directive, ACDA had been an active exponent of arms restraint, but both the Department of Defense and the Department of State had opposed many of its policy proposals.

What was new, and welcomed by the *defense* community, was the emphasis on what the government's role should be in providing direct support for arms exports. The Clinton Administration, in a first for the U.S. government, directed that once approval for an arms sale is received, American personnel overseas should be "tasked" with supporting the marketing efforts of American companies seeking defense contracts. In addition, senior U.S. officials should promote sales of particular importance, and the Department of Defense would assist U.S. arms-producing companies in participating at international air and trade exhibitions—at considerable expense, it should be noted, to taxpayers.

The latter was a reversal of the policy in the early days of the Administration, when Secretary of Defense Les Aspin ruled against Pentagon support for defense contractors at arms shows, which involved the provision and transport of tanks, missiles and military aircraft to be put on display. The new policy was, however, confirmation of what was already occurring at the cabinet level. In 1993 Commerce Secretary Ron Brown had energetically boosted the U.S. aerospace and defense industry at the Le Bourget airshow outside of Paris, the world's premier showcase for military aircraft, seeking among other sales to persuade Malaysia to buy the McDonnell Douglas F/H-18 fighter planes. Brown had also traveled to Saudi Arabia and elsewhere to encourage purchases of American arms. In 1995, Secretary of Defense William Perry attended the defense exhibit in the United Arab Emirates. With a U.S. Navy aircraft carrier and frigate sitting in the harbor, he made reference to the Iranian military threat to the Persian Gulf and pointed out to his hosts the advantages of possessing an American frigate that would be interoperable with U.S. naval forces in the region.[69]

Senior Clinton Administration officials denied that U.S. arms transfer policy was motivated by commercial incentives, and frequently cited national security justifications for a forthcoming sale. At the same time, they noted that maintaining a strong and sustainable defense/industrial base was in the national interest and that arms sales abroad were essential to this interest. According to the White House directive, the impact on U.S. industry and the defense industrial base of whether a sale was approved or not was to be a criterion in decisions on sales. A government report observed that the "Clinton policy publicly elevates the significance of domestic economic considerations in the arms transfer decision-making process to a higher degree than has been the case in previous administrations."[70]

The Clinton policy announcement of 1995 resulted in a missed opportunity. In the first major review since the end of the Cold War, there was no attempt to identify how changed world conditions might lead to a new approach. For example, the inexorable linkage between the new dangers resulting from the proliferation of weapons of mass destruction and the growing availability of sophisticated conventional weapons—which can be the means of delivery of nuclear and chemical or biological arms—was not addressed. The artificial separation between these two categories of weapons was maintained, even though the need for delivery in reality creates a continuum between the two. Little attention was given to regional tensions and conflict and to what might be done to encourage conventional arms limitations on a regional basis through negotiated ceilings on weapons transfers. Such regional limitations could be sought on either a quantitative or qualitative basis, or a combination of the two. Nor in a world of growing ethnic conflict and civil strife was attention given to the distribution of small arms, which are the ones typically used in such conflicts.

A number of specific suggestions that had been made by knowledgeable observers found no place in the Clinton statement. First, the United States could take the lead in negotiating restrictions on licensed production or co-production of military equipment, as Lora Lumpe of the Federation of American Scientists has sensibly proposed.[71] This would be important, because the current increase in the production capacity for weapons is creating further pressure for more permissive sales. Once the new producers have fulfilled their own needs for arms, they often aggressively seek exports to maintain their own industries. Second, the direct support that governments give to arms exports, including their financing, might be limited among the major suppliers. One of the major complaints of American companies has been that their European competitors receive much government assistance not available to them. This disparity has not, however, stopped them from competing very successfully, or from seeking equivalent assistance from Congress. In February 1996, after three years of lobbying by the defense industry, the defense authorization bill signed by President Clinton included a $15 billion "Defense Export Loan Guarantee" program, under which the U.S. government will insure commercial loans taken by countries to buy or lease American-made weapons.[72] Third, multilateral, or in some cases unilateral, restrictions might be sought on the transfer of specific types of weapons technologies or weapons systems, such as stealth technology, anti-ship missiles or inhumane and indiscriminate weapons such as cluster bombs, anti-personnel landmines and blinding lasers. Fourth, specific bans could be established on the introduction of top-of-the-line weaponry into unstable regions. Fifth, with respect to the American decision-making process, the executive branch could be required to make an "arms sale impact statement" that assesses the likely effect of a proposed sale upon a region or specific country. Sixth, to make Congress fully a partner in the arms transfer process, it could be agreed that no sale above a specified amount should go forward unless both Houses of

Congress approve it by a simple majority, rather than the current situation in which Congress can only engage in a blocking action. Seventh, in those cases where prospective purchasers of American arms are also recipients of U.S. foreign aid, a determination should be required as to whether additional military spending is justified. Eighth, the human rights record of a prospective recipient of arms might be explicitly taken into account. Too often American arms have been used by governments to repress their own citizens.[73]

Not all of these suggestions should necessarily have been accepted. But they do serve to illustrate that the Clinton Administration review might have led to a more creative, comprehensive and future-oriented policy than was ultimately adopted. Such a policy would, moreover, have received public support. The selling of American arms overseas has never been highly popular. Over the years, polls by the Chicago Council on Foreign Relations have found that sales of military equipment by the United States to other countries have consistently been opposed. In 1995 public opposition was higher than ever, with 77 percent against and 15 percent in favor (in comparison, in 1990 59 percent were opposed and 32 percent in favor). Leaders or "elites" were more in favor but still below 50 percent in favor.[74]

From a broader perspective, the Clinton policy review was influenced by two of the most important systemic changes affecting the international arms trade in the 1990s. The end of the high stakes, East-West political and security competition in the Third World, which came about with the end of the Cold War, has reduced concerns about an unregulated international competition in arms transfers. In the past there was anxiety that the super-powers would be drawn into a regional conflict through their arms supply relationships with their clients and friends. Second, the domestic economic considerations supporting arms sales have become far more important than in the past. Conventional weapons have not yet quite become just another item to be sold overseas, like automobiles or washing machines. However, their role as job providers, as export items in which the United States clearly has the technological edge, as balance-of-trade enhancers and as a way to achieve an improved economy of scale for America's own armed forces cannot help but weigh heavily.[75] This factor certainly pertained within an Administration whose president had achieved office with the theme of "It's the economy, stupid!"

Recognition that the Clinton Administration had moved too far in the direction of giving priority to economic considerations in making arms sales decisions was, ironically, a principal conclusion of a bipartisan,

senior-level advisory board appointed to review conventional arms proliferation issues. Reporting in June 1996, after the Clinton arms transfer
policy had been in place for sixteen months, two former directors of ACDA,
Ronald Lehman and Paul Warnke, joined others in a commission chaired by
Janne Nolan in firmly recommending that U.S. arms transfer policy "can
and should be developed separately from issues of maintenance of the
defense industrial base, which we believe are better handled by specific
DoD industrial base policy." In what could be read as an admonishment, the
members went on to make clear that

> We do not believe that arms sales that would be rejected on the basis of foreign
> policy and national security considerations should be approved simply to
> preserve jobs or keep a production line open. The Board believes it is essential
> that the U.S. government take steps to make this policy clear at home and
> abroad.[76]

Beyond Arms Restraint: Dealing with Overcapacity in Defense Production

Most of the countries that maintain defense industries are experiencing a
serious problem of overcapacity in arms production. They will need to
address this problem if the strong influence that economic pressures have
on decision-making on arms sales is to be reduced. Although many
countries have recognized the problem in recent years in a *national* context,
it has not been addressed *internationally*. Yet this may be essential if
beggar-thy-neighbor policies are not to be pursued, with the resulting
inability to cope with the powerful economic forces that buttress the arms
trade.

This overcapacity is primarily, but not solely, the result of the end of the
Cold War. A secondary factor is the escalating costs of arms, as they have
become more loaded with advanced technology. More expensive weapons
tend to be bought in smaller quantities, thereby reducing production runs,
which in turn adds to the pressure to export.

Global defense spending began to decline sharply in the late 1980s,
falling from $1,284 billion in 1988 to $840 billion in 1994 (in constant 1994
dollars), and the trend is continuing downward.[77] As the need for defense
capabilities has declined with the easing of East-West tensions, defense
budgets have been reduced throughout the Western world and in the
countries of the former Soviet Union. This has led to some slimming and

rationalization of defense industries, but the results have been uneven and, in most countries, insufficient.

Along with the decline in defense spending has come a decline in world arms exports, from $71 billion in 1988 to $22 billion in 1994 (in constant 1994 U.S. dollars), according to ACDA figures.[78] Library of Congress data show a decline from $66 billion in 1988 to $28 billion in 1995 (in constant 1995 U.S. dollars) (see table 15-2). This decline may appear surprising, given the growth in concern about arms exports. That heightened concern reflects the new emphasis on the importance of arms exports as a way of making up for the decline in sales to the indigenous military establishments of the exporting countries. Arms exports are seen as a way to compensate for declining national purchases. Companies see it as their solution to the shrinking market at home. Governments have more actively supported their defense corporations when they seek to arrange an arms deal. Ministers have been sent to the arms-buying Persian Gulf states, for example, to stress the importance their government attaches to a sale. Presidents make telephone calls to foreign potentates. Senior generals, admirals and defense ministers from prospective purchasers are extremely well-received in Paris, London and Washington, as well as at lavish international arms fairs around the world. Fierce competitiveness has become the hallmark of the international arms trade.

But arms exports will not be the panacea for the declining defense industries. American arms exports have accounted for approximately 15 to 20 percent of the nation's arms production in the 1990s. Even with their considerable success, they are unlikely to go much higher. France exported 40 percent of its weapons production in the first half of the 1980s, making it the most dependent on overseas sales among the major producers. Since then, however, its overseas sales have declined. The other European nations have been considerably less export-dependent, although not for lack of effort, especially by Britain, which outsold France in the mid-1990s. None, however, face a bright future as exporters. Nor, it is now abundantly clear, do arms exports have any chance at all of becoming the savior of the very large and inefficient Russian arms industry. In 1992–93 General Mikhail Malei, at the time President Yeltsin's senior counselor for dealing with the problems of the military-industrial sector, argued that expanding exports was the way to finance the conversion of much of the defense industry, which he estimated would cost a total of $150 billion. Total Russian exports were then no more than $2 billion annually. Even with the planned increase to $3–4 billion by 1995–96, Malei was clearly engaged in wishful thinking.

The current and prospective export markets are much too small to offset the overall decline in defense procurement.

Accordingly, the downsizing of defense industries through their restructuring is the only valid solution in the long term. The emphasis on expanding exports has had the perverse and regrettable effect of distracting from, and perhaps even delaying, the necessary conversion.

To date, the record among the arms-producing countries in defense industry restructuring is very uneven. The most dynamic changes have occurred in the United States, through the merger of a number of large defense contractors. In 1993, for example, Martin Marietta, then primarily a missile and electronics firm, purchased General Electric's defense division and General Dynamic's space division. At about the same time, the aircraft maker Lockheed purchased General Dynamic's aircraft division, while Loral, another major firm, purchased LTV, Ford Aerospace and Unisys. Then in 1994 Lockheed merged with Martin Marietta to become Lockheed Martin. A year later Lockheed Martin purchased Loral, a move that created a $30 billion giant known as Lockheed Martin Loral—now reported to account for 40 percent of the Pentagon's procurement budget. During this same period, Northrop outbid Martin Marietta for the Grumman aircraft corporation, and the new company in turn bought the defense division of Westinghouse. Brookings Institution defense analyst and former senior Department of Defense official Lawrence J. Korb has estimated that the value of such mergers and acquisitions climbed from $300 million in 1991 to $14.2 billion by 1993 and would top $20 billion in 1996.[79]

In Western Europe, on the other hand, there was only a very limited consolidation of defense companies in the first half of the 1990s. The hurdles there are greater because of the fragmentation of Europe's defense industry. Although its defense market is half the size of the United States', the EU has ten aircraft makers versus five in the United States, eleven missile manufacturers versus five, ten armored vehicle makers as opposed to two, and fourteen warship builders versus against four.[80] Within this surfeit of manufacturers, a number have been identified as "national champions"—firms to be promoted as leaders, often because they are thought to have a technological advantage. Considerations of prestige also frequently play a role. Promoting national champions inevitably can run counter to the need for consolidation and European integration.

Yet the pressures for downsizing are inexorable. The need for change is greatest in France, a country that has sought to maintain a broad-based defense industry in the belief that self-sufficiency is essential to an independent defense and foreign policy. Thus France has given bountiful state support to a large military industrial sector and avoided as long as possible the reductions being undertaken elsewhere. It fell to an heir to the Gaullist tradition, President Jacques Chirac, to undertake in 1996 the major restructuring and reductions that were essential to the economic health of the nation. Another factor was that France needed to bring its budget deficit down from 5 percent to the 3 percent of gross domestic product needed to qualify for the planned European Monetary Union. The government announced plans to eliminate at least 54,000 of the country's 256,000 defense jobs, reduce military procurement from $20.5 billion in 1996 to $16.6 billion in 1997, and cut the total size of the armed forces over six years from 483,910 to 336,100 while doing away with conscription. In addition, Chirac pledged to reorganize the defense industry by forcing the state-owned Aerospatiale to merge with the privately owned Dassault, in which the government has a large stake, so as to have a single aerospace group similar to British Aerospace and Daimler-Benz Aerospace, Europe's two other large aircraft manufacturers. In undertaking these steps, France will be belatedly following the restructuring that took place earlier in the United Kingdom and on a smaller scale in Germany.

Sweden continues to build its own indigenous fighter, the Grippen, made by Saab, for which it seeks export markets. It abandoned a main battle tank, however. The Dutch have sought to avoid closing the Fokker aircraft firm. Throughout Western Europe there remains a defense overcapacity that the creation of European collaborative production projects, such as the Eurofighter, being built jointly by Germany, Italy, Spain and the United Kingdom, has not done much to relieve.

In Russia the situation is truly grim. Many huge industrial plants remain open, conversion having had only very limited success, but without markets for their products. An estimated five million workers in 2,000 plants are mostly idle, living on meager state subsidies. There have been a few important sales overseas, notably eighteen Mig-29s to Malaysia and India, but these are not sufficient to maintain more than a small fraction of the industry. The Russian government has few funds available for re-equipping its own military services, which still have an overabundance of Soviet-era weaponry. As a result there has been a dramatic decline in Russia's procurement of arms, with estimates ranging from a 55 percent to 83 percent fall during the 1991 to 1994 period.[81] The situation is much the same in Ukraine, which had plants making advanced ballistic missiles (the SS-18 and SS-24) and space-launch vehicles. Although skilled technical teams can still be put together, once again the market is lacking. In Eastern Europe, where there once were productive arms plants supplying the Warsaw Pact countries and Soviet allies with, for example, Czech tanks and Polish helicopters, times are also very hard.[82]

A Cooperative Build-Down

Given these widespread circumstances, developing international collaboration for the downsizing of defense industries should have double appeal. On the one hand, it would help preserve some national defense industries, or parts thereof. On the other, it would reduce the pressures to export arms, some of which could easily lead to sales that run counter to the longer term interests of international security. Thus far little has been done beyond limited American assistance for selected Russian efforts at defense conversion through the U.S. Cooperative Threat Reduction Program, known as the Nunn-Lugar program.

Something much more ambitious and far-reaching is needed: a cooperative, multilateral build-down of defense production capacities. This could be sought through the creation of something akin to an international division of labor—a specialization in the manufacture of defense products. Cooperation, rather than competition, should become the approach to the structural overcapacity in defense industries.

How might this be achieved? Clearly the task is difficult, given the myriad economic interests, political sensitivities and technological stakes involved. It is important not to underestimate, however, the major benefits and incentives that multilateral cooperation can offer participating nations:

—Sharing the costs and reducing the risks of researching, developing, and manufacturing new weapons systems.

—Gaining access to innovative foreign technologies.

—Helping to achieve economies of scale in the production of increasingly expensive weapons systems.

—Developing and penetrating foreign markets that might otherwise be closed to imports from a particular country.

—Enhancing the combat efficiency and effectiveness of military alliances by eliminating wasteful duplication in arms production while promoting standardization and interoperability of battlefield weapons.

—Fostering other types of international cooperation, such as NATO or Western European Union political cohesion or Economic Union economic integration.[83]

It would probably be best initially to divide the task into a European-American one and an East-West one.

In Western Europe there is already a resurgence of interest in trans-Atlantic arms cooperation. This comes at a time of recognition that the *intra*-European arms collaboration that has been attempted in the last fifteen years often has not been successful, due to intense rivalry and the desire to protect national capabilities. Failure in collaboration has led to duplication, higher costs and an inability to reduce overcapacity. Now American technological advances in weaponry, especially since the Persian Gulf War, and the ability of the United States to produce on a larger scale, and therefore at lower unit costs, have often made its arms more attractive and less expensive for purchasers. The sheer size of the U.S. defense market, which is double the value of total European defense procurement, gives American firms an enormous advantage over their rivals. It is no wonder that, comparatively, the United States is gaining an ever larger part of the world arms market while Europe's share is shrinking.

As for trans-Atlantic trade, for a long time there has been, and still is, a serious imbalance, with the United States selling five times as much defense equipment to Europe as vice-versa. The Europeans have long complained that the "Buy America" policy for equipping U.S. forces the Pentagon follows, often at Congress' urging, has largely frozen them out of the American market. Senator Sam Nunn (D-Ga) made a commendable effort to create a true "two-way street" by initiating legislation for arms co-development projects in 1993, but it foundered in the Pentagon's labyrinthine procurement practices. For their part, the Europeans in recent years have become much more willing to cross the Atlantic for their own purchases, should there be adequate reciprocity. The reason is budgetary pressures. Faced with the choice of a purchase at home that would support the domestic industry or a cheaper buy from the United States, more than one defense minister has been tempted by the latter option—especially if it were to involve a co-production or offset deal.

The conditions are therefore more propitious than ever for a collaborative effort that leads to a rationalization of European and American defense industries at a lower level. The *quid pro quo* for this downsizing would be that the industries or companies that were selected would have their long-term viability guaranteed. This is of great importance at a time when leading European Union officials and defense executives are going so far as to express fear that Europe risks losing its aerospace and armaments industries because of American competition.[84]

The first step would undoubtedly require a candid multilateral discussion of the problem, for which NATO appears to be the best venue. Prior to entering the Clinton Administration, former U.S. Department of Defense Representative at the U.S. Mission to NATO Catherine Kelleher called for a transatlantic forum to explore ways of creating collaborative management of defense production.[85] The next step would involve the adoption of various forms of specialization. Selected countries or companies would be assigned certain R&D missions or production runs. Such activities could involve intensive multinational collaboration. They would be the building blocks for collaborative management of what could become a common Western defense industrial base.

As for Russia, it is arguably even more important to create a cooperative relationship with it, given the high stakes involved in the maintenance of good relations. Unfortunately, many Russians believe that the United States, in particular among the Western powers, has systematically sought to prevent the export of Russian arms and more generally to tear down the Russian defense establishment. Some mistakenly see this as the reason Washington has opposed arms sales to some countries such as Iran and Iraq, while allowing U.S. defense companies to be fiercely competitive in the world arms market in general. The steep rise in American weapons sales since the end of the Cold War—in proportionate rather than absolute terms—when placed alongside the decline in Russia's sales, is a matter of considerable irritation and hurt feelings.[86]

Downsizing and converting the once mammoth and still excessively large Russian defense industry are extraordinary challenges. An amazingly large part of the Soviet Union's economy had been dedicated to supporting the defense sector—approximately 60 percent of machine production and more than 80 percent of electronics manufacturing, for example. Priority was consistently given to the defense sector over the civilian and consumer economy; that sector received the best scientific brains and had first call on all resources.

No wonder conversion and downsizing have been difficult and slow. An equally important factor, however, has been the still large role of the military-industrial sector and its supporters in the ongoing political struggle in Russia between the conservatives and the reformers. Even more than in the United States and Western Europe, Russian leaders must maintain the support of the military and parts of the military-industrial complex.

Senators Nunn and Richard Lugar (R-Ind) had a good understanding of the problem and its centrality to American interests, as did Secretary of

Defense Perry. Congress, however, has given far too little financial support to the Nunn-Lugar program. Moreover, most of that funding has been spent on or been earmarked for de-nuclearization and prefabricated housing for military officers, with only a small amount left for the conversion of military factories to produce civilian goods.

As a great power Russia will continue to maintain an arms industry. Hence it is desirable for the Western nations to explore how some of the most promising parts of the Russian defense industry might be included in collaborative projects while the outdated and non-viable sectors are brought to a close. Russian scientists and defense officials believe their research and design teams and technology experts are capable of making major contributions. They also see possibilities for joint ventures, especially in space and missile technology.

Another approach that might be explored is the adoption of "market-sharing" arrangements for arms sales among the major producers, West and East. This would be based on the assumption that the maintenance of viable, smaller and not excessive defense industries in Europe, Russia (and possibly Ukraine) and the United States is in the common interest of all. Market-sharing would also reduce the unbridled competition for arms sales. Russian Foreign Minister Andrei Kozyrev—who in his earlier career wrote a book on the international arms trade—put forth this concept, as did President Yeltsin at his meeting with President Clinton in Vancouver in 1993. At the time it received short shrift and no exploration within the U.S. government.

While it is true that the concept of market-sharing is somewhat alien to American notions of unrestricted free enterprise, the reality is that partially managed trade already exists. Weapons, moreover, are not just another commercial commodity. By their nature there is large governmental involvement in their manufacture and distribution. No one is more cognizant of this than the American defense firms that make and market weapons. Arms sales are, above all, matters of foreign policy and national security. Thus there is little reason why market-sharing should not be explored as a possible instrument of policy.

Market-sharing could be structured in a number of ways. A country could be given the exclusive rights to build a particular weapons system or to use a specified technology; if not exclusive rights, it could be guaranteed a certain market share. Alternatively, a country might be given the lead in arms sales to a region. France, for example, might be assured a non-competitive environment in sales to francophone Africa or the United States to parts of Latin America. Some combination of technology and geography also could be the basis for a market-sharing arrangement.

An excellent place to start would be Central and Eastern Europe. The United States has given its defense companies a green light to sell F-16 and F/A-18 advanced fighter aircraft to Poland, Hungary and the Czech Republic. Should the expense be too large, the U.S. Air Force has offered to lease some of its older F-16s. This policy has been justified as contributing to the growth of military ties in the context of the Partnership for Peace, as well as the eventual enlargement of NATO. The United States has also approved the eventual sale of an assortment of weapons, such as tanks, to an additional seven nations, including the Baltic republics and even Albania and Bulgaria.

But for Russia, this policy is a double whammy. Not only does it threaten to move NATO's military armaments closer to its borders, but it deprives Russia of a weapons market it has traditionally held. For Washington, this policy could become deeply counter-productive. Russia is unlikely to receive sympathetically the United States' strongly stated desire that it curtail its weapons sales to Iran and China, given what it sees as an American incursion into its potential markets closer to home.

As this example suggests, collaborative management of the arms trade would no doubt require some degree of self-denial, most of all for the United States given its dominant position. But the United States is likely to be in an enviable position in any case. A Pentagon forecast of the worldwide arms trade to the end of the century and beyond concluded that the United States will continue to be preeminent in defense markets, remaining the "preferred provider . . . with little meaningful competition from other countries."[87] This position is attributed to a number of factors, including top-of-the-line products, continued R&D, outstanding after-sale service, desired interoperability with U.S. forces, and competitive prices. Moderate self-limitation therefore would be a small price to pay for the political and security benefits that would flow from the maintenance and stability of reduced but viable defense industries in a number of key countries, most of which are allied to the United States or with whom good relations are desirable. And it would greatly facilitate cooperative defense conversion—an essential step toward managing conventional arms proliferation.

Is An International Regime Feasible?

"No" must be the answer if the aim is a formal, treaty-based, international regime with inspection or enforcement powers. A classic arms control regime such as the Nuclear Non-Proliferation Treaty, or the START I agreement, goes beyond what can be achieved with conventional arms.

"Yes" can be the response if the aim is a multifaceted regime with overlapping and complementary institutions and initiatives, none of which may be fully adequate by themselves but which are mutually reinforcing. Their totality could bring about a fundamental transition in approach toward to the problem of conventional arms proliferation. This would be akin to a broad and flexible *process,* designed to *manage* the international arms trade so as to reduce the risk of dangerous, destabilizing weapons transfers and other undesirable consequences.

In the latter sense, an international regime should consist of a composite of the varied and wide-ranging proposals discussed in this chapter. Constructing such a regime to manage arms proliferation will be a long-term, complex task. It will have to involve a number of approaches: unilateral and multilateral; regional and global; and supplier-based and recipient-initiated. Among the instruments available are national regulations and legislative oversight, measures to provide transparency, and agreements for specified multilateral controls.

Some of the latter can deal with particular arms and technologies such as landmines, blinding lasers, highly advanced weapons systems or even small arms. The co-production of weapons, the licensing of their manufacture, the provision of offsets and restrictions on the end-use of armaments that have been sold are all subject to agreement and restraint. Some of these arrangements could be, even at best should be, regionally prescribed. Others will require a universal approach.

Fortunately, two significant and promising building blocks for such a regime have been put into place in recent years, although both now need to be improved upon. The U.N. Register of Conventional Arms provides a new level of official transparency, especially for the majority of countries where such military data have not been publicly available. To be useful in the international context, however—for example, to warn of a major arms accumulation that could be destabilizing within a region—nations will need to give more detailed information as to types of weapons and to expand their information to include national procurement and military holdings.

Still more important is the Wassenaar Arrangement on Export Controls for Conventional Arms and Dual-Use Goods and Technologies established in 1996. This is the closest yet to an international regime, with a thirty-three nation group of arms producers and exporters having committed themselves to maintaining effective national export controls and to engage in information exchange on arms deliveries and transfers of sensitive technologies. Here, also, key improvements need to be made, such as a requirement for mandatory *prior* notification of arms sales and a provision of "no undercutting" by one state of another state when it decides to deny a technology transfer out of proliferation concerns. The "small group" of principal suppliers within the Wassenaar Arrangement should meet regularly to discuss such matters as common approaches to potentially unstable regions such as South Asia and the Middle East, or prohibitions on the introduction of certain advanced weaponry into regions where it does not presently exist. The Wassenaar Arrangement should be seen as an embryonic and evolving regime that holds considerable promise if fully implemented and further developed.

Beyond such measures, curbing the arms trade will depend on the establishment of a stronger norm than exists today against the proliferation of conventional arms. In this sense, conventional arms are unlike nuclear weapons. There is no equivalent to the near global support for limiting the dispersion of nuclear weapons. Nor can there be, given the prevalence of weapons since Cain killed Abel. Because nations have legitimate needs for armies and armaments, normative judgments must deal with issues of quantity and quality, of the use to which arms are put and of their impact on regional balances and international security. The availability of conventional arms may induce a conflict or deter it. They may be used aggressively or defensively, in the service of evil or in the cause of good. Because everything depends on circumstances, judgments about arms transfers must be undertaken case-by-case.

Nevertheless, elements of a norm are currently falling into place. In the Cold War era, arms transfers were often justified in the comparatively simplistic context of their being instruments of policy in East-West competition. Arms were supplied to support allies and gain friends. In today's more chaotic world, there is more concern about the impact of widely available arms in so-called "failed" states such as Somalia, Rwanda or Bosnia. Ethnic and internal conflicts, border and irredentist disputes, nationalist and fundamentalist struggles—all make for an explosive brew. Add to this brew the continuation of long-established regional enmities and the rapidity with which regional balances can be unbalanced by high-technology weaponry—as occurred with Saddam Hussein's Iraq—and the arms-supplying nations have reason to be cautious in their arms transfer policies. Their concern is all the more justified given that advanced conventional arms can easily become carriers of nuclear weapons. Accordingly, many countries are increasingly cognizant of the need to manage arms transfers wisely for reasons of international security.

A relatively new consideration stems from the belief that arms received from abroad are sometimes used against their populations by regimes that abuse human rights. A Republican legislator, Senator Hatfield of Oregon, has been a critic of American arms sales on the grounds that, by his estimate, 85 percent of U.S. arms sales between 1990 and 1994 went to "non-democratic" governments. The proposed Code of Conduct for arms sales, which has advocates in both the United States and Western Europe, had sizable support in Congressional votes and is indicative of some of the fresh thinking that could contribute to a new norm.

Another consideration voiced in recent years is the contention that those developing nations with critical needs for economic, social and educational development should be discouraged from spending their scarce resources on military expenditures. Nicole Ball discusses the possibility of creating a measure of "conditionality" in bilateral and multilateral economic assistance so as to avoid wasteful defense spending.[88]

Contributing to the development of a stronger norm against conventional weapons proliferation is the new recognition that the primary—although not sole—responsibility for arms restraint lies with the supplier states rather than with the recipient, purchasing nations. This recognition, which is a major turnaround in international thinking, can be traced back to the shock created by, and lessons learned from, the Persian Gulf War. Prior to then it was widely believed that arms control undertakings in the Third World had to be initiated by nations within a region, even though not much came of the few such efforts. Only after the Persian Gulf War did all the major conventional arms supplier nations decide to come together, for the first time, to discuss such issues in the P-5 talks and the post-COCOM deliberations that led to the Wassenaar Arrangement.

The international community should move toward adopting a new culture of restraint in the proliferation of conventional arms. This will require careful multilateral management by the suppliers, working with recipient countries as appropriate.

Above all, this will demand dedicated American leadership. The United States has such a predominant share of the world's production of arms and of the global arms trade that its leadership is absolutely essential if any progress in achieving arms transfer restraints is to be made. For the same reasons, an American initiative is the sine qua non in developing international collaboration in the needed downsizing of national defense industries. Elimination, or even reduction, of the overcapacities in arms production would greatly reduce the economic pressures behind arms exports. Such a cooperative build-down would go a long way toward a build-up of world peace. American influence in the world may be declining, but not on many of the issues discussed in this volume. With enlightened American leadership, significant progress may be possible. Absent such leadership, little will be accomplished.

Notes

1. Richard F. Grimmett, *Conventional Arms Transfers to Developing Nations, 1988–1995,* Congressional Research Service, U.S. Library of Congress (Washington, D.C., 1996), 8; Richard F. Grimmett, *Conventional Arms Transfers to Developing Nations, 1987–1994,* Congressional Research Service, U.S. Library of Congress (Washington, D.C., 1995), 7.

2. See U.S. Office of Technology Assessment, *Global Arms Trade: Commerce in Advanced Military Technology and Weapons* (Washington, D.C., 1991); Richard A. Bitzinger, *The Globalization of Arms Production: Defense Markets in Transition* (Washington, D.C., 1993); Richard A. Bitzinger, *Adjusting to the Drawdown* (Washington, D.C., 1993); David C. Morrison, "Eat or Be Eaten," *National Journal* (6 March 1993), 559–563.

3. See M. Granger Morgan and Mitchell B. Wallerstein, "Controlling the High Technology Militarization of the Developing World," in W. Thomas Wander and Eric H. Arnett (eds.), *The Proliferation of Advanced Weaponry: Technology, Motivations and Responses* (Washington, D.C., 1992) 283–299.

4. Systemic factors in a historical context are very well-analyzed in Edward J. Laurance, *The International Arms Trade* (New York, 1992), and Keith Krause, *Arms and the State: Patterns of Military Production and Trade* (Cambridge, England, 1992). For additional excellent and broad analyses, see: Michael Klare, *Rogue States and Nuclear Outlaws: America's Search for a New Foreign Policy* (New York, 1995); David Mussington, *Understanding Contemporary International Arms Transfers,* Adelphi Paper 291, International Institute for Strategic Studies (IISS) (London, 1994); Robert E. Harkavy and Stephanie G. Neuman (eds.), *The Arms Trade: Problems and Prospects in the Post-Cold War World, The Annals of the American Academy of Political and Social Science* 535 (1994).

5. For an account of early attempts at international regulations see Robert E. Harkavy, *The Arms Trade and International Systems* (Cambridge, Mass., 1975), 211–225.

6. Statement to the Conference of the Committee on Disarmament by U.K. Minister Lord Goronwy-Roberts (1 July 1976) (mimeo).

7. U.S. Department of State, "Prospects for Multilateral Conventional Transfer Restraints," unpublished paper (Washington, D.C., 1976).

8. "Eight Latin American Governments Sign Declaration Aimed at Limiting Armaments," *U.N. Monthly Chronicle* (March 1975), 54–57.

9. Thomas Ohlson, *Arms Transfer Limitations and Third World Security* (London, 1988), 242.

10. Andrew J. Pierre, *The Global Politics of Arms Sales* (Princeton, 1982), 310.

11. Joanna Spear, *Carter and Arms Sales: Implementing the Carter Administration's Arms Transfer Restraint Policy* (New York, 1995).

12. See also Andrew J. Pierre, "How to Curb Mideast Arms Sales," *The New York Times* (28 June 1991).

13. *Foreign Policy Overview and Budget Requests for Fiscal Year 1992,* Hearing before the U.S. Senate, Committee on Foreign Relations, 102nd Congress, First Session (Washington, D.C., 7 February 1991), 10.

14. "U.S. Arms Transfers to the Middle East Since the Invasion of Kuwait," Arms Control Association (Washington, D.C., 13 April 1992); "Register of U.S. Government-to-Government Arms Transfers to the Middle East," Arms Control Association (Washington, D.C., August 1996).

15. Grimmett, *Conventional Arms Transfers 1988–1995,* 22.

16. Ibid., 58. See also Stockholm International Peace Research Institute (SIPRI), *SIPRI Yearbook 1995: World Armaments and Disarmament* (Oxford, 1995), 495–496.

17. The term "P-5" or "Perm-5" was widely used within governments to refer to the permanent U.N. Security Council membership. The name was misleading, however, since the major arms suppliers are not necessarily the same countries as the permanent members of the Council.

18. "Statement of the Five on Arms Transfers and Non-Proliferation" (Paris, 8–9 July 1991).

19. "Communiqué of the Five on Arms Transfers and Nonproliferation" (London, 18 October 1991).

20. See Harald Muller, Matthias Dembinski, Alexander Kelle, and Annette Schapper, *From Black Sheep to White Angel? The New German Export Control Policy* (Frankfurt, 1994).

21. Sergei Kortunov, "Arms Export Controls—Competition Among Executive Agencies," in Andrew J. Pierre and Dmitri Trenin (eds.), *Russia in the World Arms Trade* (Washington, D.C., 1997).

22. U.N. General Assembly, A/RES/46/36 (9 December 1991).

23. Malcolm Chalmers and Owen Greene, *Taking Stock: The U.N. Register After Two Years,* Bradford Arms Register Studies No. 5 (Boulder, 1995), 44; Malcolm Chalmers and Owen Greene, *The U.N. Register of Conventional Arms: Examining the Third Report,* Bradford Arms Register Studies No. 1 (Bradford, England, 1995), 2–3.

24. There are a number of excellent appraisals of the U.N. Arms Register, several of them emanating from two centers that have rapidly developed expertise on the subject, one at the University of Bradford (United Kingdom) and the other at the Monterey Institute of International Studies (California). In addition to the previous citation, see Malcolm Chalmers, Owen Greene, Edward Laurance, and Herbert Wulf, *Developing the UN Register of Conventional Arms* (Bradford, England, 1994); Malcolm Chalmers and Owen Greene, "Implementing and Developing the UN Register of Conventional Arms," Peace Research Report No. 32, University of Bradford (Bradford, England, 1993); Edward J. Laurance, "The U.N. Register of Conventional Arms: Rationales and Prospects for Compliance and Effectiveness," *Washington Quarterly* (Spring 1993), 163–172; Edward Laurance,

Simon Wezeman, and Herbert Wulf, *Arms Watch: SIPRI Report on the First Year of the U.N. Register of Conventional Arms* (Oxford, 1993); Ian Anthony, "Assessing the U.N. Register of Conventional Arms," *Survival,* XXXV (1993–94), 113–129; Natalie Goldring, *Moving Toward Transparency: An Evaluation of the United Nations Register of Conventional Arms* (Washington, D.C., 1993); *The International Control of the Arms Trade,* Oxford Research Group (Oxford, 1992); Michael Moodie, "Transparency in Armaments: A New Item for the New Security Agenda," *Washington Quarterly,* XV (1992), 75–82; Hendrik Wagenmakers, "The U.N Register of Conventional Arms: A New Instrument for Cooperative Security," *Arms Control Today,* XXIII (1993), 16–21, as well as his "The U.N. Register of Conventional Arms: The Debate on Future Issues," *Arms Control Today,* XXIV (1994), 8–13.

25. Sales to NATO members, Australia and Japan are excluded from the requirement. See *Executive-Legislative Consultations on Arms Sales,* Congressional Research Service, U.S. Library of Congress (Washington, D.C., 1982), 7; Ian Anthony, *Arms Export Regulations* (London, 1991), 192–193.

26. See Andrew J. Pierre, "Beyond the 'Plane Package': Arms and Politics in the Middle East," *International Security,* III (1978), 148–161; William D. Hartung, *And Weapons for All* (New York, 1994), 277–283.

27. OTA, *Global Arms Trade,* 30.

28. Ibid., 31.

29. As described by Julian Cooper in chapter 8, "Russia," of this volume.

30. These insights into the Russian experience are the result of the Russian-American Conventional Arms Proliferation Working Group the author has directed since 1993 at the Moscow Center of the Carnegie Endowment for International Peace. See also Pierre and Trenin, *Russia in the World Arms Trade.* For a comprehensive discussion of the condition of the Russian defense industry, see Clifford G. Gaddy, *The Price of the Past: Russia's Struggle with the Legacy of a Militarized Economy* (Washington, D.C., 1996). See also Kevin P. O'Prey, *A Farewell to Arms? Russia's Struggle with Defense Conversion* (New York, 1995).

31. "The Scott Report: Devil in the Detail," *The Economist* (24 February 1996).

32. Quoted in Elizabeth Clegg, "The Scott Report: Implications for U.K. Export Controls," *The Export Practitioner* (May 1996).

33. See Wolfgang H. Reinicke, "Arms Sales Abroad: European Community Export Controls Beyond 1992," *The Brookings Review,* X (1992), 22–25; Harald Muller, "The Export Controls Debate in the 'New' European Community," *Arms Control Today,* XXIII (1993), 10–14; "EC Nations Vote for Controls on Weapons Exports," *The Independent* (19 September 1992); Trevor Taylor, "European Cooperation on Conventional Arms Exports," a paper for the International Studies Association Conference on the New Agenda of World Politics (1994).

34. See chapter 9, "China," by Gerald Segal, in this volume; John W. Lewis, Hua Di and Xue Litai, "Beijing's Defense Establishment: Solving the Arms Export Enigma," *International Security,* XV (1991), 87–109; Karl W. Eikenberry, *Explaining and Influencing Chinese Arms Transfers,* McNair Paper No. 36 (Washington, D.C., 1995).

35. Grimmett, *Conventional Arms Transfers 1988–1995,* 58, 83.

36. SIPRI, *SIPRI Yearbook 1995,* 493.

37. This section is largely drawn from the author's analysis in "The Wassenaar Arrangement," *Strategic Comments* (International Institute for Strategic Studies) (August 1996). For background, see also Owen Greene, "Launching the Wassenaar

Arrangement: Challenges for the New Arms Export Control Regime," University of Bradford (Bradford, U.K., 1996).

38. As of July 12, 1996, the participating countries were Argentina, Australia, Austria, Belgium, Bulgaria, Canada, Czech Republic, Denmark, Finland, France, Germany, Greece, Hungary, Ireland, Italy, Japan, Luxembourg, the Netherlands, New Zealand, Norway, Poland, Portugal, Republic of Korea, Romania, Russia, Slovakia, Spain, Sweden, Switzerland, Turkey, Ukraine, the United Kingdom, and the United States. The inclusion of Russia was a particular problem that created delays. The United States insisted that, before Russia could be included in the Arrangement, it make known the details of its arms sales accords with Iran and agree that there be no further such deals. This impasse was resolved after a year's delay and several discussions at the level of the Gore-Chernomyrdin Commission. In April 1996 the launching of the Wassenaar Arrangement was postponed by Russia's objection to a key aspect of the notification procedures, and a number of other countries refused to proceed without Russia's participation. Finally, in July 1996 all the necessary agreements were reached.

39. "The Wassenaar Arrangement," address by Under Secretary of State for Arms Control and International Security Lynn E. Davis, Carnegie Endowment for International Peace (Washington, D.C., 23 January 1996).

40. "Initial Elements, the Wassenaar Arrangement on Export Controls for Conventional Arms and Dual-Use Goods and Technologies," as adopted 11–12 July 1996.

41. For the full text, see *Proliferation and Export Controls* (London, 1995), 99.

42. For a well-informed, although skeptical, appraisal, see Paul Cornish, *The Arms Trade and Europe* (London, 1995).

43. "CSCE Forum for Security Cooperation," 49th Plenary Meeting of the Special Committee, *Journal* (CSCE) (Vienna, 24 November 1993).

44. Grimmett, *Conventional Arms Transfers 1988–1995,* 11, 38, 66.

45. U.S. Department of Defense, *World-Wide Conventional Arms Trade: A Forecast and Analysis (1994–2000)* (Washington, D.C., 1994), 21.

46. Henry L. Stimson Center, *Report of the Study Group on Multilateral Arms Transfer Guidelines for the Middle East* (Washington, D.C., 1992).

47. U.S. Congressional Budget Office, *Limiting Conventional Arms Exports to the Middle East* (Washington, D.C., 1992).

48. Kenneth Watman, Marcy Agmon, and Charles Wolf, Jr., "Controlling Conventional Arms Transfers: A New Approach with Application to the Persian Gulf," a report prepared for the Under Secretary of Defense for Policy, RAND (Santa Monica, 1994).

49. For further discussion, see Geoffrey Kemp, *The Control of the Middle East Arms Race* (Washington, D.C., 1992); Geoffrey Kemp, "Cooperative Security in the Middle East," in Janne E. Nolan (ed.), *Global Engagement: Cooperation and Security in the 21st Century* (Washington, D.C., 1994), 391–418; Alan Platt (ed.), *Arms Control and Confidence Building in the Middle East* (Washington, D.C., 1992).

50. See chapter 10, "The Middle East and the Persian Gulf: An Israeli Perspective," by Gerald Steinberg, in this volume.

51. See chapter 11, "The Middle East and the Persian Gulf: An Arab Perspective," by Abdel Monem Said Aly, in this volume.

52. See chapter 13, "South Asia," Rodney W. Jones, in this volume.

53. See chapter 12, "Asia-Pacific," Andrew Mack, in this volume.

54. From 1991 to 1993 the U.S. Department of Commerce approved over 350 licenses for torture and police equipment, including thumbscrews, thumbcuffs, shackles and leg irons. See Federation of American Scientists, *Arms Sales Monitor* (20 July 1993).

55. See Center for Defense Information, "Landmines: The Real Weapons of Mass Destruction," *The Defense Monitor,* XXV (1996); Human Rights Watch Arms Project, *Landmines: A Deadly Legacy* (Washington, D.C., 1993); and International Campaign to Ban Landmines, *Second NGO Conference on Landmines: Report of Proceedings,* Geneva, May 9–11, 1994, International Campaign to Ban Landmines (Washington, D.C.).

56. Senator Patrick Leahy, "The CCW Review Conference: An Opportunity for U.S. Leadership," *Arms Control Today,* XXV (1995), 20–24; "First CCW Review Conference Ends in Discord over Landmines," *Arms Control Today,* XXV (1995), 26.

57. Human Rights Watch Arms Project, "Blinding Laser Weapons: The Need to Ban a Cruel and Inhuman Weapon" (Washington, D.C., 1995).

58. See chapter 3, "The Subterranean Arms Trade: Black-Market Sales, Covert Operations and Ethnic Warfare," by Michael T. Klare, in this volume.

59. For the best complete analysis of this, see Jeffrey Boutwell, Michael Klare, and Laura Reed (eds.), *Lethal Commerce: The Global Trade in Light Weapons* (Cambridge, Mass., 1995), especially the chapter on controlling transfers of light arms by Jo L. Husbands. For a more optimistic approach to controls see Aaron Karp, "The Arms Trade Revolution: The Major Impact of Small Arms," *Washington Quarterly,* XVII (1994), 65–77. See also Jasjit Singh (ed.), *Light Weapons and International Security* (Delhi, 1995); Edward J. Laurance, *The New Field of Micro-Disarmament: Addressing the Proliferation and Buildup of Small Arms and Light Weapons* (Bonn, 1996).

60. Office of Technology Assessment, *Global Arms Trade.*

61. U.S. General Accounting Office (GAO), *Offset Demands Continue to Grow* (Washington, D.C., 1996), 1.

62. Offsets are well-discussed by Lora Lumpe in "Sweet Deals, Stolen Jobs," *The Bulletin of Atomic Scientists,* L (1994), 30–35; "Sweet Deals and Low Politics," *Public Interest Report* (Federation of American Scientists) (January/February 1994). See also Grant T. Hammond, "The Role of Offsets in Arms Collaboration," in Ethan Kapstein (ed.), *Global Arms Production: Policy Dilemmas for the 1990s* (Boston, 1992), 205–221.

63. William W. Keller, *Arm in Arm: The Political Economy of the Global Arms Trade* (New York, 1995), 136.

64. U.S. Department of Commerce, Bureau of Export Administration, *Offsets in Defense Trade* (Washington, D.C., 1996), 70.

65. See Duncan Clarke, "Israel's Unauthorized Arms Transfers," *Foreign Policy* (Summer 1995), 89–109.

66. Grimmett, *Conventional Arms Transfers 1988–1995,* 47, 83. These figures represent a downward revision of the data. In the 1986–1993 report, Grimmett estimated that the United States made 73 percent of the agreements with the developing world. ACDA's annual report assigns 56 percent of global arms deliveries to the United States in 1994. See U.S. Arms Control and Disarmament

Agency (ACDA), *World Military Expenditures and Arms Transfers 1995* (Washington, D.C, 1996), 16.

67. Interview with the author.

68. The White House, statement by the Press Secretary, "Conventional Arms Transfer Policy" (Washington, D.C., 17 February 1995).

69. Hartung, *And Weapons for All,* and William D. Hartung, "Nixon's Children: Bill Clinton and the Permanent Arms Bazaar," *World Policy Journal,* XII (1995), 31–32.

70. Grimmett, *Conventional Arms Transfers 1987–1994,* 3.

71. Lora Lumpe, "Clinton's Conventional Arms Export Policy: So Little Change," *Arms Control Today,* XXV (1995), 9–14.

72. See "'Fiscally Conservative' 104th Congress Grants Billions in New Arms Export Subsidies," *Arms Sales Monitor* (32) (5 March 1996), 5.

73. See the list of "Twenty Measures to Implement a New U.S. Arms Transfer Policy" in *The Defense Monitor,* XXIII (1994).

74. John E. Rielly, *American Public Opinion and U.S. Foreign Policy 1995* (Chicago, 1996), 32.

75. The heightened importance of the economic dimension of arms sales is well-discussed in Keller, *Arm in Arm.*

76. Janne E. Nolan, Edward Randolph Jayne, III, Ronald F. Lehman, David E. McGiffert, and Paul C. Warnke, *Report of the Presidential Advisory Board on Arms Proliferation Policy* (1996), 15–16.

77. ACDA, *World Military Expenditures 1995,* 53.

78. Ibid., 103.

79. Lawrence J. Korb, "Merger Mania," *The Brookings Review,* XIV (1996), 22. See also, "Eat or Be Eaten," *National Journal* (6 March 1993), 559–562; Jacques S. Gansler, *Defense Conversion: Transforming the Arsenal of Democracy* (Cambridge, Mass., 1995).

80. "Defense Industries in Transition," *Strategic Survey, 1995–1996* (London, 1996), 21; see also Bernard Gray, "Europe Lags US in Alliance-Making," *Financial Times* (30 August 1996); Herbert Wulf (ed.), *Arms Industry Limited* (Stockholm, 1993); Pierre de Vestel, *Defense Markets and Industries in Transition in Europe: Time for Political Decisions?* Chaillot Papers no. 21 (Paris, 1995).

81. *SIPRI Yearbook 1995,* 474; Kevin P. O'Prey, *The Arms Export Challenge: Cooperative Approaches to Export Management and Defense Conversion,* Brookings Occasional Paper (Washington, D.C., 1995), 25.

82. See Ian Anthony (ed.), *The Future of Defense Industries in Central and Eastern Europe* (Oxford, 1994).

83. Richard A. Bitzinger, *The Globalization of Arms Production: Defense Markets in Transition* (Washington, D.C., 1993).

84. "EU Officials Fear U.S. Threat to Aerospace, Defense Base," *Defense News* (11–17 November 1996).

85. Catherine McArdle Kelleher, *The Future of European Security: An Interim Assessment* (Washington, D.C., 1995), 152–157.

86. Pierre and Trenin, *Russia in the World Arms Trade.*

87. U.S. Department of Defense, "World-Wide Conventional Arms Trade (1994–2000)," v-vi.

88. See chapter 14, "'Conditionality': Linking Development Assistance to Military Expenditures," by Nicole Ball, in this volume.

About the Authors

Ian Anthony is the Leader of the Arms Transfers and Arms Production Project of the Stockholm International Peace Research Institute. He is editor of *Arms Export Regulations* and *The Future of the Defence Industries in Central and Eastern Europe,* co-author of *West European Arms Production: Structural Changes in the New Political Environment* and author of *The Arms Trade and Medium Powers: Case Studies of India and Pakistan, 1947–90.*

Nicole Ball is a Fellow at the Overseas Development Council in Washington, D.C. where she has dealt with the role economic policies can play in promoting reform in the defense sector. She has been a consultant for the World Bank, the International Labour Office, the International Development Center of Japan and the Swedish Ministry of Foreign Affairs. Her publications include *Security and Economy in the Third World, Pressing for Peace: Can Aid Induce Reform?* and *Making Peace Work: The Role of the International Development Community.*

Julian Cooper is Director of the Center for Russian and East European Studies at the University of Birmingham (U.K.) and Professor of Russian Economic Studies. He has been a consultant on military conversion and science policy in the former Soviet Union to the North Atlantic Treaty Organization, the Organisation for Economic Co-operation and Development and the European Commission. His publications include *The Soviet Defence Industry: Conversion and Reform* and *The Conversion of the Former Soviet Defence Industry.*

Lawrence Freedman is Professor and Head of the Department of War Studies at King's College, London, and Director of the Center for Defence Studies of the University of London. He has held research positions at

Oxford University, the International Institute for Strategic Studies and the Royal Institute of International Affairs. His publications include *The Evolution of Nuclear Strategy* and *The Gulf Conflict 1990–1991: Diplomacy and War in the New World Order.*

Rodney W. Jones is President of Policy Architects International. Previously he was Executive Director of the Washington Council on Nonproliferation and worked at the Bureau of Strategic and Eurasian Affairs of the U.S. Arms Control and Disarmament Agency. He has lived in India and Pakistan and has published extensively on regional security and proliferation in Asia. Among his works are *Emerging Powers: Defense and Security in the Third World, Modern Weapons and Third World Powers* and *Small Nuclear Forces and U.S. Security Policy.*

Ethan B. Kapstein holds the Harold E. Stassen Chair for International Peace at the University of Minnesota. He has been Director of Studies at the Council on Foreign Relations, co-director of the Economics and National Security Program of the Olin Institute at Harvard University and Professor at Brandeis University and has been with the Organisation for Economic Co-operation and Development in Paris. He is the author of *The Political Economy of National Security: A Global Perspective* and editor of *Downsizing Defense* and *Global Arms Production: Policy Dilemmas for the 1990s.*

Michael T. Klare is the Five College Professor of Peace and World Security Studies (a joint appointment at Amherst, Hampshire, Mount Holyoke and Smith Colleges and the University of Massachusetts at Amherst) and Director of the Five College Program in Peace and World Security Studies. He is the author of *American Arms Supermarket* and *Rogue States and Nuclear Outlaws* and co-author of *A Source of Guns: The Diffusion of Small Arms and Light Weapons in Latin America.*

Andrew Mack is Professor of International Relations at the Australian National University and was Director of the Peace Research Centre in the Strategic and Defense Studies Centre at the University. He has been affiliated with the University of California at Berkeley, the London School of Economics and the East-West Center in Hawaii. He is the author of *Maritime Security in the Asia-Pacific in the 1990s* and *Asian Flashpoint: Security and the Korean Peninsula.*

Martin Navias teaches in the Department of War Studies, King's College, London, and specializes in the arms trade, proliferation and Southern African issues. He is author of *Nuclear Weapons and British Strategic*

Planning 1955–58 and *Going Ballistic: The Build Up of Ballistic Missiles in the Middle East.*

Janne E. Nolan is a Senior Fellow at the Brookings Institution, Adjunct Professor at Georgetown University and formerly was an official of the U.S. Arms Control and Disarmament Agency. She was Chair of the Presidential Advisory Board on Arms Proliferation Policy, which focused on conventional weapons in its 1996 report. She is the author of *Trappings of Power: Ballistic Missiles in the Third World* and editor of *Global Engagement: Cooperation and Security in the 21st Century.*

Andrew J. Pierre teaches at the School of Advanced International Studies of Johns Hopkins University and previously was a Senior Associate at the Carnegie Endowment for International Peace. Earlier he was Director-General of the Atlantic Institute for International Affairs in Paris, Senior Fellow at the Council on Foreign Relations, with the Brookings Institution and Hudson Institute, and taught at Columbia University. He also was a foreign service officer with the U.S. Department of State. He is the author of *The Global Politics of Arms Sales* and editor of *Arms Transfers and American Foreign Policy* and *Russia in the World Arms Trade.*

Andrew L. Ross is Professor of National Security Affairs at the U.S. Naval War College and an Associate at the John M. Olin Institute for Strategic Studies at Harvard University. He is the editor of *The Political Economy of Defense: Issue and Perspectives* and author of *Arms Production in Developing Countries: The Continuing Proliferation of Conventional Weapons.*

Abdel Monem Said Aly is Director of the Al-Ahram Center for Political and Strategic Studies in Cairo. He has written extensively in Arabic on the Arab-Israeli conflict and Arab regional relations and in English on Egypt's political system, national security and arms control policies, as well as on Middle East regional security. He has been a research fellow at the Brookings Institution and has participated in projects at the Stockholm International Peace Research Institute and the Institut Français des Relations Internationales.

Gerald Segal is Senior Fellow in Asian Security Studies at the International Institute for Strategic Studies and Co-Chair of the European Council for Security Cooperation in Asia-Pacific. Formerly he was a Senior Research Fellow at the Royal Institute of International Affairs. He is the author of *The Great Power Triangle, The Soviet Union and the Pacific* and *The*

Fate of Hong Kong and co-editor of *Chinese Economic Reform: The Impact on Security.*

Gerald Steinberg is a Professor at Bar-Ilan University, Ramat-Gan, Israel, and is the Research Director of the Center for Strategic Studies there. He has participated in the Track Two multilateral workshop on Middle East Arms Control and Regional Security for the Israel Ministry of Foreign Affairs. He has written widely on arms control, strategy and technology and Israeli foreign and defense policy.

The World Peace Foundation

THE WORLD PEACE FOUNDATION was created in 1910 by the imagination and fortune of Edwin Ginn, the Boston publisher, to encourage international peace and cooperation. The Foundation seeks to advance the cause of world peace through study, analysis, and the advocacy of wise action. As an operating, not a grant-giving foundation, it provides financial support only for projects which it has initiated itself.

Edwin Ginn shared the hope of many of his contemporaries that permanent peace could be achieved. That dream was denied by the outbreak of World War I, but the Foundation has continued to attempt to overcome obstacles to international peace and cooperation, drawing for its funding on the endowment bequeathed by the founder. In its early years, the Foundation focused its attention on building the peacekeeping capacity of the League of Nations, and then on the development of world order through the United Nations. The Foundation established and nurtured the premier scholarly journal in its field, *International Organization,* now in its forty-ninth year.

From the 1950s to the early 1990s, mostly a period of bipolar conflict when universal collective security remained unattainable, the Foundation concentrated its activities on improving the working of world order mechanisms, regional security, transnational relations, and the impact of public opinion on American foreign policy. From 1980 to 1993 the Foundation published nineteen books and seven reports on Third World security; on South Africa and other states of southern Africa; on Latin America, the Caribbean, and Puerto Rico; on migration; and on the international aspects of traffic in narcotics. In 1994 and 1995, the Foundation published books on Europe after the Cold War; on the United States, southern Europe, and the countries of the Mediterranean basin; and on reducing the world traffic in conventional arms.

441

The Foundation is now focusing its energies and resources on a series of interrelated projects entitled Preventing Intercommunal Conflict and Humanitarian Crises. These projects proceed from the assumption that large-scale human suffering, wherever it occurs, is a serious and continuing threat to the peace of the world, both engendering and resulting from ethnic, religious, and other intrastate and cross-border conflicts. The Foundation is examining how the forces of world order may most effectively engage in preventive diplomacy, create early warning systems leading to early preventive action, achieve regional conflict avoidance, and eradicate the underlying causes of intergroup enmity and warfare.

Index